ROBOTICS INTEGRATED DESIGN AND PRACTICE

ROBOTICS INTEGRATED DESIGN AND PRACTICE

Zeming Fan
Xiaojun Yu
Zhongkui Sun
Zhunga Liu
Wenquan Liu
Fei Ning

Northwestern Polytechnical University, China

World Scientific

NEW JERSEY • LONDON • SINGAPORE • BEIJING • SHANGHAI • HONG KONG • TAIPEI • CHENNAI • TOKYO

Published by

World Scientific Publishing Co. Pte. Ltd.
5 Toh Tuck Link, Singapore 596224
USA office: 27 Warren Street, Suite 401-402, Hackensack, NJ 07601
UK office: 57 Shelton Street, Covent Garden, London WC2H 9HE

Library of Congress Cataloging-in-Publication Data
Names: Fan, Zeming author | Yu, Xiaojun (Associate professor) author
Title: Robotics integrated design and practice / Zeming Fan, Xiaojun Yu, Zhongkui Sun, Zhunga Liu, Wenquan Liu, Fei Ning, Northwestern Polytechnical University, China.
Description: Hackensack, NJ : World Scientific, [2026]
Identifiers: LCCN 2024061784 | ISBN 9789819810246 hardcover | ISBN 9789819810253 ebook | ISBN 9789819810260 ebook other
Subjects: LCSH: Robots--Design and construction | Robots--Programming | Robotics in education
Classification: LCC TJ211 .F36 2026
LC record available at https://lccn.loc.gov/2024061784

British Library Cataloguing-in-Publication Data
A catalogue record for this book is available from the British Library.

For any available supplementary material, please visit
https://www.worldscientific.com/worldscibooks/10.1142/14223#t=suppl

Desk Editors: Soundararajan Raghuraman/Julio Hong/Amanda Yun

Typeset by Stallion Press
Email: enquiries@stallionpress.com

Foreword by Lihua Xie

Robotics is a branch of engineering and computer science that involves the conception, design, manufacturing, and operation of robots that replicate or substitute human actions. Robotics has been an active research field in both academic and industrial labs since the invention of the first robot in the 1960s. Currently, robotics is advancing rapidly in the modern world and has achieved spectacular success in various areas.

While robotics describes the field of study focused on developing robots and automation, a robot is a programmable machine that can complete a task. A robot might resemble a human or be capable of performing specific tasks like human beings. To function, a combination of computer programs and algorithms, actuators, control systems, as well as real-time sensors is implemented to enable robots to perceive the environment and carry out tasks. Typically, this is a huge undertaking, which requires a multitude of ideas from various "classical" fields.

As the field of robotics and exploration of the potential uses and functionality of robots have grown substantially in the 21st century, different aspects of robotics research are carried out by experts in various fields. Although it is usually not the case that any single individual has the entire area of robotics in his or her grasp, the main functionalities of a robot are quite similar. Therefore, a category of robot according to its main functionalities is natural. At a relatively high level of abstraction, splitting robot functionalities into four main areas seems reasonable, i.e., mechanical manipulation, locomotion, computer vision, and artificial intelligence.

This book introduces mechanical manipulation for both serial robots and parallel robots with typical hands-on examples. Specifically, Chapters 1 and 2 cover the basics of a robot, focusing mainly on its constructions. Chapters 3 and 4 introduce mechanical engineering and mathematics, including the forward kinematics and inverse kinematics of a robot. Chapters 5 and 6 present robot visual perceptual and object recognition, while Chapters 7 and 8 introduce robot path planning and trajectory planning, respectively. Finally, Chapters 9 and 10 introduce control, a practical robot system, and hands-on practice of its main functions. This book emphasizes practical issues of robotic systems, from design and construction to its control and functional control in detail. Additionally, at the end of each chapter, there is a set of exercises, including some practical examples related to the main contents of the chapter.

This book originated from both class notes and the practical experiences of authors in teaching the robot major curriculum at Northwestern Polytechnical University, Xi'an, China, since 2005. This book is characterized by the combination of theory, engineering design, and practice, and it is especially suitable for both senior undergraduate and postgraduate students majored in robotics, and it is also suitable for those scientific and technological researchers engaged in the fields of robot research, development, and application.

The overall quality of this book is excellent, emphasizing practical implementation step by step. To further help students in this process, this book also provides a large amount of supplementary materials, which can be found by scanning the QR code within the book at each chapter. I like this book and would like to recommend it for publication.

Lihua Xie
Nanyang Technological University
April, 2024

Foreword by Yaonan Wang

Robotics, as an interdisciplinary field bridging engineering and computer science, encompasses the theoretical conception, systematic design, precision manufacturing, and operational implementation of automated systems capable of replicating or augmenting human capabilities. Since the landmark development of the first industrial robot in the 1960s, this discipline has evolved into a vital research domain spanning academic institutions and industrial laboratories. Today's technological landscape witnesses unprecedented advancements in robotics, demonstrating remarkable success across manufacturing, healthcare, space exploration, and service industries.

While the term "robotics" refers to the scientific discipline itself, a robot constitutes an autonomous or semi-autonomous system that executes programmable operations through integrated sensing, computation, and actuation mechanisms. Modern robotic systems may adopt anthropomorphic configurations or specialized architectures tailored for specific operational environments. Their functionality relies on sophisticated integration of computational algorithms, sensor fusion techniques, actuator control systems, and real-time decision-making processes — a technological synthesis demanding multidisciplinary expertise from mechanics to artificial intelligence. The 21st century has witnessed exponential growth in robotic applications, prompting specialized research trajectories within the field. Though individual mastery of the entire robotics spectrum remains impractical, core functional components maintain conceptual consistency across platforms. We therefore propose a systematic classification of robotic capabilities into four foundational pillars: mechanical manipulation, dynamic locomotion, computer vision, and cognitive intelligence.

This comprehensive volume focuses specifically on mechanical manipulation systems, providing rigorous theoretical frameworks and practical implementations for both serial and parallel robotic architectures. The book's structure follows a progressive pedagogical approach. Chapters 1 and 2 establish fundamental concepts of robotic system design and mechanical configurations. Chapters 3 and 4 develop mathematical foundations through forward and inverse kinematic analyses. Chapters 5 and 6 integrate visual perception systems with object recognition algorithms. Chapters 7 and 8 address path planning methodologies and trajectory optimization techniques. Chapters 9 and 10 synthesize control system design with practical implementation strategies.

Distinguishing features of this book include the following: industry-aligned practicality — detailed case studies spanning design conception to functional implementation; multidisciplinary integration — seamless combination of mechanical engineering principles with computational techniques; progressive learning system — chapter-end exercises ranging from theoretical derivations to real-world simulation projects; digital resources — QR-code accessible supplementary materials including CAD models and MATLAB/ Python codebases.

Originating from two decades of instructional refinement at Northwestern Polytechnical University's robot developments, this work uniquely synthesizes theoretical rigor with industrial applicability. Its systematic presentation makes it particularly valuable for senior undergraduates and graduate students in robotics engineering, while advanced researchers will appreciate its comprehensive treatment of contemporary manipulation technologies.

As a practitioner who has witnessed the evolution of robotic technologies, I find this volume exceptional in its balanced approach to academic depth and practical relevance. The authors' meticulous organization of complex concepts into implementable knowledge frameworks deserves commendation. Without reservation, I endorse this text as an essential resource for the next generation of robotics professionals.

Yaonan Wang
Academician of the Chinese Academy of Engineering,
Hunan University
April, 2024

Preface

Six decades have elapsed since the birth of the first industrial robot, marking the emergence of robotic technology as a dynamic interdisciplinary field that continues to attract significant academic and industrial investment. The remarkable progress achieved in this domain has been particularly accelerated since the advent of the 21st century, with the global robotics industry demonstrating an annual growth rate exceeding 30%. Notably, Asian markets have emerged as principal drivers of this expansion, with China projected to dominate the global robotics marketplace in forthcoming decades. This sustained growth trajectory has catalyzed intensified research and development initiatives among leading robotics nations, creating a fertile environment for technological innovation and practical implementation.

This monograph presents a comprehensive exploration of integrated design methodologies and practical implementations across three primary robotic architectures: serial manipulators, parallel mechanisms, and mobile platforms. Through systematic examination of core robotic competencies — including intelligent sensory systems, autonomous navigation protocols, kinematic modeling, and advanced control architectures — the text provides a structured approach to developing fully functional robotic systems with integrated obstacle avoidance capabilities.

Distinctive in its integration of theoretical foundations with practical engineering applications, this work serves dual pedagogical purposes: (1) bridging the gap between academic robotics theory and real-world implementation challenges and (2) cultivating essential competencies in mechatronic system design, critical analysis, and collaborative problem-solving. The progressive knowledge architecture is designed to equip readers with both conceptual understanding and practical skills in holistic robotic system development, thereby laying a robust foundation for professional practice in robotics engineering.

Target readership encompasses senior undergraduates and graduate students specializing in robotics engineering, automation, and mechatronic systems. Additionally, the text offers substantial value to researchers and industrial practitioners engaged in robotic system development, particularly those working on integrated hardware–software solutions for autonomous operations. The balanced treatment of theoretical principles and practical design considerations ensures relevance across academic and industrial contexts, while numerous case studies and implementation examples facilitate knowledge transfer to real-world engineering challenges.

Key Features: Unified treatment of serial, parallel, and mobile robotic architectures; Systematic integration of sensing, navigation, and control subsystems; Project-based learning approach to mechatronic system integration; Emphasis on industry-relevant design methodologies; Balanced presentation of theoretical models and practical implementations.

About the Editors

Zeming Fan received his B.S., M.S., and Ph.D. degrees all in engineering in 1993, 1999, and 2002, respectively. After obtaining his Ph.D. from Xi'an University of Technology, China, he continued his research work as a post-doctoral researcher, focusing mainly on robot design and its applications. In August 2005, he joined Northwestern Polytechnical University, China as an associate professor with the School of Automation. Currently, he is a full professor with his research focus mainly on artificial intelligence and intelligent autonomous humanoid robots, and he has published more than 60 peer-reviewed top journal papers and three books. Prof. Fan is also an active peer reviewer for several international journals. Prof. Fan has been PI for more than 20 research funds and has also won several excellent national awards for both competitions and teaching, e.g., the first prize in Shaanxi Teaching Achievement, the second prize of Shaanxi Provincial University Science and Technology Award, and several first-class teaching achievement awards of the university. Prof. Fan is also a chair professor of the national first-class major of "Robot Engineering" and the head of one national and two provincial first-class courses.

Xiaojun Yu received his Ph.D. degree from Nanyang Technological University, Singapore, in 2015. From January 2015 to August 2017, he worked as a postdoctoral research fellow at the same university. In 2017, he joined Northwestern Polytechnical University (NWPU), China, and currently, he is a tenured associate professor at NWPU's School of Automation. Dr. Yu is an IEEE senior member, and his main research interests include robotic control and biomedical signal/image processing. Dr. Yu has published more than 100 peer-reviewed journal papers and 40 international conference papers and has received the Outstanding Paper Award of SPIE BISC 2018 and IEEE ICICN 2021. He is also an active peer reviewer for several international journals.

About the Associate Editors

Zhongkui Sun is a full professor and doctoral supervisor at Northwestern Polytechnical University, China. He served as a postdoctoral fellow at the University of Potsdam, Germany, as well as a visiting scholar at the University of South Florida, USA. He was awarded the Shaanxi Provincial Youth Science and Technology Rising Star Talent Support Program, the Shaanxi Provincial 'Special Support Program' for leading talents in science and technology innovation, and the Shaanxi Provincial 'Sanqin Talent Allowance'. Dr. Sun is mainly engaged in the fields of robot dynamics and control, stochastic dynamics and control, and nonlinear vibration analysis and control, and has published more than 100 SCI papers in leading academic journals at home and abroad, such as *Physical Review E, Nonlinear Dynamics, Journal of Sound and Vibration, and Science China*. He has presided over six projects for the National Natural Science Foundation of China and two provincial and ministerial-level scientific research projects, and won the second prize in National Teaching Achievement, the first prize in Shaanxi Teaching Achievement, the second prize of Shaanxi Natural Science Award, the first prize of Shaanxi Science and Technology Award, and the first prize of Shaanxi Provincial University Science and Technology Award.

Dr. Sun has been invited to serve as a Topic Editor of *Symmetry* (IF: 2.713), the Young Editorial Board member of *Applied Mathematics* and Mechanics, and the Editorial Board member of *Studies in Advanced Mathematics.* He acts as a member of the 8th Youth Working Committee of the Chinese Society of Mechanics, a member of the Random Vibration Committee of the Chinese Society of Vibration Engineering, and a member of the Nonlinear Vibration Committee of the Chinese Society of Vibration Engineering.

Zhunga Liu received his B.S., M.S., and Ph.D. degrees in Control Science and Engineering from Northwestern Polytechnical University (NWPU), China, in 2007, 2010, and 2014, respectively. After obtaining his Ph.D., he joined NWPU's School of Automation as a faculty, where he is currently a full professor and the associate dean. Prof. Liu's research interests focus mainly on intelligent information fusion, target tracking, detection, and recognition, on which he has published more than 50 peer-reviewed journal papers. Prof. Liu has been awarded the first prize in Natural Science from the Chinese Society of Automation, the second prize in Technological Invention by the Ministry of Education, the first prize in Natural Science of Shaanxi Province, and the Youth Science and Technology Award of the Chinese Society of Aeronautics, amongst others. He is currently serving as an associate editor of several international journals, including the *International Journal of Approximate Reasoning.*

Wenquan Liu is currently an engineer with the School of Automation at Northwestern Polytechnical University, China and the Deputy Director of the School's Experimental Teaching Center. Dr. Liu has been engaged in experimental practice teaching. His research focuses mainly on motor control, power supply design, and intelligent network. Dr. Liu has been the Principal Investigator (PI) for more than 20 scientific research projects to develop multi-functional charge and discharge equipment, data message network communication protocol software, and to design control module detectors. Dr. Liu has extensive hands-on experience in the field of robot design, control, and applications.

Fei Ning is currently an engineer at the School of Automation, Northwestern Polytechnical University, China. Dr. Ning has been Principal Investigator (PI) or co-PI of several research grants. He has published several books with extensive hands-on experience in the field of robot design, control, and applications.

Contents

Chapter 1

Introduction

Since the beginning of the 21st century, humanity has not only focused on its own development but also on the development of robots. Today, the name "robot" is familiar to most people. From ancient myths and legends to modern science fiction novels, movies, and TV shows, there are numerous captivating depictions of robots. Despite the significant achievements in robotics, the majority of real-world robots are neither as intelligent or brave as those portrayed in myths and literature nor as versatile as some entrepreneurs advertise. Currently, the capabilities of robots remain relatively limited, but their rapid development is increasingly impacting industrial production, space exploration, ocean exploration, and various aspects of human life.

1.1 Origin and Development of Robots

1.1.1 *Origin of robots*

Following the onset of the modern era, humans became increasingly fixated on the idea of inventing various mechanical tools and power machines to alleviate the burden of physical labor. The 18th century witnessed a pivotal moment with the advent of the steam engine, which heralded the era of machinery replacing human effort.

This transition marked the onset of the first industrial and scientific revolution, introducing a plethora of automated machines and power systems that transformed the concept of robots from mere fantasies into tangible manifestations of automatic machinery. This period saw the emergence of various mechanical control robots, including intricate machine toys and handicrafts.

The term "robot" was initially coined in 1920 by Czech writer Karel Čapek in his science-fiction play, *R.U.R.* (Rossum's Universal Robots), drawing inspiration from the Slovak word *robota*, meaning slave or laborer. Čapek's play garnered significant attention for its portrayal of the ominous implications of robot development for society. Within the narrative, robots produced by the Rossum Company initially function as mindless laborers, devoid of emotion or autonomy. However, a pivotal breakthrough in robotics technology imbues these robots with intelligence and emotions, leading to their integration as indispensable members of human society in both industrial and domestic settings.

Čapek's work also addresses concerns regarding the safety, intelligence, and autonomy of robots, highlighting the potential consequences of unchecked advancements in robotics technology. While the realm of science fiction remains speculative, there exists a palpable apprehension among humans regarding the potential realization of such scenarios. In response to societal unease surrounding the impending proliferation of robots, renowned science-fiction writer Isaac Asimov introduced his famous "Three Laws of Robotics" in his 1950 novel, *I, Robot*:

(1) A robot shall not harm a human being or stand by and watch a human being harmed;
(2) Robots must be absolutely obedient to humans unless that obedience is harmful to humans;
(3) A robot must protect itself from harm unless it is sacrificed to protect a human or the human orders it to do so.

These three codes gave a new ethical dimension to the robotics community and popularized the concept of robotics, making it more acceptable to human society. To this day, the Three Codes of Robotics continue to provide meaningful guidelines for robotics researchers, designers, manufacturers, and users.

1.1.2 *Development of robots*

During the early 1960s up to the early 1970s, following the introduction of industrial robots, progress in robotics was sluggish, with numerous research organizations and companies facing setbacks. Notable advancements during this period included the creation of a mobile intelligent robot, Shakai, by the Stanford Research Institute in 1968 and the commercialization of the first robot, T3, by the Cincinnati-Milacron Corporation in 1973.

The 1970s witnessed a growing interest in robotics within the artificial intelligence (AI) community. It became evident that the emergence of robots revitalized AI development, offering a fertile ground for testing and application, which held promise for significant advancements in AI. This realization garnered recognition from scientific, technological, industrial, and governmental sectors globally. With advancements in automatic control theory, electronic computers, and aerospace technology, robotics entered a new phase of development by the mid-1970s. This momentum continued into the 1980s, with the robotics manufacturing industry emerging as one of the most rapidly growing sectors of the economy by the mid-1980s.

However, by the late 1980s, saturation of industrial robot users led to a surplus of industrial robots, prompting closures and mergers among robot manufacturers. This downturn caused a slump in international robotics research and the industry. The early 1990s saw signs of recovery and continued development in the robotics industry, yet this resurgence was short-lived, as another downturn was experienced in 1993–1994. Despite a consistent increase in the number of industrial robots globally each year, market fluctuations characterized the period from the 1980s to the end of the 20th century, marked by three distinct phases of growth and decline. The momentum of robot development gained traction by 1995, paving the way for a promising future into the 21st century.

Since the beginning of the 21st century, the industrial robotics industry has accelerated, with an annual growth rate of about 30%. In particular, the growth rate of industrial robots in Asia is up to 43%, which is the most prominent.

In recent years, the worldwide robotics industry has been developing rapidly, and the total sales volume of the worldwide robotics industry increased by 10% in 2007 compared to 2006. Humanization, heavy-duty capabilities, and intelligence are becoming the main trends of the future robot industry.[1] The total number of industrial robots in service worldwide is now over one million. In addition, there are millions of service robots in operation. However, more than 90% of these robots are not intelligent.

With the rapid growth in the number of industrial robots and the development of industrial production, human beings have put forward higher requirements for the working ability of robots, namely that these robots should be able to utilize not only a variety of feedback sensors but also a variety of learning, reasoning and decision-making techniques in AI. Intelligent robots also apply many of the latest intelligent technologies, such as presence sensing technology, virtual reality technology, multi-agent technology, artificial neural network technology, genetic algorithms and genetic programming, bionic technology, multi-sensor integration, fusion technology, and nanotechnology.

Intelligent robots are a class of robotic systems capable of sensing the environment and their own state through sensors, realizing goal-oriented autonomous movement in the environment with obstacles, so as to complete certain operational functions.[2] Mobile robots are a class of robots with high intelligence, which constitutes one of the frontiers and key fields of intelligent robot research. The difference between mobile robots and other robots lies in their "mobile" characteristics. Mobile robots can not only play a bigger and bigger role in production and life but also serve as an effective tool and experimental platform for researching the generation of complex intelligent behaviors and exploring human thought patterns. The intelligence level of 21st-century robots will rise to a marvelous higher level.

1.2 Categories of Robots

Numerous methods exist for classifying robots. This chapter organizes robots based on their connectivity, mobility, control mechanisms, geometric structure, level of intelligence, and intended applications.

1.2.1 *Categories by connection*

1.2.1.1 *Tandem robots*

Tandem robots represent an earlier iteration of robotic systems utilized in various industrial fields. Illustrated in Figure 1.1 is a tandem-type robot that functions as an open kinematic chain employing an open-loop mechanism as its primary prototype. The tandem robot's design, characterized by its open-loop tandem mechanism, enables the end-effector to maneuver across a broad workspace, facilitating flexible operations with simpler control systems and structural designs. This design has seen widespread application across diverse fields, such as machine tools and assembly plants, owing to its well-established research and successful implementation.

However, the continuous connection of tandem robot joints presents limitations. Due to the continuity of the connection of a tandem robot, when any end effector of the robot is subjected to any force, not only do the joints share the load, but they also have to bear the superimposed weight, and each joint has to bear the sum of the forces from itself to the end effector joint. Consequently, the load capacity and positional accuracy of tandem robots are significantly lower than those of multi-axis machinery. Additionally, the placement of motors within joint parts results in substantial rotational inertia during movement, diminishing the robot's dynamics. Furthermore, cumulative errors at tandem joints pose a significant challenge, impairing the working accuracy of tandem robots.

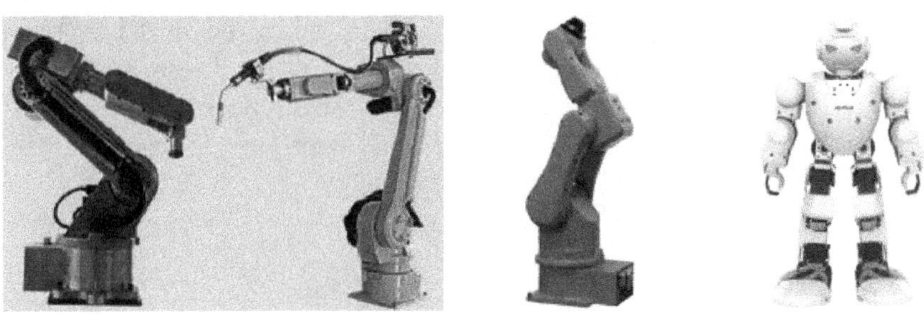

Figure 1.1. Tandem robots.

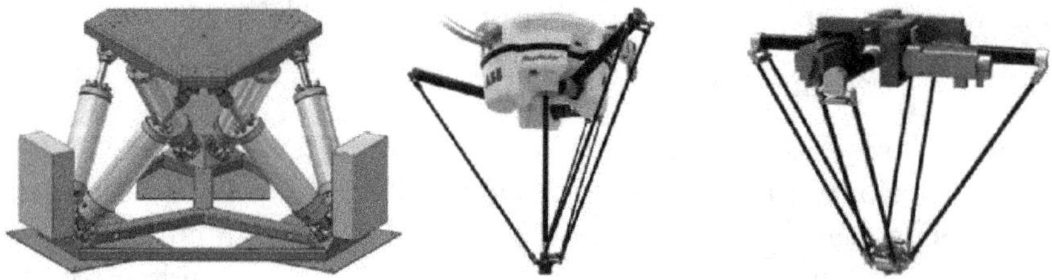

Figure 1.2. Parallel robots.

1.2.1.2 *Parallel robots*

A parallel robot is a closed kinematic chain. As shown Figure 1.2, a parallel-type robot generally consists of an upper and lower platform, as well as two or more kinematic branch chains. Compared to series-connected robots, parallel-connected robots are characterized by small motion inertia and thermal deformation and are less prone to dynamic and cumulative errors because the coordinates of joints are interconnected by one or several closed loops; in addition, they have the advantages of higher precision, high machine rigidity, compact and stable structure, high load-bearing capacity, and easy inversion.[4] Based on these characteristics, parallel robots have been a research hotspot in the field of robotics for the past 30 years. Although they appeared later than tandem robots and there are still many theoretical problems that remain unsolved, research on the structural design, dynamics, and control strategies of the parallel multi-motion arm, as well as the workspace and station singularity of the spindle motors, has matured and has been widely applied in the fields that require high stiffness, high precision, high load capacity, and compact workspace. Motion simulators and delta robots are successful examples of parallel-type robots.

1.2.1.3 *Hybrid robots*

The hybrid robot represents an innovative robot design, tailored to meet specific needs within industrial settings, emphasizing workspace versatility and operational adaptability. This novel structure integrates elements of both parallel and tandem mechanisms, forming a sophisticated hybrid system characterized by intricate design and optimization strategies. Combining the robustness, load-bearing capacity, speed, and precision typical of parallel robots with the expansive motion capabilities, ease of control, and flexibility akin to serial robots, the hybrid robot offers a comprehensive solution for applications demanding precise motion control.

However, the presence of parallel mechanisms necessitates consideration of motion decoupling in structural design, making the strategic setup of these parallel mechanisms a crucial area of research for hybrid robots. Moreover, hybrid robots, augmented with parallel mechanisms, exhibit characteristics such as micro-motion and high precision, offering promising prospects in industries demanding precision machining. These robots find applications not only in conventional sectors, such as food, medicine, 3C, daily chemicals, and logistics for tasks like material handling, sorting, and transfer, but also extend their usability through multi-angle pickup capabilities, thereby broadening their application scope.

The emergence of hybrid tandem-parallel robots has expanded application scenarios in robotics, enabling more effective alignment with market demands and fulfilling personalized customization requirements. This integrated approach facilitates the establishment of efficient automated solutions that enhance the core competitiveness and profitability of China's manufacturing enterprises while accelerating their transformation and upgrading. By organically integrating serial and parallel robotic architectures, hybrid configurations combine the positional flexibility of serial manipulators with the high stiffness and dynamic performance of parallel mechanisms, thereby creating comprehensive robotic systems that synergize the advantages of both structural paradigms. Figure 1.3 shows the multi-functional and multi-purpose intelligent robot organically integrated with series, parallel, and mobile characteristics, as researched by Prof. Fan Zeming of Northwestern Polytechnical University; this also serves as the experimental and innovative practice platform for this book.

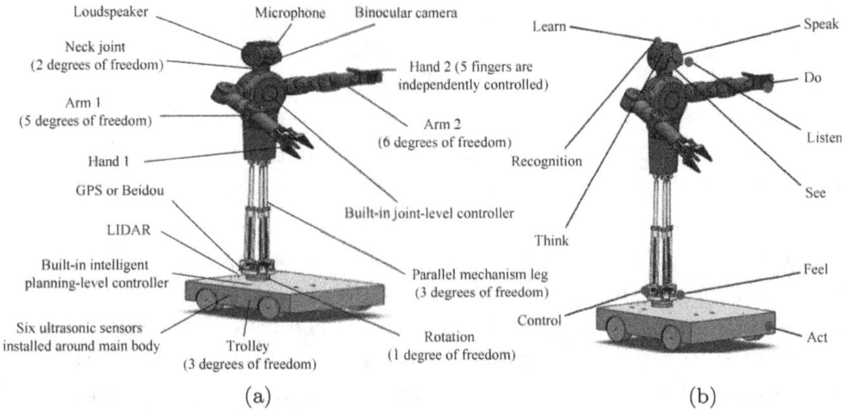

Figure 1.3. An intelligent robot: (a) structure and its composition diagram and (b) functions.

1.2.2 *Categories by mobility*

1.2.2.1 *Stationary robots*

A stationary robot is fixed to a certain base, and the entire robot or manipulator cannot be moved, but only the individual joints, as shown in Figure 1.4.

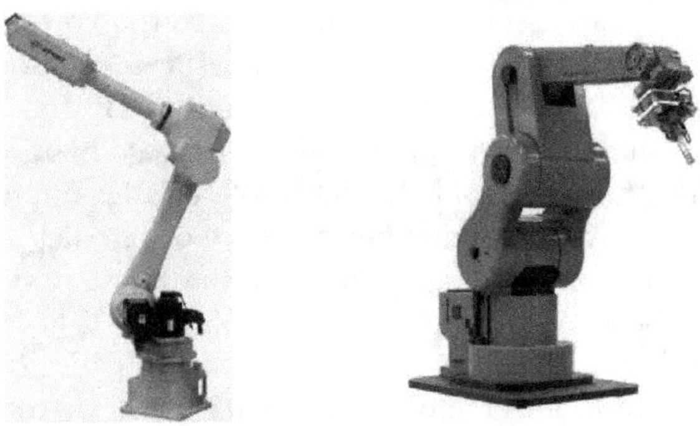

Figure 1.4. Stationary robots.

1.2.2.2 *Automated guided vehicles (AGVs)*

AGVs are primarily used to move materials indoors and outdoors in factories, warehouses, and transportation areas (an application known as material handling). As shown in Figure 1.5, AGVs may be

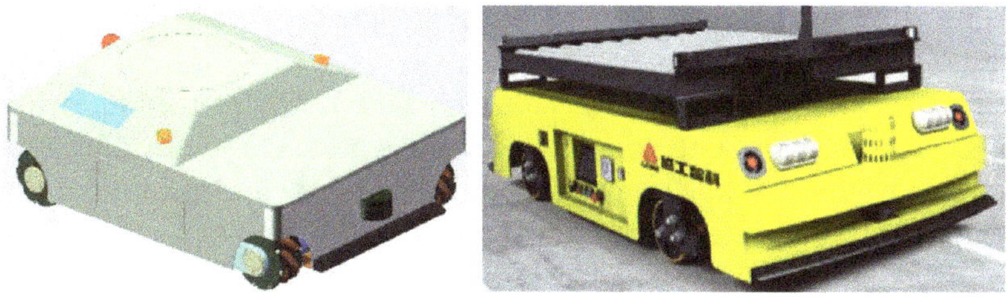

Figure 1.5. AGVs.

transporting automotive parts at a manufacturing site, newsprint at a publishing company, or waste material at a nuclear power plant.

Modern transportation systems typically use wireless communications to connect all vehicles to a central computer responsible for controlling traffic flow. Vehicles are further categorized on the basis of whether they pull trailers loaded with materials (towed AGVs), are loaded and unloaded with forked AGVs (fork-open AGVs), or are transported on platforms on the roof of the vehicle (unit-loaded AGVs). AGVs are probably the most developed market for mobile robots. In addition, moving materials as well as loading and unloading of trucks, trains, ships, and airplanes are potential applications for future generations of transportation.

1.2.2.3 *Service robots*

Service robots perform tasks that would be considered service industry jobs if they were performed by humans. As shown in Figure 1.6, some service tasks, such as mail, food, and medicine delivery, are considered "light" material handling, similar to the work of

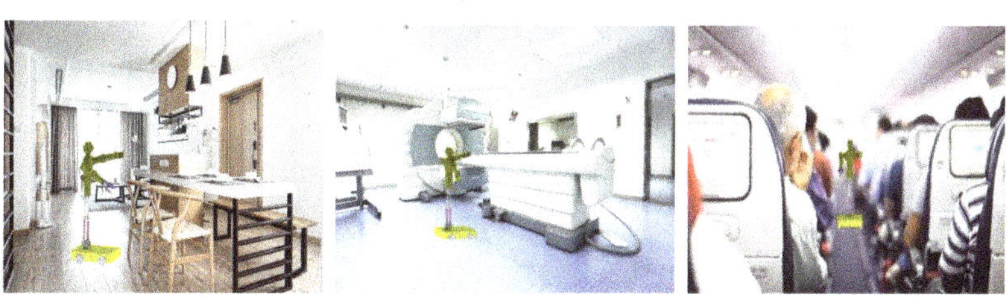

(a) (b) (c)

Figure 1.6. Intelligent robots used in service areas: (a) home service robots, (b) medical service robots, and (c) aviation service robots.

AGVs. However, many service tasks are characterized by a higher degree of human intimacy.

Healthcare service robots can be used to deliver food, water, medicines, and reading materials to patients. They can also move biological samples and waste, medical records, and administrative reports from one location to another.

Surveillance robots are like automated security guards, and in some cases, their automated ability to competently move around an area and simply detect intruders is valuable. This application is one of the early interests of mobile robot manufacturers.

Other service robots include those designed for floor cleaning and lawn maintenance in institutional and household settings. Cleaning robots are deployed in public spaces including airports, supermarkets, shopping malls, and industrial facilities, and their operational capabilities extend to surface sanitization, debris removal, vacuum-based particulate extraction, and refuse collection. These devices are distinguished not by their capacity for targeted navigation or material transport, but rather by their systematic coverage of spatial domains with at least single-pass completeness.

1.2.2.4 *Social robots*

Social robots are service robots that are specifically designed to interact with humans, and their main purpose is usually to deliver information or provide entertainment. While only stationary to deliver information, social robots also need to be mobile for one reason or another.

Some potential applications include answering questions about the location of products in retail stores (such as grocery and hardware stores), delivering burgers to children in restaurants, and robotic assistants for the elderly and sick to help them see (robotic guide dogs), move around, or remember their medications.

In recent years, Sony has produced and marketed some impressive entertainment robots, the earliest of which were packaged as "pets". Automated guides for museums and expositions can direct patrons to specific exhibits.

1.2.2.5 *Field robots*

Field robots perform tasks in challenging outdoor natural terrain "the field" conditions. Almost any type of vehicle that has to move and work in an outdoor environment has the potential to be automated. Most things are harder to do outdoors: It's hard to see in bad weather, it's hard to decide how to navigate through complex natural terrain, and it's easy to get stuck in a rut. Field robots are usually weapons and/or tools mounted on mobile pedestals, which are capable of not only moving to different locations but also physically interacting with the environment in some way.

In agriculture, potential applications for robots include planting, weeding, chemical (herbicides, pesticides, and fertilizers) application, pruning, harvesting, and grasping fruits and vegetables. Unlike home mowing, large-scale mowing is necessary in parks, golf courses, and highway medians. Human-powered vehicles specialized for mowing are good candidates for automation. In forestry, tending nurseries and harvesting mature trees are potential applications. The robot shown in Figures 1.7 was researched by Prof. Zeming Fan at Northwestern Polytechnical University.

Robots have multiple applications in mining and excavation. Above ground, excavators, loaders, and rock trucks have been automated in open-pit mines. Underground, drilling rigs, anchor machines, continuous coal miners, and dump trucks have been automated.

(a) (b)

Figure 1.7. Intelligent robots used in the field of fruits and vegetables grasping: (a) schematic diagram of fruit grasping and (b) the fruit grasping process.

1.2.2.6 *Inspection, reconnaissance, surveillance, and detection robots*

Robotic devices designed for inspection, reconnaissance, surveillance, and detection purposes are deployed on mobile platforms to examine specific areas or identify targets within them. The primary motivation for utilizing robots in such tasks is often the inherent danger that would be posed to humans undertaking them. Environments characterized by high levels of radiation, such as the interiors of nuclear power plants, certain military operations like reconnaissance and bomb disposal, and space exploration, exemplify scenarios where robots are preferred. For instance, the humanoid robot depicted in Figures 1.8 represents an advanced research endeavor led by Prof. Zeming Fan at Northwestern Polytechnical University.

Within the energy sector, robots have found applications in inspecting various components of nuclear reactors, including steam generators, heating tubes, and waste storage tanks. Moreover, prototypes or operational robots have been developed for the inspection of high-voltage power lines, as well as gas and oil pipelines. The scope of robotic intervention extends to underwater environments, where remotely controlled vehicles are increasingly utilized to scrutinize

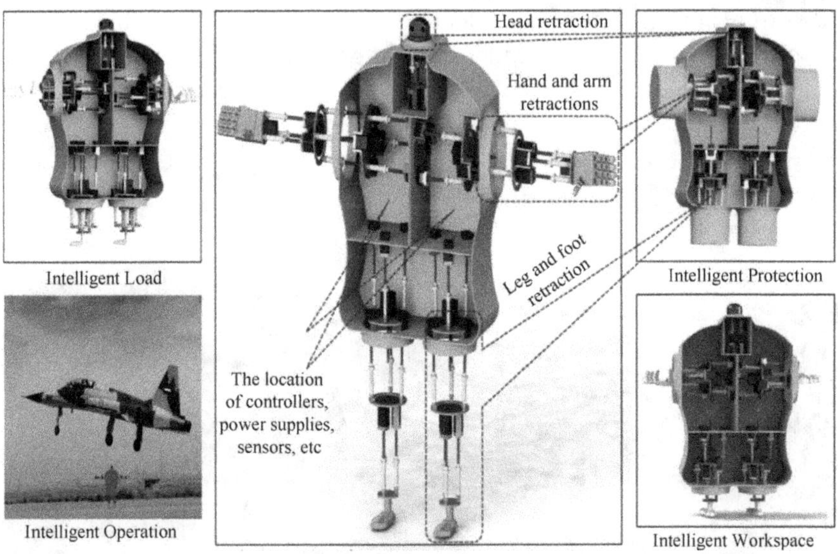

Figure 1.8. Schematic diagram of intelligent humanoid robot structure.

structures such as oil rigs, undersea communication cables, and even assist in locating shipwrecks.

Research and development of robotic soldiers have been particularly intense in recent years. As shown in Figure 1.9, robots are used in warfare, such as for reconnaissance and surveillance, troop supply, minefield mapping and clearance, and ambulance services. And military vehicle manufacturers are already working to incorporate a wide variety of robotic technologies into their products. Bomb disposal is already a mature market.

Figure 1.9. Intelligent humanoid robots used in the field of warfare.

In space, a number of robotic vehicles have already traveled kilometers autonomously on the surface of Mars, and the concept of a vehicle flying around a space station propelled by thrusters has been on the agenda for some time.

1.2.3 *Categories by control mode*

Robots can be categorized into non-servo robots and servo-controlled robots according to the type of control.[6]

(1) **Non-servo robots:** These robots are more limited in their ability to work, and they tend to involve what are called "endpoint", "grab-and-go", or "switch" robots and particularly "finite sequence" robots. These robots work in a pre-programmed sequence and use end-limit switches, brakes, pin boards, and sequencers to control the movement of the manipulator.

(2) **Servo-controlled robots:** These robots have a greater working capacity than non-servo robots, and are therefore more expensive and, in some cases, less reliable than simple robots. The feedback signal obtained by the feedback sensor and the integrated signal from the given device are compared with a comparator, and the error signal obtained is amplified and used to stimulate the robot's drive device, which in turn drives the end-effector device to move according to a certain law to reach the specified position, speed, etc. This is obviously a feedback control system. Servo-controlled robots can be divided into point servo-controlled and continuous path servocontrolled robots.

1.2.4 *Categories by geometry*

Robots have a variety of mechanical structure forms, and the most commonly used structure forms are described by the coordinate characteristics shown in Figure 1.10, such as the Cartesian coordinate structure, cylindrical coordinate structure, polar coordinate

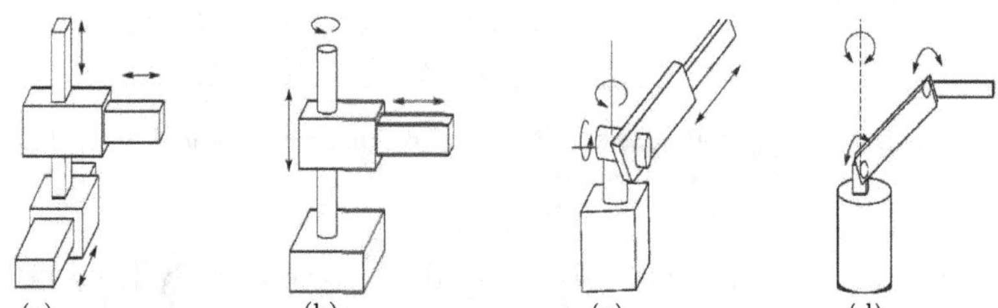

Figure 1.10. Geometric structure of a robot. (a) Cartesian coordinate structure, (b) Cylindrical coordinate structure, (c) Polar coordinate structure and (d) Articulated coordinate structure.

structure, spherical coordinate structure, and articulated coordinate structure.

1.2.5 *Categories by intelligence levels*

(1) A general robot is one that is not intelligent but only has general programming capabilities and operational functions, as shown in Figure 1.1.
(2) Intelligent robots, with varying degrees of intelligence, can be subdivided into sensing robots, interactive robots, and self-reliant robots, as shown in Figures 1.3 and 1.8.

1.2.6 *Categories by use*

(1) Industrial robots are mainly used in industrial and agricultural production or in manufacturing for welding, painting, assembly, handling, and processing of agricultural products, as shown in Figure 1.7.
(2) Exploration robots, which are used to conduct space or ocean exploration, can also be used for surface and underground exploration, as shown in Figure 1.11.
(3) Service robots, which are robots that work semi- or fully autonomously and perform services that contribute to improving

Figure 1.11. Exploration robot.

human lives or improving the functionality of equipment outside of manufacturing, as shown in Figure 1.6.

(4) Military robotics are used for military purposes, either offensive or defensive. They can be subdivided into aerial military robots, maritime military robots, and ground military robots, or simply air force robots, navy robots, and army robots, as shown in Figure 1.9.

1.3 Overview of the Book

Successful robots encompass diverse applications and markets, yet they all share a common attribute: Their design embodies an amalgamation of various fields of expertise, rendering robotics inherently cross-disciplinary. Addressing challenges in motion necessitates comprehension of mechanical structures, kinematics, dynamics, and control theory. Similarly, constructing resilient perceptual systems mandates integration of signal analysis and specialized domains such as computer vision to effectively utilize diverse sensor processing technologies. Localization, navigation, and target recognition, on the other hand, require knowledge of computer algorithms, information theory, AI, and probability theory.

Figure 1.12 shows a topological map of thc rclationship among the chapters of this book, which identifies many of the intellectual topics related to robotics. As undergraduate and graduate students studying robotics, as well as researchers who are passionate about the field, a familiarity with matrix algebra, calculus, probability theory, and computer programming is highly beneficial for studying this book. At the same time, the intelligent robot developed by Prof. Fan Zeming of Northwestern Polytechnical University (see Figure 1.3) is an experimental and innovative practical teaching platform developed for the study of this book, which can complete most of the tasks discussed in this book for experimental and innovative practical training. Learning based on this platform will have a multiplier effect.

1.3.1 *Design and practice of robot kinematics*

The analysis of a robotic mechanical structure's motion, known as kinematic analysis, involves understanding how the robot moves

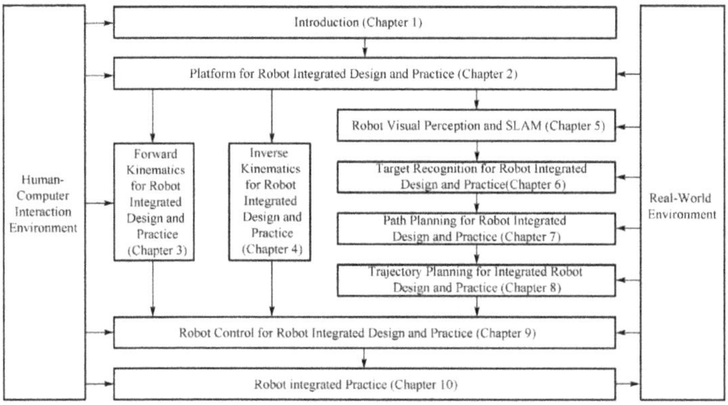

Figure 1.12. Topology of the relationship among the chapters of the book.

relative to a fixed Cartesian coordinate system, without considering the forces or moments causing the movement. It's essential to differentiate between kinematics and differential kinematics. In the context of a robotic manipulator, kinematics examines the mathematical relationship between joint positions and end-effector orientations, while differential kinematics explores the relationship between joint movements and end-effector motions in terms of velocities using Jacobi matrices.

Expressing kinematic relationships in formulas enables the investigation of two crucial robotics problems: the positive kinematics problem and the inverse kinematics problem. Positive kinematics employs linear algebra tools to establish a systematic and comprehensive approach for describing how the end-effector moves based on joint motions. Inverse kinematics considers the inverse of the previous problem, and the essential role of its solution is to transform the desired motion formulated for the end-effector in the workspace into the motion of the corresponding joint.[7]

Developing a kinematic model of the robot is also useful in determining the relationship between the forces and moments acting on the joints when in a static equilibrium position and the forces and moments acting on the end-effector.

1.3.1.1 *Design and practice of positive kinematics of the operating arm*

Kinematics studies the motion of an object without considering the forces that cause that motion. In kinematics, we study position, velocity, acceleration, and higher-order differentiation of position variables with respect to time or other variables. In this way, the study of manipulator arm kinematics is all about the geometric and temporal properties of motion.

Almost all operator arms are composed of rigid links, with neighboring links connected by joints that allow relative motion. These joints are usually equipped with position sensors that measure the relative position of the neighboring links. In the case of rotating joints, this displacement is referred to as the joint angle. If some of the operating arms contain sliding (or moving) joints, then the displacement of the two neighboring links is a linear motion, which is sometimes referred to as joint offset.

The number of degrees of freedom of the operating arm is the number of operating arms with independent position variables that define the position of all components in the mechanism. The end-effector is mounted at the free end of the manipulator arm. Depending on the application of the robot, the end-effector can be a fixture, a torch, a solenoid, or some other device. We typically describe the position of the operator arm in terms of a tool coordinate system attached to the end-effector, which corresponds to a base coordinate system attached to the fixed base of the operator arm.

A typical problem in the study of operator arm kinematics is positive kinematics. Computing the position and orientation of the end-effector of an operator arm is a static geometric problem. Specifically, given a set of values for the joint angles, the positive kinematics problem involves computing the position and orientation of the tool coordinate system with respect to the base coordinate system.[8] In general, we refer to this process as the representation of the position of the actuator arm from a joint space description to a Cartesian space description.

1.3.1.2 *Design and practice of inverse kinematics for operating arms*

In later chapters, we will discuss manipulator arm inverse kinematics, which is the process of calculating, given the position and orientation of an end-effector of the manipulator arm, all the joint angles that can be reached for that given position and orientation.[9] This is a fundamental problem in the practical application of operating arms.

Operator arm inverse kinematics is a fairly complex geometric problem, yet humans and other biological systems perform thousands of such solutions every day. For an artificially intelligent system such as a robot, we need to design an algorithm in the control computer to realize this inverse computation. In a way, the solution of the inverse kinematics problem is the most important part of the manipulator arm system.[10]

We consider this to be a "localization" mapping problem, a mapping of the robot's position and orientation from a 3D Cartesian space to an internal joint space. This mapping is required when the robot's target position and orientation is expressed in external 3D spatial coordinates. Some early robots did not have such algorithms, and they could simply be moved (sometimes manually taught to move) to a desired position and orientation while a series of joint variables (e.g., position and orientation in each joint space) were recorded to reproduce the motion. Obviously, if the robot were simply recording and reproducing the its joint position and motions, then there would be no need for any transformation algorithm from the joint space to the Cartesian space. However, it has become difficult to design a robot without such inverse kinematics algorithms.

Inverse kinematics is not as simple as positive kinematics. Because the equations of kinematics are nonlinear, it is difficult to obtain closed solutions, and sometimes there are no solutions,[11] while in the case of existence of solutions, the problem of multiple solutions may arise.

The study of the above problem gives a proper explanation for the phenomenon that the human brain and nervous system unconsciously guide the movement of the arm and hand, as well as the manipulation of objects. The presence or absence of a solution to the kinematic equations limits the workspace of the manipulating arm. The absence of a solution indicates that the target point is outside the workspace and the manipulator arm is unable to reach that desired position and orientation.[12]

1.3.2 *Design and practice of robot perception and object recognition*

Robots need to recognize objects in various engineering applications.[13] For example, robots need to recognize part drawings during machining; they need to recognize the shapes of workpieces during assembly; they need to recognize the objects being carried during handling; they need to recognize the shapes and colors of fruits, branches, trunks, and so on during grasping of fruits; and they need to recognize the shapes of obstacles and environments when they are moving in an environment. Therefore, object recognition is ubiquitous in the field of robotics.

1.3.3 *Design and practice of robot localization and map construction*

Because robots have to accomplish a given task in an unknown environment, localization and map construction play an important role.[14] In robotics, localization and mapping are usually interdependent. For mapping, we can only accurately describe the map information of the surrounding environment if we know the location of the robot; and for localization, we can only make more accurate localization based on the features in the environment depicted by mapping.

1.3.4 *Design and practice of robot planning and obstacle avoidance*

Cognition in robotics is discussed through robot sensors, robot mapping techniques, and localization techniques.[15] In general, cognitive representation systems utilize purposeful decisions and execute them to achieve the highest level of goals.

1.3.5 *Design and practice of robot control*

Control characteristics and control techniques for robots are analyzed, and robot joint space control, workspace control, and force control are discussed. Strictly, linear control techniques are only applicable to systems that can be mathematically modeled with linear differential equations. For the control of the manipulator arm, this linear approach is essentially an approximation because, as we will see in Chapter 5, the equations describing the dynamics of the manipulator arm are generally nonlinear differential equations. However, usually, this approximation is feasible, and these linear methods are the most commonly used in current engineering practice. Workspace control involves direct and decoupled control methods and adaptive control methods. When the robot is tracking a trajectory motion in space, position control can be used. Still, when the end-effector collides with the working environment, as in the case of a grinding robot, not only position control but also force control should be considered.

1.4 Summary

This chapter describes the origin and development of robots, the classification of robots, and the structure of this book. The section on the classification of robots describes six aspects of classification: connection, mobility, control, geometry, intelligence, and use; and the section on the structure of the book presents an overview of robot kinematics, perception and object recognition, localization and map construction, planning and obstacle avoidance, and the design and practice of control.

1.5 Exercises

1.1 What characterizes the development of domestic and foreign robotics?

1.2 Present a chronology of the main events in the development of robotics over the past 40 years.

1.3 Briefly describe the positive kinematics of a robot.

1.4 Briefly describe the inverse kinematics of a robot.

1.5 Briefly describe the speed and statics of a robot.

1.6 Briefly describe the dynamics of a robot.

1.7 Briefly describe the types of control for a robot.

Chapter 2

Platform for Robot Integrated Design and Practice

2.1 Introduction

Robotics represents a fusion of various technologies including information technology, electronics, computer science, control systems, electrical engineering, mechanical engineering, and bio-cognitive engineering. It employs principles from natural sciences, social sciences, humanities, and related fields to explore intelligent perception, optimal control, and system design of robots, as well as human–robot interaction across diverse domains. This book serves as a resource, offering theoretical insights, design principles, and practical guidance, aimed at nurturing high-caliber, globally oriented professionals. These individuals are envisioned to possess a robust foundation, extensive knowledge, practical skills, and a mindset conducive to innovation, enabling them to contribute effectively to both national and local advancements in the robotics sector.

This book is structured in accordance with the "International Engineering Education Accreditation System" (hereafter referred to as the International Accreditation System), which advocates elevating design, experimentation, and innovative practice to a new standard in talent development. The corresponding roadmap for engineering discipline development aligned with this advanced educational framework is illustrated in Figure 2.1. It outlines a blueprint for engineering specialization, emphasizing the integration of practical experience and innovative endeavors, with experimental and

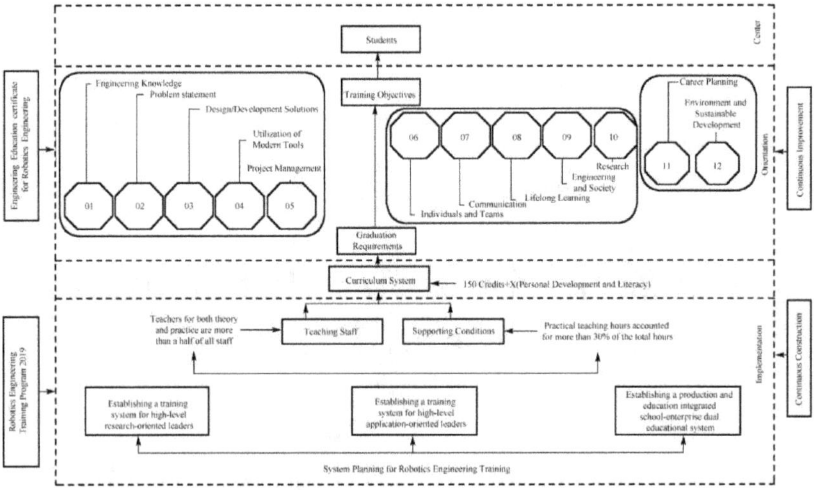

Figure 2.1. Planning for the development of engineering majors.

innovative practice courses constituting over 30% of the total credits. Among these, comprehensive practical courses should make up at least 60% of the experimental and innovative practice curriculum.

Consequently, establishing an integrated innovative practice platform for the Robotics Design and Innovation Practice course becomes imperative.

According to the high standard and cutting-edge requirements, based on the talent cultivation goal of "Solid foundation, broad coverage, rigorous practice, and pursuit of innovation" of the curriculum system for engineering majors proposed by the international accreditation system, the main courses are categorized into several major course modules, and the courses support and connect with each other to highlight the characteristics of the majors and meet the requirements for the cultivation of academic, interdisciplinary, employment, and entrepreneurial talents. In order to meet the requirements of training such talents, and considering the possibility of browser-based synchronous teaching of theory, experiment, and innovative practice integration at home, dormitory, or any other place with a network, the experimental and innovative practice platform shown in Figure 2.2 is designed and constructed.

The platform integrates the equipment group, server group, teacher group, and student group organically through the campus

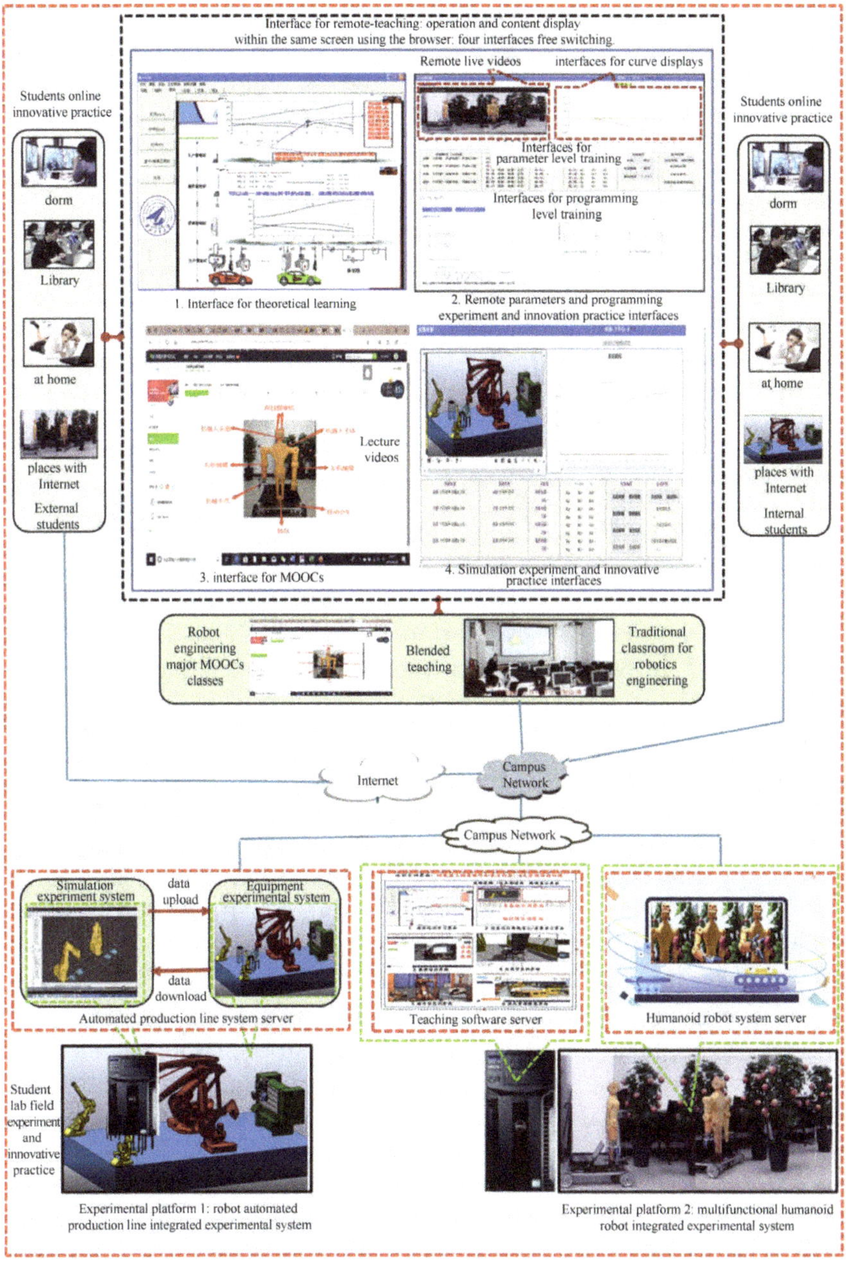

Figure 2.2. Browser-based platform for integrated and synchronized teaching of theory, experiments, and innovative practices.

network and the Internet. Whether it is a college classroom or a catechism classroom, the teacher can complete the synchronous teaching of the integration of theory, experiments, and innovative practice in the classroom, and the students can also carry out synchronous after-school learning of the integration of theory, experiment, and innovative practice in the dormitories, libraries, and other places where there is a network.

Upon accessing the system, users can follow the instructional content to access relevant sections such as "courseware", "video", and "experiment." Within the "courseware" section, they engage with theoretical concepts, while the "video" section enables real-time monitoring of experimental setups. The "experiment" section facilitates the completion of practical tasks aligned with the theoretical learning.

The "Experiment" interface comprises three main components: "Live Video", "Data Collection Display", and "Operational Interface". The "Live Video" component allows for continuous monitoring of laboratory equipment, while the "Data Collection Display" presents sensor-generated data in dynamic visualizations. The "Operational Interface" offers functionalities such as experiment queuing and parameter configuration.

Prior to commencing experiments, users log in, activate the camera, queue up experiments, and proceed if the relevant equipment is available. Upon initiating the default parameters, they execute the "Start Connection" and "Send Data" commands.

During experimentation, users can monitor equipment status via the "Live Video" feed and assess data feedback through the "Data Collection Display". They can seamlessly switch between the "courseware" and "experiment" interfaces, ensuring synchronous integration of theoretical learning, practical experimentation, and innovative application. This seamless transition supports blended learning, both online and offline, fostering a comprehensive understanding of theory and its practical implications. In the "Online Programming" panel, it is also possible to write experimental and innovative practice programs in the underlying language. Through the running state of the equipment and feedback curve display, users can analyze and verify the correctness of the experiment and innovative practice algorithms. At the end of the experiment, the generated data report is analyzed to achieve the purpose of summarizing, correcting errors, and optimizing the control algorithm.

The platform is characterized as follows:

(1) **Innovative:** Organic integration of theory, experiment, and innovative practice; blending of university classroom and catechism classroom; organic integration of teacher teaching and student learning.
(2) **Pedagogical:** Through an integrated virtual teaching platform, teachers can conduct instruction while students autonomously access diverse experimental apparatuses via portable devices, constituting an exclusive pedagogical ecosystem with ubiquitous accessibility. This intelligent platform provides students with comprehensive academic support throughout the entire educational journey from matriculation to graduation.
(3) **Scientific:** The use of "Internet + education technology" connects laboratory equipment, teachers, and students into one, realizing highly efficient teaching for teachers, highly efficient learning for students, and highly efficient use of equipment.
(4) **Enlightenment:** Teachers or students condense theories through experiments, guide experiments through theories, inspire innovative practices through experiments, and enhance the level of theories through innovative practices, realizing the three to complement each other.
(5) **Practicality:** Both teachers and students can conduct synchronous integration of theoretical instruction, experimental operations, and innovative practical activities with web browsers simply using a laptop, enabling unified platform-based teaching and autonomous learning. Experimental devices can be seamlessly integrated into the system with minimal technical requirements.

2.2 Hardware System

To enable robots to mimic human flexibility and undertake tasks in challenging environments, including those that are hazardous, repetitive, or mundane, much of their design and functionality draw inspiration from the human body. This book introduces a novel

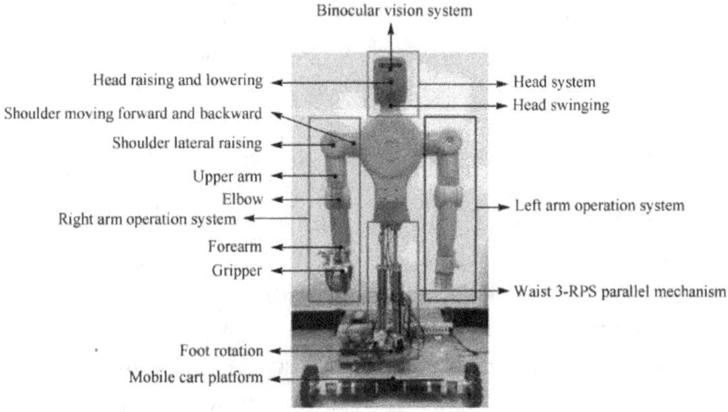

Figure 2.3. Robot hardware.

hybrid-connected humanoid robot platform with 25 degrees of freedom. The upper body of this robot closely resembles human anatomy, encompassing the head, arms, torso, and waist.

Moreover, the robot's upper limbs are linked to a mobile cart via a parallel mechanism and a bottom turntable. The mobile cart facilitates the robot's movement in various directions, while the parallel mechanism enables squatting and standing motions. Utilizing a vision system embedded in the robot's head, it identifies and locates target objects. Furthermore, autonomous navigation, walking, and object manipulation tasks are accomplished through a combination of vehicle-mounted LiDAR and head vision systems.

The robot system is delineated into two main components: hardware and software systems, both of which are extensively detailed in this chapter. The hardware components of the robot platform include features like a binocular vision system, a dual-arm operation system, a waist 3-RPS parallel mechanism, and a mobile cart platform. The overall structure is depicted in Figure 2.3.

As shown in Figure 2.3, the hardware system of each part of the robot is assembled and built as a whole, and coordinated and controlled by the software system to realize the function of autonomous robot operation.

2.2.1 *Binocular vision system*

This book employs the Intel Realsense D435 camera as the visual system for the robot. This camera integrates a wide field of view

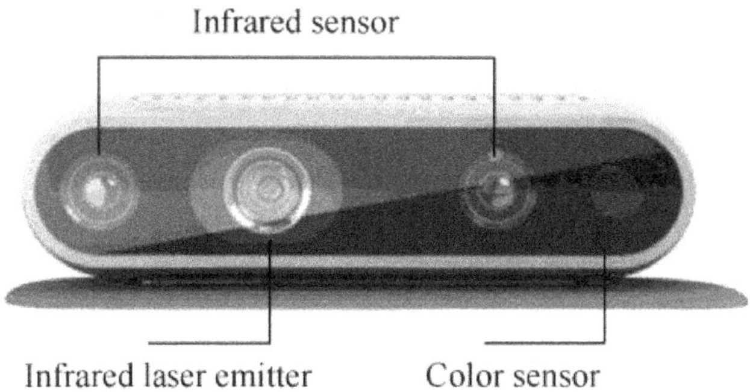

Figure 2.4. Robot vision system.

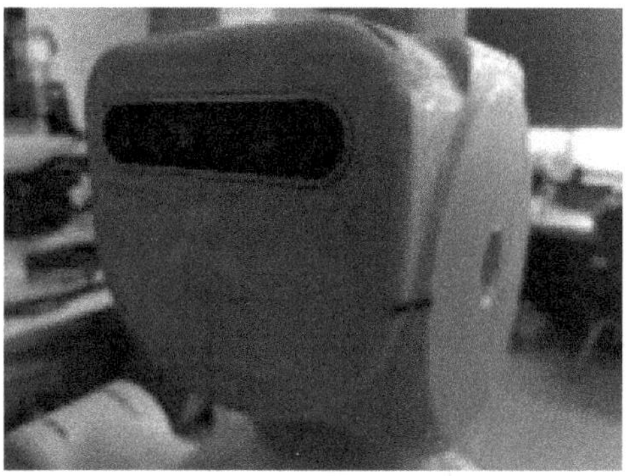

Figure 2.5. Robot head.

and a global shutter sensor, making it suitable for both indoor and outdoor use. Positioned from left to right on the front of the camera are the infrared sensor, infrared laser emitter, infrared sensor, and color sensor, as depicted in Figure 2.4.

The depth-sensing capabilities of this camera are utilized in the book to detect and locate targets, ensuring precision in the grasping experimental process. The camera has a minimum depth distance of approximately 0.105 m and can capture objects up to a distance of 10 m, with accuracy influenced by calibration, environmental factors, and lighting conditions. The binocular camera is compact enough to be mounted on the robot's head and scans in different directions with the movement of the robot's neck in two directions: pitch and rotation, as shown in Figure 2.5.

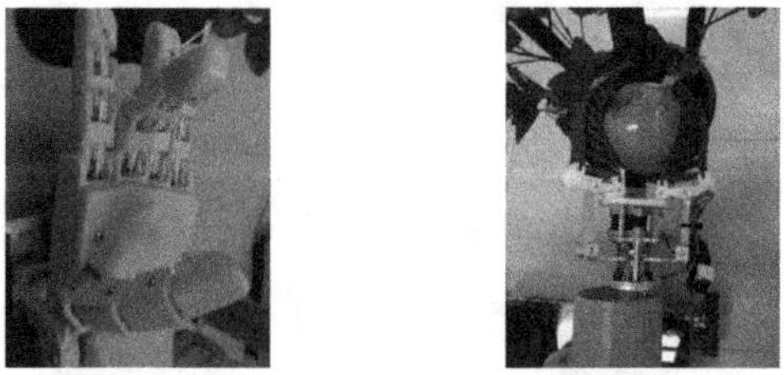

(a) (b)

Figure 2.6. Dual-arm end actuator. (a) Auxiliary boom end-effector and (b) Main boom end-effector.

2.2.2 *Dual-arm operating system*

Modeled after the human body's double-arm structure, the outer layer and support structure of the robot system's double arms are designed in a modular metal structure, and the shell is made by 3D printing using plastic materials with less density and high strength, which can greatly reduce the overall weight of the robot. The main and auxiliary operating arms of the robot platform have five and six degrees of freedom, respectively, and the main and auxiliary operating arms include the shoulder forward and backward rotating joints, shoulder left and right rotating joints, big arm rotating joints, elbow joints, and small arm rotating joints, and thc auxiliary operating arm has a higher number of joints at the wrist compared to the main operating arm. At the same time, in order to meet the operational task requirements of the main operating arm, its end-effector adopts a three-finger flexible robotic claw, and the dual-arm end-effector is shown in Figure 2.6.

The auxiliary arm end-effector has five finger joints: thumb, index finger, middle finger, ring finger, and pinky finger, each of which mimics a human finger and performs the corresponding movements. More importantly, the end-effector can be used to remove obstacles that block the target object and assist the main operating arm in grasping the target object. The end-effector of the main operating arm is a three-finger flexible robotic gripper, which is made of flexible material to ensure that the epidermis of the target is protected from damage.

The robotic platform's dual-arm system boasts a total of 16 degrees of freedom. Achieving human-like flexibility requires precise management of the robot's joints. This book proposes the utilization of a serial bus intelligent servo for this purpose. This servo employs an ARM 32-bit microcontroller as its primary control unit. Position sensing is accomplished through a 360-degree, 12-bit precision magnetic induction angle program. Communication is facilitated via RS485, known for its robust resistance to interference. Communication occurs in a serial asynchronous mode, with a question-and-answer exchange between the controller and the servo. Within a bus control network, multiple servos can operate, each assigned a unique ID. Control commands transmitted by the controller contain ID information, ensuring that only servos with matching IDs execute the commands and return responses accordingly. The drive servo and servo control board are shown in Figures 2.7 and 2.8, respectively.

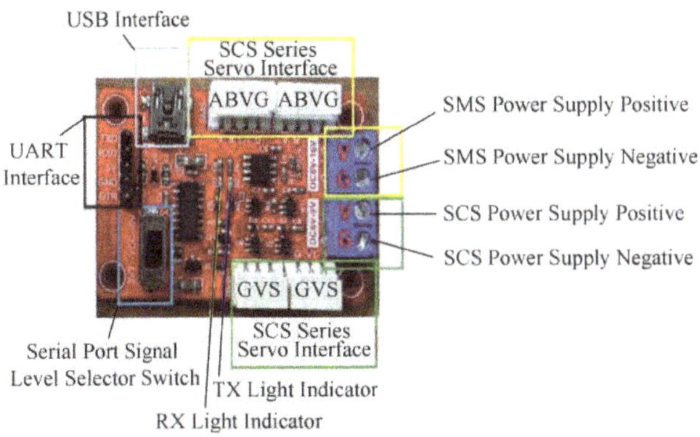

Figure 2.7. Driving servo.

Figure 2.8. Servo control board.

2.2.3 *Waist 3-RPS parallel mechanism*

Generally, when the upper body of a humanoid robot is heavy, a high-strength, high-stiffness, and high-carrying-capacity mechanism is required if a conventional tandem structure is used to connect it to the lower body. This is because the tandem mechanism is a cantilever beam structure, which is structurally unstable, has low stiffness, and has low load carrying capacity. However, compared to the traditional series mechanism, the parallel mechanism has the advantages of structural stability, high stiffness, higher load carrying capacity, high output accuracy, and low motion load. Therefore, this book adopts the 3-RPS parallel mechanism as the "waist" of the robot, connecting the upper body of the humanoid robot to the mobile cart, as shown in Figure 2.9.

The 3-RPS parallel mechanism consists of a static platform, a moving platform, and three tandem branches. The bottom of the tandem branch is fixed to the static stage, and the top of the tandem branch is connected to the moving stage through a ball–hinge joint. At the same time, the tandem branches are designed as a ball screw nut structure with 120° symmetric distribution and are driven and controlled by a digital servo. The robotic platform, employing its symmetrical structural design, enables the end-effector moving platform of the parallel mechanism to be considered as a

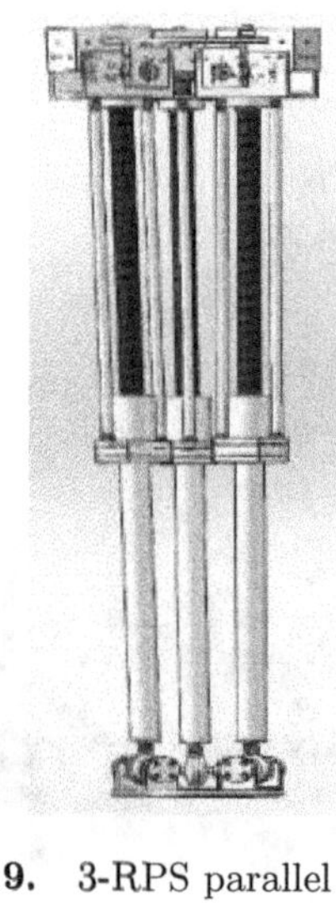

Figure 2.9. 3-RPS parallel mechanism.

3-degree-of-freedom (3-DOF) spatial motion mechanism with two rotational and one translational degree of freedom. In other words, the 3-RPS parallel mechanism allows the robot to play the role of rotation and movement in its joint space, thus realizing the forward and backward bending, left and right bending, and lifting and lowering functions of the robot. This waist 3-RPS parallel mechanism not only increases the workspace of the hybrid humanoid robot but also improves the stiffness and load-bearing capacity of the robot's waist.

2.2.4 *Mobile trolley platform*

To guarantee the adaptability, dependability, and precision of the entire robotic system, we opted for a four-wheeled mobile cart utilizing McNamee wheels as the foundation for the robot's movement. The McNamee wheel comprises a hub and a series of drum rollers evenly arranged around it, forming the core of the support mechanism. Typically, the hub axis and roller axis are set at a specific angle, commonly 45 degrees, as illustrated in Figure 2.10.

In the direction of non-active motion, the small rollers are free to rotate around their own axis, enabling omnidirectional movement of the robot. The free rotation of the passive small rollers provides the individual McNamee wheels with three degrees of freedom independent of each other. McNamee wheel-based mobile robots usually consist of four McNamee wheels divided into left and right rotations, each equipped with a motor and an encoder, allowing both synergistic

Figure 2.10. Mobile cart platform.

and independent control between the four wheel sets. Compared with other types of mobile carts, the McNamee wheel-based mobile cart has the following characteristics: (1) lightweight chassis, high structural strength, and high load-bearing capacity; (2) motion control is convenient and simple, and the control system is easy to realize; (3) the wheel mechanism is simple in design, and the wheelset has a certain width, which makes it smoother when moving. When moving the mobile cart in any required direction, the force of each wheel will eventually be superimposed to synthesize a combined force vector, so that the mobile cart is free to move without affecting the inherent direction of the wheels. Based on the motion characteristics of the four McNamee wheels, the robot platform accomplishes forward, backward, left, right, diagonal, and spinning motion patterns.

2.3 Software System

After constructing the robot hardware system, the subsequent step involves developing the software system to manage its operations. However, allocating significant hardware resources to build the robot and then writing numerous underlying drivers can substantially escalate the workload and development expenses. The Robot Operating System (ROS), an open-source platform introduced by Willow Garage, facilitates the implementation of various software functionalities such as drawing, localization, sensing, remote control, simulation, and action planning. By leveraging ROS, users can easily control robots by selecting appropriate functions from the available software packages. Over the years, ROS has undergone continuous enhancement, and developers have dedicated considerable effort to conducting in-depth research within related fields. Its compatibility and extensive capabilities make it particularly suitable for fostering innovative practical training among students, hence the platform's preference for ROS as the operating system.

2.3.1 *Robot model*

The ROS package for robot modeling is the URDF package, which contains a C++ parser for a uniform robot description format. In the URDF package, a file with the extension .urdf must be created to hold the connection relationships between the robot links and joints. In this case, the link tag denotes the robot's link and the joint

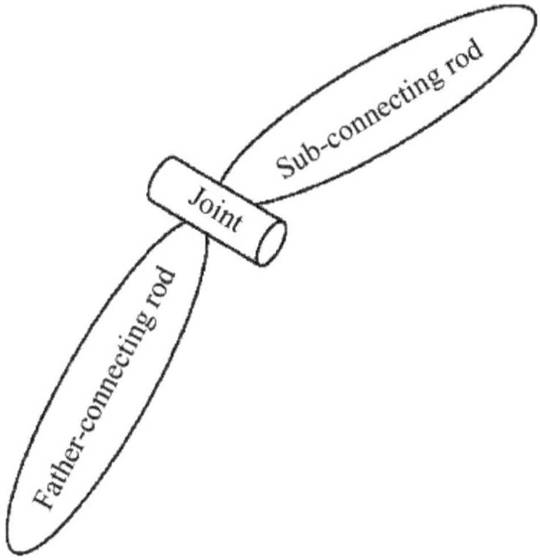

Figure 2.11. Joints and links.

tag denotes the joint. The two links are connected through the joints, and the former link is referred to as the parent link and the latter link is referred to as the child link, as shown in Figure 2.11.

In this book, the robot model in SolidWorks is exported as a URDF model file, and it is configured in the RVIZ simulation plug-in so that the simulation model exactly resembles the real robot by configuring its parameters and other operations, as shown in Figure 2.12.

In the RVIZ simulation plugin, debugging was performed on the robot, and the Setup Assistant was applied to configure the planning groups of the robot. Simultaneously, a communication interface between the "MoveIt!" module and the physical robot was established to validate the correctness of the robot model. In this model, to ensure proper motion of the waist parallel mechanism, it was equivalently modeled as a spherical joint in the URDF (Unified Robot Description Format) file. This representation enables simultaneous forward-backward and lateral bending capabilities while maintaining kinematic consistency.

When a set of angle values $(-100°, -10°, 30°, -60°, 24°, 0°)$ is randomly given to the joints of the robot's auxiliary operating arm, the motion states of the simulation model and the real robot are as shown in Figure 2.13.

Ine the RVIZ simulation plug-in, the randomly given end-effector target point is planned, and through the inverse kinematics

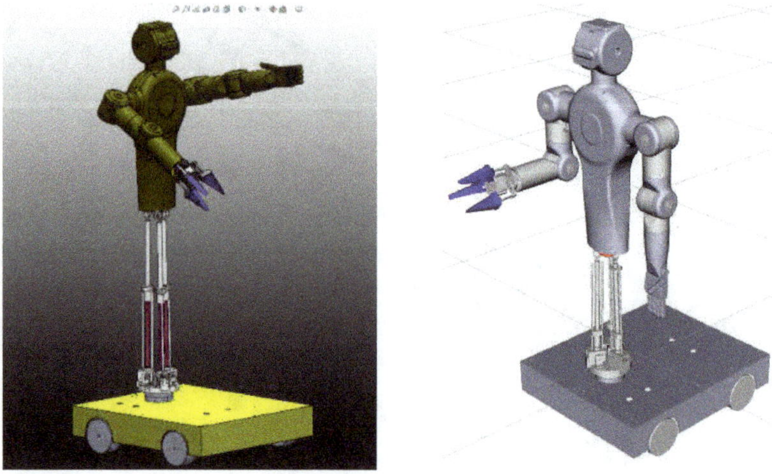

(a) (b)

Figure 2.12. Robot model: (a) SolidWorks and (b) URDF models.

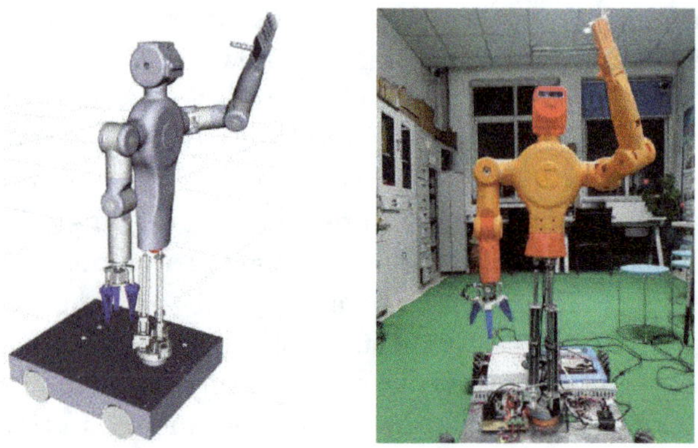

(a) (b)

Figure 2.13. Motion of the simulation model and the real robot: (a) motion state of the simulation model and (b) motion state of the real robot.

algorithm, the operation arm end-effector in the simulation model can be made to arrive at the target point and communicate with the bottom layer to drive the real operation arm to move toward the target point. When the target position and orientation of the main operating arm end-effector are known, planning is performed using the main operating arm planning group to obtain the values

(a) (b)

Figure 2.14. Inverse kinematics test: (a) simulation model and (b) real robot.

of each joint angle of the main operating arm and send them down to the bottom layer to drive the operating arm motion, as shown in Figure 2.14.

2.3.2 *Robot communication*

ROS packages typically have multiple executables working as processes, also known as nodes. Due to the complexity of the robot's functional modules, it is not usual to centralize all the functions in a single node; instead, a distributed approach is used, where different modules are placed in different nodes, which in turn communicate with each other using topics, services, and actions. The communication between some of the nodes in the control program used in this book is shown in Figure 2.15.

Rational provisioning between nodes is managed through the node manager (Master), and nodes are registered with the node manager before they are incorporated into the overall ROS program. Communication between nodes is also paired by the node manager in order to realize point-to-point communication. The communication methods of ROS mainly include the following.

Topic: A topic is a unidirectional peer-to-peer communication method. After establishing a contact once, a topic can be published by multiple nodes at the same time or subscribed to by multiple nodes at the same time. Each topic has a unique name, and any

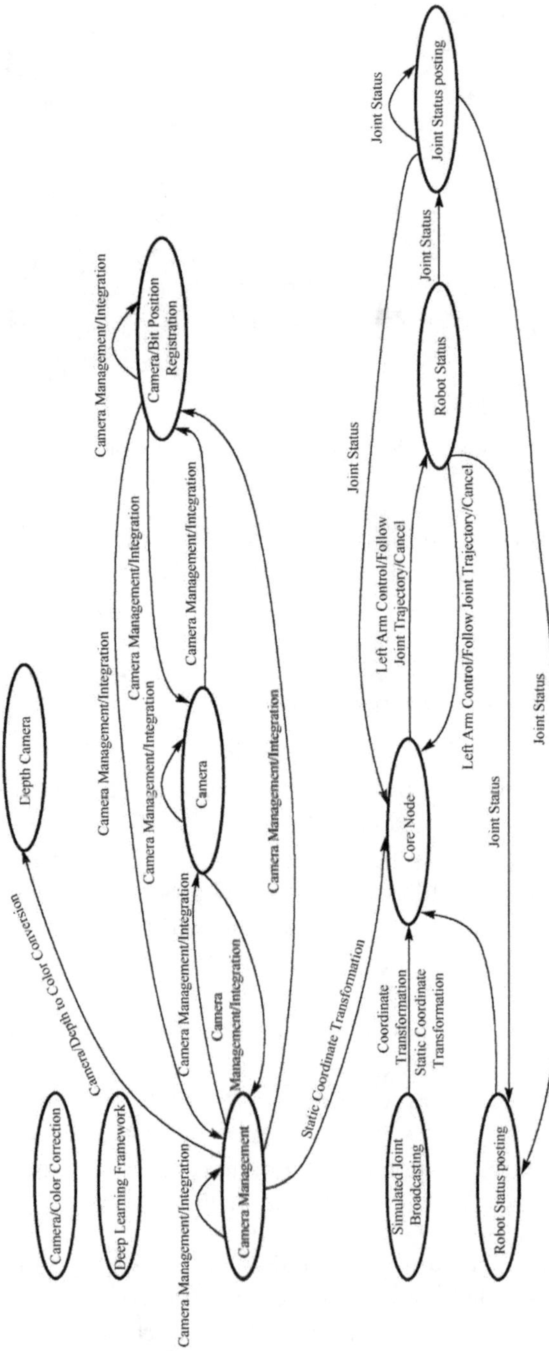

Figure 2.15. Communication between some nodes.

node can access this topic and send data through it. The data communicated between nodes has a certain format standard; this data format contains the message (Message). Usually, the messages that need to be published are placed in a folder named msg, with the file extension .msg.

Service: The communication of services is bidirectional, enabling not only message transmission but also feedback, thereby forming a request/response interaction pattern. This paradigm specifies that one node requests another node to execute a particular task. Similar to ROS message files (msg files), the data format for service communication is defined within the srv directory of functional packages, utilizing the .srv file extension. Each service definition comprises two distinct components, i.e., a request structure and a corresponding response structure.

Action: The communication method of action is similar to service communication and, at the same time, makes up for the shortcomings of service communication. When the robot performs a long-duration task, the requesting party of service communication will not receive feedback for a long time, and the communication process will be hindered. Here, action communication can check the progress at any time and can also terminate the request. Action messages also have a certain format standard, which is placed in a file with the extension .action in the action folder of the function package.

The communication method is shown in Figure 2.16.

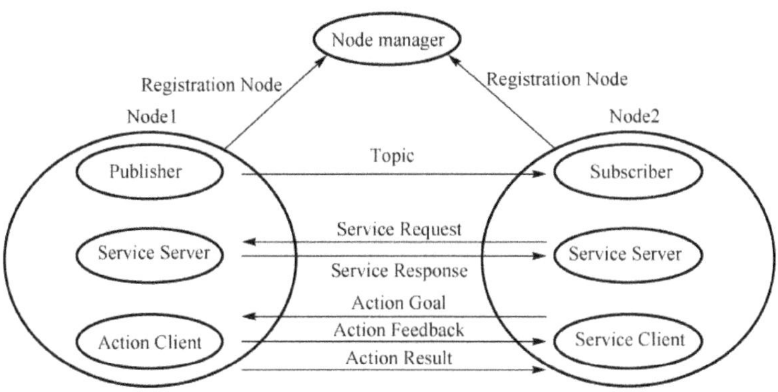

Figure 2.16. Communication method.

2.4 Summary

This chapter primarily introduces the hardware and software components of the complete robot platform. The hardware system encompasses elements such as the binocular vision system, dual-arm operation system, waist 3-RPS parallel mechanism, and mobile cart platform. On the other hand, the software system employs the ROS, where control programs are written based on algorithms to manage the robot and execute predetermined functions. During the operation of the robot platform, the hardware and software systems work in tandem. The software system governs the hardware system, with the host computer orchestrating the robot's operational movements based on the acquired information about external target objects. Ultimately, this integrated approach ensures the successful execution of the robot's tasks.

2.5 Exercises

2.1 Discuss the components of the waist 3-RPS parallel mechanism.
2.2 Discuss the characteristics of McNamee wheel-based mobile carts.
2.3 Discuss the function and performance of binocular vision systems.
2.4 Discuss the features and performance of ROS.
2.5 Explain what is the meaning of degrees of freedom?

Chapter 3

Design and Practice for Robot Forward Kinematics

3.1 Theory of Robot Forward Kinematics

Analyzing the kinematics of robots involves setting up a coordinate system. While a general method can be used to establish the robot's coordinate system, there is no consistent pattern in the positioning and orientation relationship between the coordinate systems of each link. This lack of regularity complicates the transformation matrix between the links and makes calculating the position and orientation matrix of the entire robot burdensome.

In 1956, Denavit and Hartenberg (DH) introduced a method that addressed this issue. Their method precisely defines the coordinate axes of each system, thereby simplifying the positional and orientational relationships between them. Consequently, the transformation matrix method[7] between the links becomes simpler, streamlining the analysis and computation of the robot's motion. This method comprises two variations: the standard-DH coordinate system establishment method and the later evolved modified-DH coordinate system establishment method, both of which are detailed in the following.

3.1.1 *Standard-DH coordinate system for robots*

3.1.1.1 *Serial number calibration of the standard-DH coordinate system*

The links of the robot are connected together by joints, and the joints are divided into moving and rotating parts. The serial numbers of the standard-DH coordinate system are assigned sequentially from the machine base to the end-effector, and the numbers of the links, joints, and coordinate system are shown in Figure 3.1.

Link numbering: The seat is link 0, and so on, corresponding to links $1, 2, \ldots, n$.

Joint numbering: Between the seat and the link 1 is joint 1, and so on, corresponding to joints $2, 3, \ldots, n$.

Coordinate system numbering: The coordinate system of the link i moves with the joint i together with the link i. We establish the coordinate system of the link i on the joint $i + 1$ of its end-effector, i.e., the coordinate system of the link 0 (base coordinate system) is established on joint 1, the coordinate system of the link 1 is established on joint 2, and so on. The end-effector link n coordinate system is established on the hand.

For rotating joints, the z axis direction of each link coordinate system coincides with the joint axis; for moving joints, the z axis direction is the direction of movement of the joint.

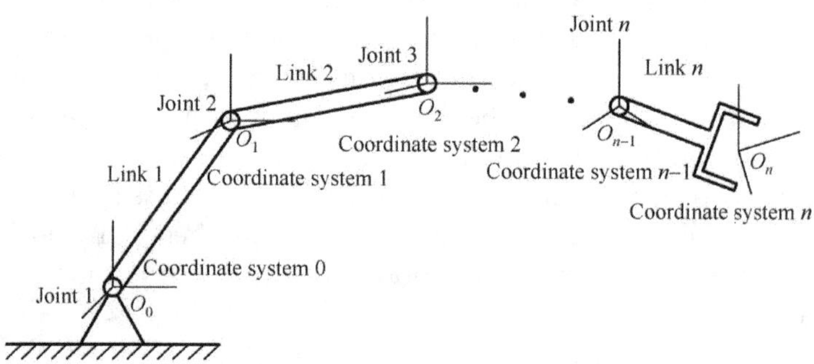

Figure 3.1. Assignment of the standard-DH coordinate system.

3.1.1.2 *Establishment of robot coordinate system*

For the rotating joints, the standard-DH coordinate system is shown in Figure 3.2 and described as follows:

(1) The i coordinate system of the link is established on the $i+1$ joint with the z_i axis on the axis of the $i+1$ joint and the z_i axis pointing as specified.
(2) The x_i axis of the coordinate system of the link i is located on the common perpendicular of the joint axes of the ends of the link i, and the direction points to the link $i+1$. If the axes of the joints i and $i+1$ intersect, it is specified that the x_i axis passes through the point of intersection of the two axes and is perpendicular to the plane in which they are located, and that the direction of the x_i axis is specified by itself.
(3) The origin of the coordinate system of the link i is the intersection of the x_i and z_i axes, and the y_i axis is determined by the x_i and z_i axes according to the right-hand rule.

The rotation joint of the standard-DH coordinate system can be described by four parameters, as shown in Figure 3.2, and the specific meanings are shown in Table 3.1.

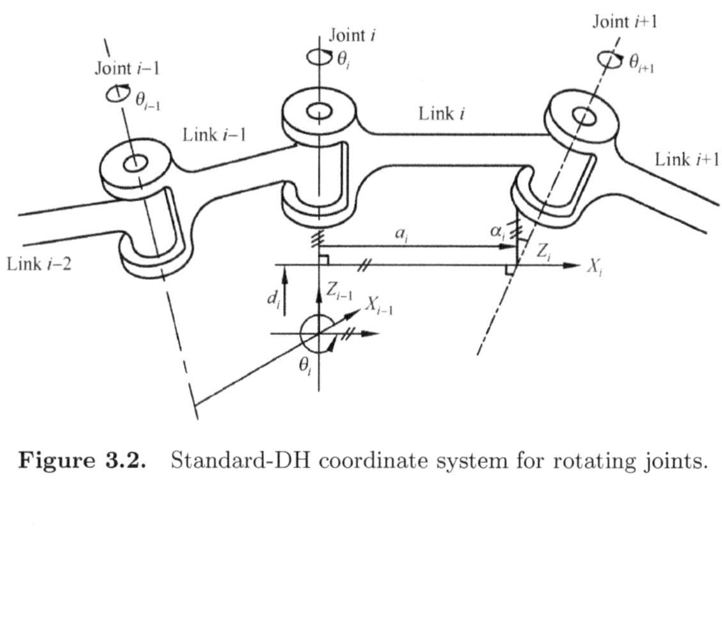

Figure 3.2. Standard-DH coordinate system for rotating joints.

Table 3.1. Meanings of the standard-DH parameters for links i of the coordinate system (joints i are rotating joints).

Hallmark	Parameters describing the relationship between two neighboring links		Parameters describing the link	
Parameters	Angle θ_i of the two links	Distance d_i between two links	Length a_i of link	Torsion angle α of link
Definition	Angle between two plumb lines in the plane perpendicular to the axis of joint i (z_{i-1} axis) (angle of rotation around the z_{i-1} axis from the x_{i-1} axis to the x_i axis)	Distance between two plumb lines along the axis of joint i (z_{i-1} axis) (displacement along the z_{i-1} axis moving from the x_{i-1} axis to the x_i axis)	Length of the common plumb line of the joint axis at each end of the link i (displacement along the x_i axis moving from the z_{i-1} axis to the z_i axis)	The angle between the axes of the two joints in the plane perpendicular to the x_i axis (the angle of rotation around the x_i axis from the z_{i-1} axis to the z_i axis)
Characterization	Joint variable	Constant	Constant	Constant

Notes: (1) Due to the two options for the pointing of the z_i axis and, in the case of intersection of the two joint axes (in this case, $a_i = 0$), the pointing of the x_i axis, the standard-DH coordinate system is not created uniquely.
(2) If the joint axis of the two ends of the link i is parallel, because of the existence of multiple values of the common perpendicular of its parallel axes, the common perpendicular alone cannot be used to determine the origin of the coordinate system of the link i; the value of d_{i+1} also needs to be taken into account in order to determine the location of the origin, choose the origin as usual, and make it meet $d_{i+1} = 0$.

3.1.2 *Modified-DH coordinate system for robots*

3.1.2.1 *Serial number calibration of the modified-DH coordinate system*

The standard-DH method fixes the coordinate system of the link i on the joint $i+1$, i.e., the coordinate system is established at the end of

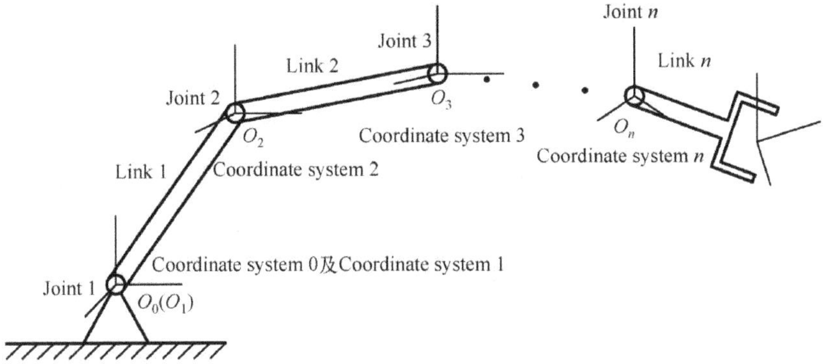

Figure 3.3. Assignment of the modified-DH coordinate system.

the link i, while the modified-DH method fixes the coordinate system of the link i on the joint i, i.e., the coordinate system is established at the beginning of the link i. Therefore, according to the order from the base to the end of the actuator, the modified-DH method is the same as the standard-DH method in that the numbers of each link and each joint remain the same, but the difference is that the number of each coordinate system is changed. As shown in Figure 3.3, the coordinate systems of link i and link i move together with the joint i, and we establish the coordinate system of link i on the joint i at its beginning. In this way, the coordinate systems of link 0 (base coordinate system) and link 1 are established on joint 1, link 2 on joint 2, and so on, and the coordinate system of the end-effector link n is established on joint n. Similarly, for rotating joints, the z axis direction of each link coordinate system coincides with the joint axis; for moving joints, the z axis direction is the direction of movement of the joint.

3.1.2.2 *Establishment of the modified-DH coordinate system for robots*

The rotating joint in the modified-DH coordinate system is shown in Figure 3.4.

(1) The coordinate system of the link i is established on the joint i, whose z_i axis is located on the axis of the joint i, and the direction of the z_i axis is defined.

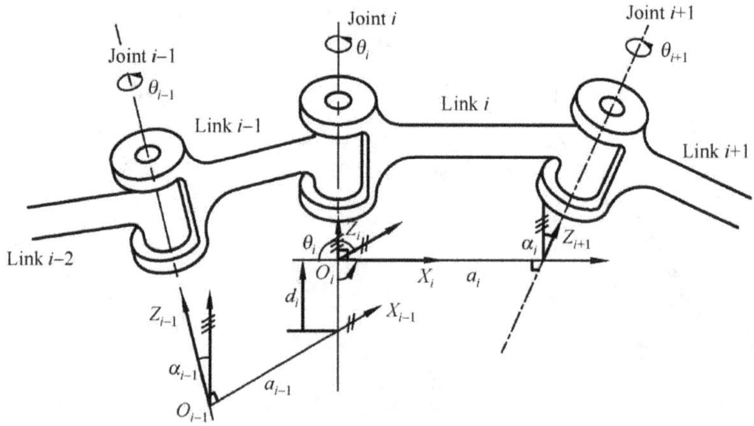

Figure 3.4. Modified-DH coordinate system for rotating joints.

(2) The x_i axis of the coordinate system of the link i is located on the common perpendicular of the joint axes at the ends of the link i and points in the direction of the link $i+1$; if the joint axes i and $i+1$ intersect, the x_i axis is specified to be perpendicular to the plane in which the joints i and $i+1$ are located, and points in the direction of your own specification.

(3) The origin of the coordinate system of the link i is the intersection of the x_i and z_i axes, and the y_i axis is determined by the x_i and z_i axes according to the right-hand rule.

(4) The coordinate systems $\{0\}$ and $\{1\}$ are established on joint 1, but the coordinate system $\{0\}$ is a fixed coordinate system (base coordinate system) while the coordinate system $\{1\}$ is a moving coordinate system, which stipulates that when the joint variable of the moving coordinate system $\{1\}$ is 0, the origin and orientation of the coordinate systems $\{0\}$ and $\{1\}$ coincide exactly.

(5) The link n coordinate system $\{n\}$ is established on the axis of the joint n. The specific position and orientation of the origin on the axis of the joint and the direction of the axis x_n can be chosen arbitrarily, and the parameters of the link are set to zero whenever possible when choosing. For example, the position and orientation of the origin of the coordinate system $\{n\}$ can be chosen so that $d_n = 0$.

(6) The robot hand is in the coordinate system $\{n\}$, and the center of the hand is at a position that is not the origin of the coordinate system $\{n\}$.

The rotational joint in the modified-DH coordinate system can also be described by four parameters, as shown in Figure 3.4, with specific meanings as shown in Table 3.2.

Table 3.2. Meanings of the modified-DH parameters for the coordinate system i of the link (joint i is a rotating joint).

Hallmark	Parameters describing the relationship between two neighboring links		Parameters describing the link	
Parameters	Angle θ_i of the two links	Distance d_i between two links	Link length a_i	Torsion angle α_i of link
Definition	Angle between two plumb lines in the plane perpendicular to the axis of joint i (z_i axis) (angle of rotation around the z_i axis from the x_{i-1} axis to the x_i axis)	Distance between two plumb lines along the axis of joint i (z_i axis) (displacement along the z_i axis moving from the x_{i-1} axis to the x_i axis)	Length of the common plumb line of the joint axis at each end of the link i (displacement along the x_i axis moving from the z_i axis to the z_{i+1} axis)	The angle between the two articulated axes z_i and z_{i+1} in a plane perpendicular to the x_i axis, (the angle of rotation around the x_i axis from the z_i axis to the z_{i+1} axis)
Characterization	Joint variable	Constant	Constant	Constant

Notes: (1) Due to the fact that there are two options for the pointing of the z_i axis, as well as two options for the pointing of the x_i axis in the case of intersection of the two joint axes (in this case, $a_i = 0$), the modified-DH coordinate system is not uniquely established.
(2) If the joint axes of the two ends of the link i are parallel, because of the existence of multiple values of the common perpendicular of its parallel axes, the common perpendicular alone cannot be used to determine the origin of the coordinate system of the link i. Usually. the origin is selected to satisfy $d_i = 0$.

3.1.3 *Coordinate pose expression of two adjacent links*

The position and orientation matrix of the robot link i coordinate system with respect to the link $i-1$ coordinate system is the coordinate transformation matrix of the link i coordinate system with respect to the link $i-1$ coordinate system, denoted by $\boldsymbol{A}_i^{i-1}$. For a robot with n joints, the position and orientational relationship between two neighboring link coordinate systems is expressed as follows:

$$\boldsymbol{A}_n^{n-1}, \ldots, \boldsymbol{A}_i^{i-1}, \ldots, \boldsymbol{A}_1^0 (i = 1, 2, \ldots, n)$$

The superscript is usually omitted and expressed as $\boldsymbol{A}_i$, i.e.,

$$\boldsymbol{A}_n, \ldots, \boldsymbol{A}_i, \ldots, \boldsymbol{A}_1 (i = 1, 2, \ldots, n)$$

Also, we often refer to $\boldsymbol{A}_i^{i-1}$ ($\boldsymbol{A}_i$) as $i^{i-1}\boldsymbol{A}$.

3.1.4 *Coordinate pose determination of two adjacent links*

3.1.4.1 *Standard-DH coordinate system position and orientation matrix*

As shown in Figures 3.1 and 3.2, after determining all the links in accordance with the standard-DH coordinate system establishment method, proceed with the following steps to establish the transformation relationship between the coordinate systems of $i-1$ and i for two adjacent links:

(1) Rotate the θ_i angle around the z_{i-1} axis so that the x_{i-1} axis is in the same direction as the x_i axis (both axes are parallel and in the same direction).
(2) Translate the distance d_i along the z_{i-1} axis so that the x_{i-1} axis is colinear and in the same direction as the x_i axis.
(3) Translate along the x_{i-1} axis (or the x_i axis line) at this moment by a distance of a_i so that the origin of the coordinate system of the link $i-1$ coincides with the origin of the coordinate system of the link i, the x_{i-1} axis coincides with the x_i axis, and the z_{i-1} axis is coplanar with the z_i axis.

(4) Rotate the α_i angle around the x_{i-1} axis (or the x_i axis) at this moment so that the z_{i-1} axis coincides with the z_i axis.

After the above-mentioned transformations, the coordinate system of the link $i-1$ coincides with the coordinate system of the link i. According to these transformation steps, using the left-multiplication rule, the coordinate transformation matrix $\boldsymbol{A}_i$ of the link i coordinate system with respect to the link $i-1$ coordinate system (i.e., the position and orientation matrix of the link i coordinate system with respect to the link $i-1$ coordinate system) in the standard-DH coordinate system can be derived as follows:

$$
\begin{aligned}
\boldsymbol{A}_i &= \mathbf{Rot}(z_{i-1},\theta_i)\mathbf{Trans}(0,0,d_i)\mathbf{Trans}(a_i,0,0)\mathbf{Rot}(x_i,\alpha_i)\\
&= \begin{bmatrix} c\theta_i & -s\theta_i & 0 & 0\\ s\theta_i & c\theta_i & 0 & 0\\ 0 & 0 & 1 & 0\\ 0 & 0 & 0 & 1 \end{bmatrix}
\begin{bmatrix} 1 & 0 & 0 & 0\\ 0 & 1 & 0 & 0\\ 0 & 0 & 1 & d_i\\ 0 & 0 & 0 & 1 \end{bmatrix}
\begin{bmatrix} 1 & 0 & 0 & a_i\\ 0 & 1 & 0 & 0\\ 0 & 0 & 1 & 0\\ 0 & 0 & 0 & 1 \end{bmatrix}\\
&\quad\times \begin{bmatrix} 1 & 0 & 0 & 0\\ 0 & c\alpha_i & -s\alpha_i & 0\\ 0 & s\alpha_i & c\alpha_i & 0\\ 0 & 0 & 0 & 1 \end{bmatrix}\\
&= \begin{bmatrix} c\theta_i & -s\theta_i c\alpha_i & s\theta_i s\alpha_i & a_i c\theta_i\\ s\theta_i & c\theta_i c\alpha_i & -c\theta_i s\alpha_i & a_i s\theta_i\\ 0 & s\alpha_i & c\alpha_i & d_i\\ 0 & 0 & 0 & 1 \end{bmatrix}
\end{aligned}
\tag{3.1}
$$

The following points should be noted when using Eq. (3.1):

(1) When the joint i is a rotary joint, θ_i in Eq. (3.1) is a joint variable, and α_i, d_i, and a_i are constants.
(2) When the joint i is a moving joint, d_i in Eq. (3.1) is a joint variable, and α_i, θ_i, and a_i are constants.
(3) For the moving joint i, if the coordinate system is established using the method shown in Figure 3.2, let the length of the link

$a_i = 0$, so the coordinate transformation matrix $\boldsymbol{A}_i$ of Eq. (3.1) becomes

$$\boldsymbol{A}_i = \mathbf{Rot}(z_{i-1}, \theta_i)\mathbf{Trans}(0, 0, d_i)\mathbf{Trans}(a_i, 0, 0)\mathbf{Rot}(x_i, \alpha_i)$$
$$= \begin{bmatrix} c\theta_i & -s\theta_i c\alpha_i & s\theta_i s\alpha_i & 0 \\ s\theta_i & c\theta_i c\alpha_i & -c\theta_i s\alpha_i & 0 \\ 0 & s\alpha_i & c\alpha_i & d_i \\ 0 & 0 & 0 & 1 \end{bmatrix} \tag{3.2}$$

3.1.4.2 *Modified-DH coordinate system position and orientation matrix*

As shown in Figures 3.3 and 3.4, all the links are determined according to the modified-DH coordinate system establishment method, and the following steps are carried out to establish the transformation relationship between the coordinate systems of two neighboring links $i-1$ and i:

(1) Rotate the α_{i-1} angle around the x_{i-1} axis so that the z_{i-1} axis is in the same direction as the z_i axis (both axes are parallel and in the same direction).
(2) Translate the distance a_{i-1} along the x_{i-1} axis so that the z_{i-1} axis is colinear and in the same direction as the z_i axis.
(3) Translate along the z_{i-1} axis (or the z_i axis line) at this moment by a distance of d_i so that the origin of the coordinate system of the link $i-1$ coincides with the origin of the coordinate system of the link i, the z_{i-1} axis coincides with the z_i axis, and the x_{i-1} axis is coplanar with the x_i axis.
(4) Rotate the θ_i angle around the z_{i-1} axis (or the z_i axis) at this moment so that the x_{i-1} axis coincides with the x_i axis.

After the above-mentioned transformations, the coordinate system of the link $i-1$ coincides with the coordinate system of the link i. According to these transformation steps, the transformation matrix of the link i coordinate system with respect to the link $i-1$ coordinate system $\boldsymbol{A}_i$ (i.e., the position and orientation matrix of the link i coordinate system with respect to the link $i-1$ coordinate system) can be deduced by applying the rule of left-multiplication in the modified-DH coordinate system, namely

$$
\begin{aligned}
\boldsymbol{A}_i &= \mathbf{Rot}(x_{i-1}, \alpha_{i-1})\mathbf{Trans}(a_{i-1}, 0, 0)\mathbf{Trans}(0, 0, d_i)\mathbf{Rot}(z_i, \theta_i) \\
&= \begin{bmatrix} 1 & 0 & 0 & 0 \\ 0 & c\alpha_{i-1} & -s\alpha_{i-1} & 0 \\ 0 & s\alpha_{i-1} & c\alpha_{i-1} & 0 \\ 0 & 0 & 0 & 1 \end{bmatrix} \begin{bmatrix} 1 & 0 & 0 & a_{i-1} \\ 0 & 1 & 0 & 0 \\ 0 & 0 & 1 & 0 \\ 0 & 0 & 0 & 1 \end{bmatrix} \begin{bmatrix} 1 & 0 & 0 & 0 \\ 0 & 1 & 0 & 0 \\ 0 & 0 & 1 & d_i \\ 0 & 0 & 0 & 1 \end{bmatrix} \\
&\quad \times \begin{bmatrix} c\theta_i & -s\theta_i & 0 & 0 \\ s\theta_i & c\theta_i & 0 & 0 \\ 0 & 0 & 1 & 0 \\ 0 & 0 & 0 & 1 \end{bmatrix} \\
&= \begin{bmatrix} c\theta_i & -s\theta_i & 0 & a_{i-1} \\ s\theta_i c\alpha_{i-1} & c\theta_i c\alpha_{i-1} & -s\alpha_{i-1} & -d_i s\alpha_{i-1} \\ s\theta_i s\alpha_{i-1} & c\theta_i s\alpha_{i-1} & c\alpha_{i-1} & d_i c\alpha_{i-1} \\ 0 & 0 & 0 & 1 \end{bmatrix}
\end{aligned}
\tag{3.3}
$$

The following points should be noted when using Eq. (3.3):

(1) In the position and orientation matrix in Eq. (3.3), there are parameters α_{i-1}, a_{i-1} for link $i-1$, and θ_i and d_i for link i.
(2) When the joint i is a rotation joint, θ_i in Eq. (3.3) is a joint variable, and α_{i-1}, d_i, and a_{i-1} are constants.
(3) When the joint i is a moving joint, d_i in Eq. (3.3) is a joint variable, and α_{i-1}, θ_i, and a_{i-1} are constants.

3.1.5 *Robot forward kinematics*

From Section 3.1.4, we set the position and orientation matrix of the latter link coordinate system with respect to the former link coordinate system $\boldsymbol{A}_i$; then working forward from the end-effector link coordinate system $\{n\}$, we can get the following: the position and orientation matrix of the nth coordinate system with respect to the $n-1$th coordinate system$\boldsymbol{A}_n$,... the position and orientation matrix of the link 2 coordinate system with respect to the link 1 coordinate system $\boldsymbol{A}_2$, and the position and orientation matrix of the link 1 coordinate system with respect to the base coordinate system $\{0\}$ $\boldsymbol{A}_1$. Assuming that there is a point $\boldsymbol{P}$ in the end-effector link coordinate system $\{n\}$ whose position in the coordinate system $\{n\}$ is ${}^n\boldsymbol{P} = [P_x \ P_y \ P_z \ 1]$, then its position in the base coordinate system ${}^0\boldsymbol{P}$ can be found using the right-multiplication rule, i.e.,

$$ {}^0\boldsymbol{P} = \boldsymbol{A}_1\boldsymbol{A}_2\boldsymbol{A}_3\cdots\boldsymbol{A}_{n-1}\boldsymbol{A}_n \cdot {}^n\boldsymbol{P} $$

Therefore, the position and orientation matrix of the end-effector link coordinate system $\{n\}$ with respect to the base coordinate system is

$$ {}_n^0\boldsymbol{T} = \boldsymbol{A}_1\boldsymbol{A}_2\boldsymbol{A}_3\cdots\boldsymbol{A}_{n-1}\boldsymbol{A}_n = \begin{bmatrix} n_x & o_x & a_x & p_x \\ n_y & o_y & a_y & p_y \\ n_z & o_z & a_z & p_z \\ 0 & 0 & 0 & 1 \end{bmatrix} \tag{3.4} $$

Equation (3.4) is the equation of motion of the robot, and ${}_n^0\boldsymbol{T}$ in Eq. (3.4) is also often expressed as $\boldsymbol{T}_n^0$ or ${}^0\boldsymbol{T}_n$.

It is further deduced that the position and orientation matrix of the robot end-effector link coordinate system $\{n\}$ with respect to the link $i-1$ coordinate system is

$$ {}_6^{i-1}\boldsymbol{T} = \boldsymbol{A}_i\boldsymbol{A}_{i+1}\cdots\boldsymbol{A}_6 \tag{3.5} $$

For a six-link robot, the position and orientation matrix of its end-effector with respect to the base coordinate system can then be expressed as

$$ {}_6^0\boldsymbol{T} = \boldsymbol{A}_1\boldsymbol{A}_2\cdots\boldsymbol{A}_6 \tag{3.6} $$

where ${}_6^0\boldsymbol{T}$ is often abbreviated to $\boldsymbol{T}_6$. If the robot parameters and the values of each joint variable (angular or linear displacement) are known, the position and orientation of the robot's end-effector (hand) can be calculated using Eq. (3.6), which is the robot positive kinematics, i.e., the robot positive solution problem.

3.2 Integrated Design of Robot Forward Kinematics

This chapter does not provide in-depth elaborations on the process of acquiring the pose of target objects using binocular cameras. Instead, it organizes the robot platform into three subsystems: perception, main operation, and auxiliary operation. In the perception phase, the robot utilizes a nine-degree-of-freedom system, encompassing the mobile cart (three degrees), robot foot rotary joint (one degree), waist 3-RPS parallel mechanism (three degrees), and robot head pitch and rotary joint (two degrees). The main operation system involves 12 degrees of freedom, including the mobile cart (three degrees), robot foot rotary joint (one degree), waist 3-RPS parallel mechanism (three degrees), and main operation system (one degree). Moreover, the main operation system encompasses the waist 3-RPS parallel mechanism (three degrees) and robot main operating arm (five degrees). The auxiliary operation system comprises the six degrees of freedom of the robot's left operating arm.

For the kinematic analysis, this book employs the modified-DH method to establish kinematic models for each subsystem, as illustrated in Figure 3.5.

In Figure 3.5, the coordinate system $\{A\}$ is the coordinate system of the target A. The binocular camera used by the robot platform has its own base coordinate system, denoted as the coordinate system $\{C\}$, and the origin of this coordinate system is located at the left eye position of the robot. In addition, the selected geodetic coordinate system $\{G\}$ is used as the base coordinate system for the entire sensing system and the main operating system, and the central point of the moving cart coincides with the origin of the coordinate system $\{G\}$ before the robot starts moving. At the same time, the auxiliary operation system only needs the auxiliary operation arm of the robot system to move when it coordinates with the main operation system to complete the operation task, so the base coordinate system of the auxiliary operation system is selected at the center of the robot's chest, which is recorded as the coordinate system $\{O\}$. The coordinate system $\{W\}$ and the coordinate system $\{H\}$ represent the waist coordinate system and the head center coordinate system of the robot, respectively.

In the following, the perception system, the main operation system, and the auxiliary operation system are analyzed separately. In order to distinguish the different coordinate systems of the three systems, the coordinate system of each joint of the perception system and its corresponding joint angle are defined as $\{i\}, \theta_i (i = 1, 2, \ldots, 9)$; the coordinate system of each joint of the main operation system and its corresponding joint angle are defined as $\{i_{\mathrm{m}}\}, \theta_{i_{\mathrm{m}}} (i = 1, 2, \ldots, 12)$; and the coordinate system of each joint of the auxiliary operation

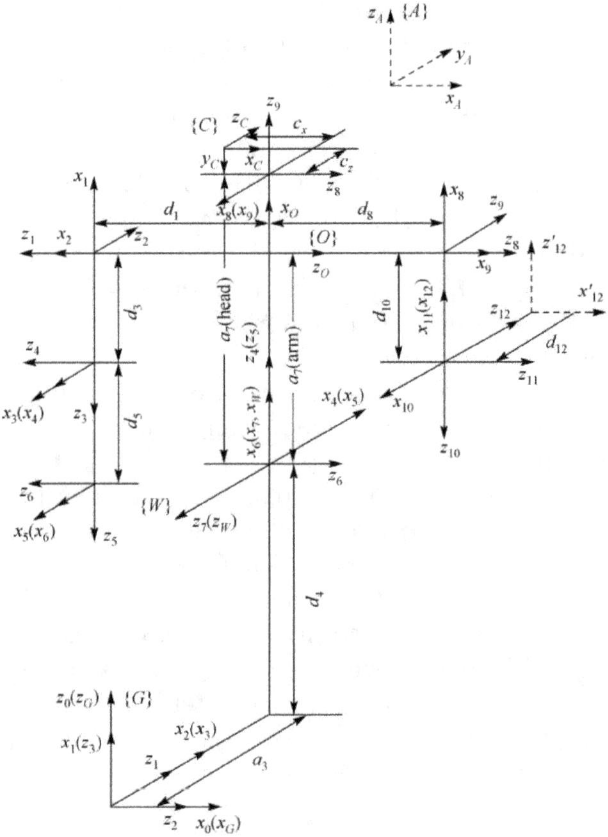

Figure 3.5. Robot system overall model.

system and its corresponding joint angle are defined as $\{i_a\}$ and $\theta_{i_a}(i = 1, 2, \ldots, 6)$, respectively.

3.2.1 *Forward kinematics of the perceptual system*

The main function of the sensing system is to determine the position and orientation information of the robot's "eyes" when the joints of the whole robot system are in different motion states, so as to obtain the position and orientation of the target. However, the position and orientation of the target A obtained by the camera is relative to the camera base coordinate system $\{C\}$. Therefore, it is necessary to convert the position and orientation obtained by the binocular camera into the geodetic coordinate system $\{G\}$ through the sensing system, so that the position and orientation information of the target in the environment can be determined. In addition, in order to assist the end-effector of the working arm to successfully reach the target point, it is also necessary to convert the position and orientation of the target acquired by the binocular camera into the base coordinate system $\{O\}$.

3.2.1.1 *Transformation of the position and orientation of target A into the geodetic coordinate system $\{G\}$*

According to Figure 3.5, the kinematic model of the perceptual system is separated as shown in Figure 3.6.

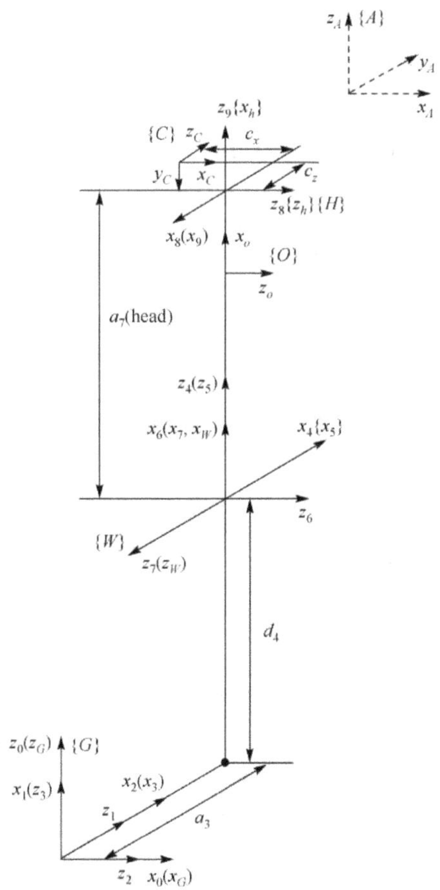

Figure 3.6. Perception system model.

Table 3.3. DH parameters of the perceptual system.

Link i	Joint angle θ_i	Torsion angle α_{i-1}	Link length a_{i-1}	Link offset d_i	θ_i angular range
1	$-90°$	$-90°$	0	$d_1 - d_y$ (variables)	A constant (math.)
2	$-90°$	$-90°$	0	$d_2{-}d_y$ $d_2{-}d_y$ (variables)	A constant (math.)
3	θ_3	$-90°$	0	0	$-180° \sim 180°$
4	θ_4	$0°$	$a_3 = 186\,\text{mm}$	$d_4 = 454\,\text{mm}$	$-180° \sim 180°$
5	$0°$	$0°$	0	$d_5 = h_z$ (variables)	A constant (math.)
6	$\theta_6 + 90°$	$90°$	0	0	$-60° \sim 60°$
7	θ_7	$-90°$	0	0	$-60° \sim 60°$
8	$\theta_8 + 90°$	$90°$	$a_7 = 505\,\text{mm}$	0	$-20° \sim 60°$
9	θ_9	$90°$	0	0	$-90° \sim 90°$

Based on the model in Figure 3.6, the DH parameters of the perceptual system obtained are shown in Table 3.3.

Among them, d_x represents the distance moved by the cart along the axes of the base coordinate system x, d_y represents the distance moved by the cart along the axes of the base coordinate system y, h_z represents the distance moved by the parallel mechanism in the vertical direction, a_3 represents the distance from the center of the cart to the center of the robot's feet, d_4 represents the distance from the center of the robot's feet to the center of its waist, a_7 represents the distance from the center of the waist to the center of the robot's head. According to the established parameter table, the transformation matrix of each link ${}^{i-1}_{i}\boldsymbol{T}$ can be obtained as shown in Eq. (3.7):

$$
{}^0_1\boldsymbol{T} = \begin{bmatrix} 0 & 1 & 0 & 0 \\ 0 & 0 & 1 & d_y \\ 1 & 0 & 0 & 0 \\ 0 & 0 & 0 & 1 \end{bmatrix} \quad
{}^1_2\boldsymbol{T} = \begin{bmatrix} 0 & 1 & 0 & 0 \\ 0 & 0 & 1 & d_x \\ 1 & 0 & 0 & 0 \\ 0 & 0 & 0 & 1 \end{bmatrix} \quad
{}^2_3\boldsymbol{T} = \begin{bmatrix} c_3 & -s_3 & 0 & 0 \\ 0 & 0 & 1 & 0 \\ -s_3 & -c_3 & 0 & 0 \\ 0 & 0 & 0 & 1 \end{bmatrix}
$$

$$
{}^3_4\boldsymbol{T} = \begin{bmatrix} c_4 & -s_4 & 0 & a_3 \\ s_4 & c_4 & 0 & 0 \\ 0 & 0 & 1 & d_4 \\ 0 & 0 & 0 & 1 \end{bmatrix} \quad
{}^4_5\boldsymbol{T} = \begin{bmatrix} 1 & 0 & 0 & 0 \\ 0 & 1 & 0 & 0 \\ 0 & 0 & 1 & h_z \\ 0 & 0 & 0 & 1 \end{bmatrix} \quad
{}^5_6\boldsymbol{T} = \begin{bmatrix} -s_6 & -c_6 & 0 & 0 \\ 0 & 0 & -1 & 0 \\ c_6 & -s_6 & 0 & 0 \\ 0 & 0 & 0 & 1 \end{bmatrix}
$$

$$
{}^{6}_{7}\boldsymbol{T} = \begin{bmatrix} c_7 & -s_7 & 0 & 0 \\ 0 & 0 & 1 & 0 \\ -s_7 & -c_7 & 0 & 0 \\ 0 & 0 & 0 & 1 \end{bmatrix} \quad {}^{7}_{8}\boldsymbol{T} = \begin{bmatrix} -s_8 & -c_8 & 0 & a_7 \\ 0 & 0 & -1 & 0 \\ c_8 & -s_8 & 0 & 0 \\ 0 & 0 & 0 & 1 \end{bmatrix}
$$

$$
{}^{8}_{9}\boldsymbol{T} = \begin{bmatrix} c_9 & -s_9 & 0 & 0 \\ 0 & 0 & -1 & 0 \\ s_9 & c_9 & 0 & 0 \\ 0 & 0 & 0 & 1 \end{bmatrix} \tag{3.7}
$$

Therefore, according to Eq. (3.7), the position and orientation of the robot's head center relative to the geodetic coordinate system ${}^{0}_{9}\boldsymbol{T}$ can be obtained, as shown in Eq. (3.8):

$$
{}^{0}_{9}\boldsymbol{T} = {}^{0}_{1}\boldsymbol{T}{}^{1}_{2}\boldsymbol{T}{}^{2}_{3}\boldsymbol{T}{}^{3}_{4}\boldsymbol{T}{}^{4}_{5}\boldsymbol{T}{}^{5}_{6}\boldsymbol{T}{}^{6}_{7}\boldsymbol{T}{}^{7}_{8}\boldsymbol{T}{}^{8}_{9}\boldsymbol{T} \tag{3.8}
$$

The sensing system has a total of nine degrees of freedom and its end-effector is located at the center of the robot head. The coordinate system $\{C\}$ is the base coordinate system of the camera itself, and the coordinate axes of x_C, y_C, and z_C are shown in Figure 3.6. Through the model, the relationship between the camera coordinate system $\{C\}$ and the ninth joint coordinate system ${}^{9}_{C}\boldsymbol{T}$ can be obtained as shown in Eq. (3.9):

$$
\begin{aligned}
{}^{9}_{C}\boldsymbol{T} &= \boldsymbol{R}_Z(90°) \times \boldsymbol{R}_X(-90°) \times \boldsymbol{D}_X(-c_x) \times \boldsymbol{D}_Z(c_z) \\
&= \begin{bmatrix} 0 & -1 & 0 & 0 \\ 1 & 0 & 0 & 0 \\ 0 & 0 & 1 & 0 \\ 0 & 0 & 0 & 1 \end{bmatrix} \begin{bmatrix} 1 & 0 & 0 & 0 \\ 0 & 0 & 1 & 0 \\ 0 & -1 & 0 & 0 \\ 0 & 0 & 0 & 1 \end{bmatrix} \begin{bmatrix} 1 & 0 & 0 & -c_x \\ 0 & 1 & 0 & 0 \\ 0 & 0 & 1 & c_z \\ 0 & 0 & 0 & 1 \end{bmatrix} \\
&\quad \times \begin{bmatrix} 1 & 0 & 0 & 0 \\ 0 & 1 & 0 & 0 \\ 0 & 0 & 1 & c_z \\ 0 & 0 & 0 & 1 \end{bmatrix} \\
&= \begin{bmatrix} 0 & 0 & -1 & -c_z \\ 1 & 0 & 0 & -c_x \\ 0 & -1 & 0 & 0 \\ 0 & 0 & 0 & 1 \end{bmatrix}
\end{aligned} \tag{3.9}
$$

In the formula, $c_x = 32.5\,\text{mm}$ indicates the distance between the origin of the camera coordinate system (located in the left eye of the robot) and the left–right difference of the binocular center, and $c_z = 68.5\,\text{mm}$ indicates the distance from the center of the binocular to the front and back of the center of the head. Then, the position and orientation of the binocular camera with respect to the geodetic coordinate system can be obtained ${}^{G}_{C}\boldsymbol{T}$, as shown in Eq. (3.10):

$$ {}^{G}_{C}\boldsymbol{T} = {}^{0}_{9}\boldsymbol{T}{}^{9}_{C}\boldsymbol{T} = {}^{0}_{1}\boldsymbol{T}{}^{1}_{2}\boldsymbol{T}{}^{2}_{3}\boldsymbol{T}{}^{3}_{4}\boldsymbol{T}{}^{4}_{5}\boldsymbol{T}{}^{5}_{6}\boldsymbol{T}{}^{6}_{7}\boldsymbol{T}{}^{7}_{8}\boldsymbol{T}{}^{8}_{9}\boldsymbol{T}{}^{9}_{C}\boldsymbol{T} \tag{3.10} $$

According to Eq. (3.10), the position and orientation of the robot's head in the geodetic coordinate system ${}^{0}_{9}\boldsymbol{T}$ and the position and orientation of the binocular camera in the geodetic coordinate system ${}^{G}_{C}\boldsymbol{T}$ can be found when the angle of rotation of the joints in the robot's sensing system and the distance moved are given. Based on this, the problem of "where the target is" can be solved. In this book, the position and orientation of target A acquired by the binocular camera is recorded as ${}^{C}_{A}\boldsymbol{T}$, and then the position and orientation of target A in the geodetic coordinate system $\{G\}$ can be obtained as ${}^{G}_{A}\boldsymbol{T}$, as shown in Eq. (3.11):

$$ {}^{G}_{A}\boldsymbol{T} = {}^{G}_{C}\boldsymbol{T}{}^{C}_{A}\boldsymbol{T} \tag{3.11} $$

3.2.1.2 *Transformation of the position and orientation of target A in the chest center coordinate system* $\{O\}$.

The robot's head has two joints, up and down pitch and left and right rotation, which can move the camera to increase its scanning spatial range. If the position and orientation of the target acquired by the camera is recorded as $\boldsymbol{T}_{\text{camera}}$, it needs to pass through the two rotational joints of the head in order to transform $\boldsymbol{T}_{\text{camera}}$ into the chest center coordinate system $\{O\}$. According to the model, the robot head can rotate around the x_h and z_h axes of the coordinate system $\{H\}$. Figure 3.7 shows the coordinate relationship between the camera coordinate system $\{C\}$, the head center coordinate system $\{H\}$, and the chest center coordinate system $\{O\}$.

According to Figure 3.7, the following transformation steps can be obtained:

Step 1: The head has two joints of pitch and left/right deflection around the coordinate system $\{H\}$; rotate α_2 around the x_h axis

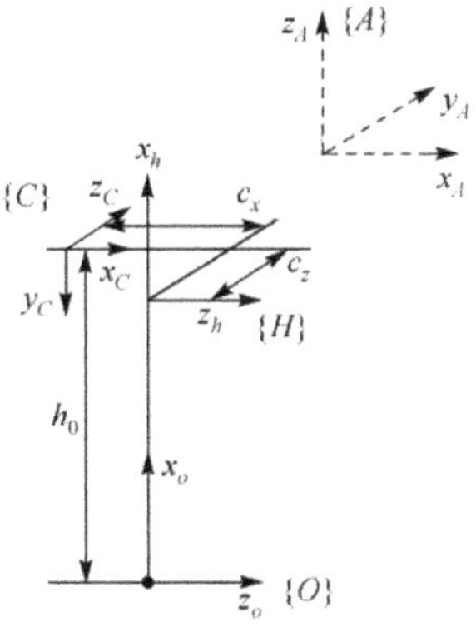

Figure 3.7. Conversion.

(counterclockwise head rotation is positive), and then rotate α_1 around the z_h axis (tilting the head is positive), and the transformation relationship is recorded as $\boldsymbol{T}_h$, as shown in Eq. (3.12):

$$\boldsymbol{T}_h = \mathbf{Rot}(z_h, \alpha_1)\mathbf{Rot}(x_h, \alpha_2) = \begin{bmatrix} c\alpha_1 & -s\alpha_1 c\alpha_2 & s\alpha_1 s\alpha_2 & 0 \\ s\alpha_1 & c\alpha_1 c\alpha_2 & -c\alpha_1 s\alpha_2 & 0 \\ 0 & s\alpha_2 & c\alpha_2 & 0 \\ 0 & 0 & 0 & 1 \end{bmatrix} \tag{3.12}$$

Step 2: From Figure 3.7, we can get that between the coordinate system $\{H\}$ and the coordinate system $\{C\}$, the direction of each coordinate axis is the same, and the origin is in the direction of the same x axis and the difference is a certain distance, which is recorded as $h_0 = 230\,\text{mm}$. Therefore, we can get the position and orientation ${}^{O}_{H}\boldsymbol{T}$ of the head center coordinate system $\{H\}$ relative to the chest center coordinate system $\{O\}$, i.e., we can directly add h_0 to the direction of the x axis in $\boldsymbol{T}_h$, as shown in Eq. (3.13):

$${}^{O}_{H}\boldsymbol{T} = \begin{bmatrix} c\alpha_1 & -s\alpha_1 c\alpha_2 & s\alpha_1 s\alpha_2 & h_0 \\ s\alpha_1 & c\alpha_1 c\alpha_2 & -c\alpha_1 s\alpha_2 & 0 \\ 0 & s\alpha_2 & c\alpha_2 & 0 \\ 0 & 0 & 0 & 1 \end{bmatrix} \tag{3.13}$$

Step 3: Transform the camera coordinate system $\{C\}$ into the head center coordinate system $\{H\}$, where the transformation relation is denoted as ${}^{H}_{C}\boldsymbol{T}$, as shown in Eq. (3.14):

$$
\begin{aligned}
{}^{H}_{C}\boldsymbol{T} &= \boldsymbol{R}_Z(90^\circ) \times \boldsymbol{R}_Y(-90^\circ) \times \boldsymbol{D}_X(-c_x) \times \boldsymbol{D}_Z(c_z) \\
&= \begin{bmatrix} 0 & -1 & 0 & 0 \\ 0 & 0 & -1 & -c_z \\ 1 & 0 & 0 & -c_x \\ 0 & 0 & 0 & 1 \end{bmatrix}
\end{aligned} \tag{3.14}
$$

c_x and c_z in Eq. (3.14) have the same meanings as c_x and c_z, respectively, in Eq. (3.9).

Step 4: According to the above-mentioned three steps, get the position and orientation relation of the camera transformed to the chest center coordinate system $\{O\}$ through the two rotary joints of the head, denoted as ${}^{O}_{C}\boldsymbol{T}$, as shown in Eq. (3.15):

$$
\begin{aligned}
{}^{O}_{C}\boldsymbol{T} &= {}^{O}_{H}\boldsymbol{T}{}^{H}_{C}\boldsymbol{T} \\
&= \begin{bmatrix} s\alpha_1 s\alpha_2 & -c\alpha_1 & s\alpha_1 c\alpha_2 & h_0 + c_z s\alpha_1 c\alpha_2 - c_x s\alpha_1 s\alpha_2 \\ -c\alpha_1 s\alpha_2 & -s\alpha_1 & -c\alpha_1 c\alpha_2 & -c_z c\alpha_1 c\alpha_2 + c_x c\alpha_1 s\alpha_2 \\ c\alpha_2 & 0 & -s\alpha_2 & -c_z s\alpha_2 - c_x c\alpha_2 \\ 0 & 0 & 0 & 1 \end{bmatrix}
\end{aligned} \tag{3.15}
$$

Step 5: Combining all the above steps, the position and orientation of the target A relative to the chest center coordinate system $\{O\}$ can be obtained as ${}^{O}_{A}\boldsymbol{T}$, as shown in Eq. (3.16):

$$
{}^{O}_{A}\boldsymbol{T} = {}^{O}_{C}\boldsymbol{T} \times \boldsymbol{T}_{\text{camera}} \tag{3.16}
$$

3.2.2 *Forward kinematics of the host operating system*

Based on Figure 3.5, the kinematic model of the main operating system is separated out, as shown in Figure 3.8.

Based on Figure 3.8, the DH parameters of the main operating system can be obtained, as shown in Table 3.4.

Among them, d_x, d_y, and h_z have the same meanings as d_x, d_y, and h_z in the perception system model, respectively; a_{3_m} and d_{4_m} have

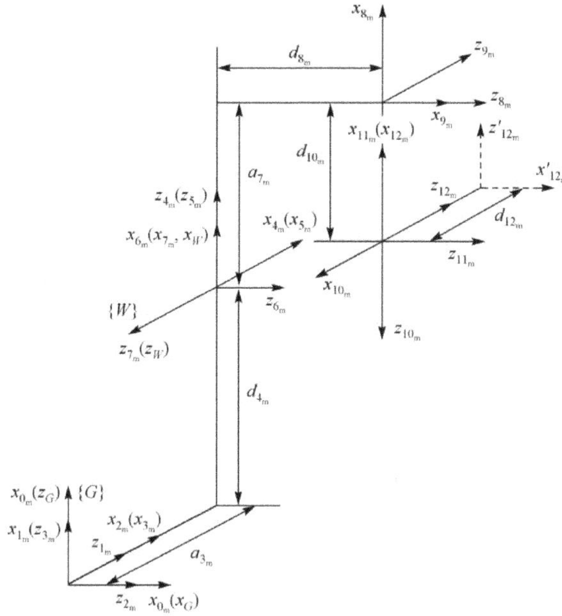

Figure 3.8. Kinematic model of the main operating system.

Table 3.4. DH parameters for the main operating system.

Link i_{m}	Joint angle $\theta_{i\mathrm{m}}$	Torsion angle $\alpha_{(i-1)\mathrm{m}}$	Link length $a_{(i-1)\mathrm{m}}$	Link offset $d_{i\mathrm{m}}$	$\theta_{i\mathrm{m}}$ angular range
1_{m}	$-90°$	$-90°$	0	$d_{1\mathrm{m}} = d_y$ (variables)	A constant (math.)
2_{m}	$-90°$	$-90°$	0	$d_{2\mathrm{m}} = d_x$ (variables)	A constant (math.)
3_{m}	$\theta_{3\mathrm{m}}$	$-90°$	0	0	$-180° \sim 180°$
4_{m}	$\theta_{4\mathrm{m}}$	$0°$	$a_{3\mathrm{m}} = 186\,\mathrm{mm}$	$d_{4\mathrm{m}} = 545\,\mathrm{mm}$	$-180° \sim 180°$
5_{m}	$0°$	$0°$	0	$d_{5\mathrm{m}} = h_z$ (variables)	A constant (math.)
6_{m}	$\theta_{6\mathrm{m}} + 90°$	$90°$	0	0	$-60° \sim 60°$
7_{m}	$\theta_{7\mathrm{m}}$	$-90°$	0	0	$-60° \sim 60°$
8_{m}	$\theta_{8\mathrm{m}}$	$90°$	$a_{7\mathrm{m}} = 275\,\mathrm{mm}$	$d_{8\mathrm{m}} = 200\,\mathrm{mm}$	$-120° \sim 120°$
9_{m}	$\theta_{9\mathrm{m}} + 90°$	$90°$	0	0	$-120° \sim 10°$
10_{m}	$\theta_{10\mathrm{m}} + 90°$	$-90°$	0	$d_{10\mathrm{m}} =$ 183 mm	$-90° \sim 90°$
11_{m}	$\theta_{11\mathrm{m}} - 90°$	$90°$	0	0	$-120° \sim 120°$
12_{m}	$\theta_{12\mathrm{m}}$	$90°$	0	0	$-90° \sim 90°$

the same meanings as a_3 and d_4 in the perception system model, respectively; a_{7_m} indicates the distance from the center of the waist to the center of the chest of the robot; d_{8_m} indicates the distance from the center of the chest to the center of the shoulder of the main operating arm of the robot; and d_{10_m} indicates the distance from the center of the shoulder of the main operating arm to the elbow of the main operating arm. In addition, d_{12_m} in Figure 3.8 indicates the distance from the center of the elbow to the center of the end-effector. According to Table 3.4, the homogeneous coordinate transformation matrix of each link of the main operating system can be obtained as ${}^{(i-1)_m}_{i_m}\boldsymbol{T}$, as shown in Eq. (3.17):

$$
{}^{0_m}_{1_m}\boldsymbol{T} = \begin{bmatrix} 0 & 1 & 0 & 0 \\ 0 & 0 & 1 & d_y \\ 1 & 0 & 0 & 0 \\ 0 & 0 & 0 & 1 \end{bmatrix} \quad
{}^{1_m}_{2_m}\boldsymbol{T} = \begin{bmatrix} 0 & 1 & 0 & 0 \\ 0 & 0 & 1 & d_x \\ 1 & 0 & 0 & 0 \\ 0 & 0 & 0 & 1 \end{bmatrix}
$$

$$
{}^{2_m}_{3_m}\boldsymbol{T} = \begin{bmatrix} c_{3_m} & -s_{3_m} & 0 & 0 \\ 0 & 0 & 1 & 0 \\ -s_{3_m} & -c_{3_m} & 0 & 0 \\ 0 & 0 & 0 & 1 \end{bmatrix} \quad
{}^{3_m}_{4_m}\boldsymbol{T} = \begin{bmatrix} c_{4_m} & -s_{4_m} & 0 & a_{3_m} \\ s_{4_m} & c_{4_m} & 0 & 0 \\ 0 & 0 & 1 & d_{4_m} \\ 0 & 0 & 0 & 1 \end{bmatrix}
$$

$$
{}^{4_m}_{5_m}\boldsymbol{T} = \begin{bmatrix} 1 & 0 & 0 & 0 \\ 0 & 1 & 0 & 0 \\ 0 & 0 & 1 & h_z \\ 0 & 0 & 0 & 1 \end{bmatrix} \quad
{}^{5_m}_{6_m}\boldsymbol{T} = \begin{bmatrix} -s_{6_m} & -c_{6_m} & 0 & 0 \\ 0 & 0 & -1 & 0 \\ c_{6_m} & -s_{6_m} & 0 & 0 \\ 0 & 0 & 0 & 1 \end{bmatrix}
$$

$$
{}^{6_m}_{7_m}\boldsymbol{T} = \begin{bmatrix} c_{7_m} & -s_{7_m} & 0 & 0 \\ 0 & 0 & 1 & 0 \\ -s_{7_m} & -c_{7_m} & 0 & 0 \\ 0 & 0 & 0 & 1 \end{bmatrix} \quad
{}^{7_m}_{8_m}\boldsymbol{T} = \begin{bmatrix} c_{8_m} & -s_{8_m} & 0 & a_{7_m} \\ 0 & 0 & -1 & -d_{8_m} \\ s_{8_m} & c_{8_m} & 0 & 0 \\ 0 & 0 & 0 & 1 \end{bmatrix}
$$

$$
{}^{8_m}_{9_m}\boldsymbol{T} = \begin{bmatrix} -s_{9_m} & -c_{9_m} & 0 & 0 \\ 0 & 0 & -1 & 0 \\ c_{9_m} & -s_{9_m} & 0 & 0 \\ 0 & 0 & 0 & 1 \end{bmatrix} \quad
{}^{9_m}_{10_m}\boldsymbol{T} = \begin{bmatrix} -s_{10_m} & -c_{10_m} & 0 & 0 \\ 0 & 0 & 1 & d_{10_m} \\ -c_{10_m} & s_{10_m} & 0 & 0 \\ 0 & 0 & 0 & 1 \end{bmatrix}
$$

$$
{}^{10_\mathrm{m}}_{11_\mathrm{m}}\boldsymbol{T}=\begin{bmatrix} s_{11_\mathrm{m}} & c_{11_\mathrm{m}} & 0 & 0 \\ 0 & 0 & -1 & 0 \\ -c_{11_\mathrm{m}} & s_{11_\mathrm{m}} & 0 & 0 \\ 0 & 0 & 0 & 1 \end{bmatrix} \quad {}^{11_\mathrm{m}}_{12_\mathrm{m}}\boldsymbol{T}=\begin{bmatrix} c_{12_\mathrm{m}} & -s_{12_\mathrm{m}} & 0 & 0 \\ 0 & 0 & -1 & 0 \\ s_{12_\mathrm{m}} & c_{12_\mathrm{m}} & 0 & 0 \\ 0 & 0 & 0 & 1 \end{bmatrix} \tag{3.17}
$$

From the model in Figure 3.8, the origin of the end-effector coordinate system of the main operating system is located at the center of the elbow, so it is necessary to transform the end-effector coordinate system to coincide with the end-effector center coordinate system $\{12'_\mathrm{m}\}$. The transformation process is as follows: Rotate the coordinate system $\{12_\mathrm{m}\}$ around the z axis by 90°, then rotate 90° around the x axis, and finally translate d_{12_m} along the y axis of the transformed coordinate system to the $\{12'_\mathrm{m}\}$ coordinate system ${}^{0_\mathrm{m}}_{12'_\mathrm{m}}\boldsymbol{T}$. Therefore, the position and orientation of the end-effector center with respect to the base coordinate system $\{0_\mathrm{m}\}$ (geodetic coordinate system $\{G\}$) is obtained as follows:

$$
{}^{0_\mathrm{m}}_{12'_\mathrm{m}}\boldsymbol{T}={}^{0_\mathrm{m}}_{12_\mathrm{m}}\boldsymbol{T}\times\boldsymbol{R}_Z(90^\circ)\times\boldsymbol{R}_X(90^\circ)\times\boldsymbol{D}_Y(d_{12_\mathrm{m}}) \tag{3.18}
$$

$$
{}^{0_\mathrm{m}}_{12_\mathrm{m}}\boldsymbol{T}={}^{0_\mathrm{m}}_{1_\mathrm{m}}\boldsymbol{T}\,{}^{1_\mathrm{m}}_{2_\mathrm{m}}\boldsymbol{T}\,{}^{2_\mathrm{m}}_{3_\mathrm{m}}\boldsymbol{T}\,{}^{3_\mathrm{m}}_{4_\mathrm{m}}\boldsymbol{T}\,{}^{4_\mathrm{m}}_{5_\mathrm{m}}\boldsymbol{T}\,{}^{5_\mathrm{m}}_{6_\mathrm{m}}\boldsymbol{T}\,{}^{6_\mathrm{m}}_{7_\mathrm{m}}\boldsymbol{T}\,{}^{7_\mathrm{m}}_{8_\mathrm{m}}\boldsymbol{T}\,{}^{8_\mathrm{m}}_{9_\mathrm{m}}\boldsymbol{T}\,{}^{9_\mathrm{m}}_{10_\mathrm{m}}\boldsymbol{T}\,{}^{10_\mathrm{m}}_{11_\mathrm{m}}\boldsymbol{T}\,{}^{11_\mathrm{m}}_{12_\mathrm{m}}\boldsymbol{T} \tag{3.19}
$$

where $d_{12_\mathrm{m}}=300\,\mathrm{mm}$. When the angular values of the movements of the individual joints of the main operating system are known, the position and orientation information of its end-effector relative to the base coordinate system can be determined according to Eqs. (3.18) and (3.19).

3.2.3 *Forward kinematics of auxiliary operating system*

The auxiliary operating system is the robot auxiliary operating arm mechanism, which contains a total of six degrees of freedom, and the base coordinate system $\{0_\mathrm{a}\}$ of its model is located in the robot chest center coordinate system $\{O\}$. According to Figure 3.5, the kinematic model of the assisted operation system is separated as shown in Figure 3.9.

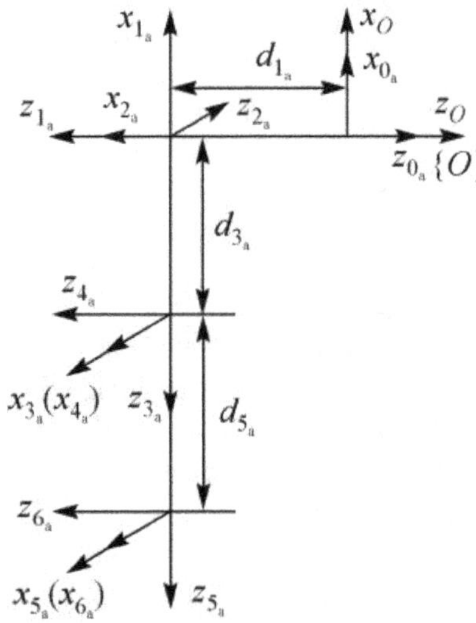

Figure 3.9. Kinematic model of the auxiliary operating system.

Table 3.5. DH Parameters for the auxiliary operating system.

Link i_a	Joint angle θ_{i_a}	Torsion angle $\alpha_{(i-1)_a}$	Link length $a_{(i-1)_a}$	Link offset d_{i_a}	θ_{i_a} angular range
1_a	θ_{1_a}	180°	0	$d_{1_a} = 200$ mm	−120°∼120°
2_a	$\theta_{2_a} - 90°$	−90°	0	0	−10°∼120°
3_a	$\theta_{3_a} - 90°$	90°	0	$d_{3_a} = 183$ mm	−90°∼90°
4_a	θ_{4_a}	−90°	0	0	−120°∼20°
5_a	θ_{5_a}	90°	0	$d_{5_a} = 300$ mm	−90°∼90°
6_a	θ_{6_a}	−90°	0	0	−15°∼15°

Based on Figure 3.9, the DH parameters of the auxiliary operating system can be obtained, as shown in Table 3.5.

Here, d_{1_a} denotes the distance from the center of the chest to the shoulder of the robot's auxiliary operation arm, d_{3_a} denotes the distance from the center of the shoulder of the auxiliary operation arm to the elbow of the auxiliary operation arm, and d_{5_a} denotes the distance from the center of the elbow to the wrist of the end-effector. According to Table 3.5, the homogeneous coordinate transformation matrix of each link of the assisted operation system can be obtained as ${}^{(i-1)_a}_{i_a}\boldsymbol{T}$, as shown in Eq. (3.20):

$$
{}^{0_a}_{1_a}\boldsymbol{T} = \begin{bmatrix} c_{1_a} & -s_{1_a} & 0 & 0 \\ -s_{1_a} & -c_{1_a} & 0 & 0 \\ 0 & 0 & -1 & -d_{1_a} \\ 0 & 0 & 0 & 1 \end{bmatrix} \quad {}^{1_a}_{2_a}\boldsymbol{T} = \begin{bmatrix} s_{2_a} & c_{2_a} & 0 & 0 \\ 0 & 0 & 1 & 0 \\ c_{2_a} & -s_{2_a} & 0 & 0 \\ 0 & 0 & 0 & 1 \end{bmatrix}
$$

$$
{}^{2_a}_{3_a}\boldsymbol{T} = \begin{bmatrix} s_{3_a} & c_{3_a} & 0 & 0 \\ 0 & 0 & -1 & -d_{3_a} \\ -c_{3_a} & s_{3_a} & 0 & 0 \\ 0 & 0 & 0 & 1 \end{bmatrix} \quad {}^{3_a}_{4_a}\boldsymbol{T} = \begin{bmatrix} c_{4_a} & -s_{4_a} & 0 & 0 \\ 0 & 0 & 1 & 0 \\ -s_{4_a} & -c_{4_a} & 0 & 0 \\ 0 & 0 & 0 & 1 \end{bmatrix}
$$

$$
{}^{4_a}_{5_a}\boldsymbol{T} = \begin{bmatrix} c_{5_a} & -s_{5_a} & 0 & 0 \\ 0 & 0 & -1 & -d_{5_a} \\ s_{5_a} & c_{5_a} & 0 & 0 \\ 0 & 0 & 0 & 1 \end{bmatrix} \quad {}^{5_a}_{6_a}\boldsymbol{T} = \begin{bmatrix} c_{6_a} & -s_{6_a} & 0 & 0 \\ 0 & 0 & 1 & 0 \\ -s_{6_a} & -c_{6_a} & 0 & 0 \\ 0 & 0 & 0 & 1 \end{bmatrix} \tag{3.20}
$$

If the robot auxiliary arm wrist is translated to the center of the end-effector along the negative direction of the y axis of its fixed coordinate system by $d = 30\,\text{mm}$, the position and orientation of the center of the end-effector with respect to the coordinate system of the center of the robot's chest ${}^{0_a}_{p_a}\boldsymbol{T}$ is shown in Eq. (3.21):

$$
\begin{aligned}
{}^{0_a}_{p_a}\boldsymbol{T} &= {}^{0_a}_{6_a}\boldsymbol{T} \times \boldsymbol{D}_Y(-d) \\
&= {}^{0_a}_{1_a}\boldsymbol{T}{}^{1_a}_{2_a}\boldsymbol{T}{}^{2_a}_{3_a}\boldsymbol{T}{}^{3_a}_{4_a}\boldsymbol{T}{}^{4_a}_{5_a}\boldsymbol{T}{}^{5_a}_{6_a}\boldsymbol{T} \times \boldsymbol{D}_Y(-d)
\end{aligned} \tag{3.21}
$$

Hence, by determining the angular measurement of motion for every joint in the robotic assistive operating system, we can derive the position and orientation data for the assistive operating arm's end-effector using the formula provided in Eq. (3.21)

3.2.4 *Forward kinematics of 3-RPS parallel mechanism*

There are three representations for robot position and orientation description: actuator space description, joint space description, and workspace description.[16] Previously, we assumed that the motion joints of the robot's waist structure are directly driven by actuators. However, the 3-RPS parallel mechanism makes the kinematic platform move by driving each of the three series-connected branches by three motors, which can be equated to two rotational joints and one moving joint in the joint space. However, the motion state of the moving platform cannot be derived directly from the motors of the parallel mechanism driving the three tandem branches, so it is necessary to equate the motion of the tandem branches to the motion state of the moving platform, which is also known as the positive kinematics of the parallel mechanism. The simplified structure of the 3-RPS parallel mechanism is shown in Figure 3.10.

As illustrated in Figure 3.10, the lower layer of the parallel mechanism is a fixed platform, where motors mounted on this platform are uniformly distributed, and those connections between kinematic pairs form an equilateral triangle. The upper level of the moving platform through the ball hinges is connected to the tandem branch, while tandem branch is driven by the upper level of the moving platform motion. The fixed platform and the moving platform planes are circular, and their radii are $R = 45\,\text{mm}$. At the same time, the initial

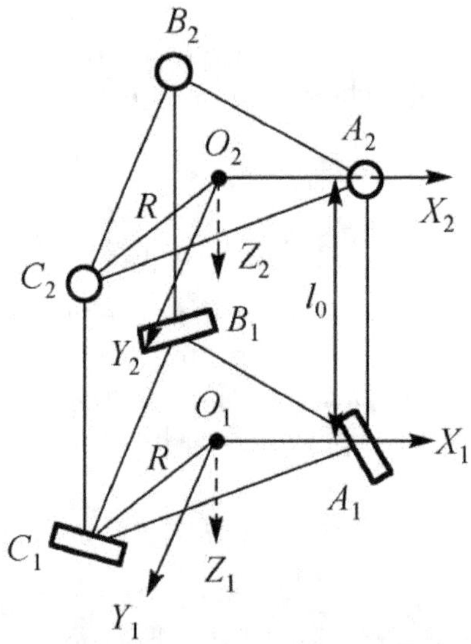

Figure 3.10. Simplified structure of the 3-RPS parallel mechanism.

length of the three tandem branches firmly connected to the fixed platform and the moving platform is $l_0 = 425\,\text{mm}$.

According to the mounting structure of the parallel mechanism of the robot, the coordinate systems $O_1 - X_1Y_1Z_1$ and $O_2 - X_2Y_2Z_2$ are established on the moving and fixed platforms with the geometric centers of the equilateral triangles as the coordinate origins, as shown in Figure 3.10. Among them, the X_i axis direction is the direction of the line from the point O_i to the vertex A_i, the Y_i axis direction is the direction parallel to the vector $\overrightarrow{B_iC_i}$ from the point O_i, and the Z_i axis direction is the direction perpendicular to the plane of the equilateral triangle downward, where $i = 1, 2$.

Therefore, the position coordinates of the vertices of the moving platform and fixed platform internal equilateral triangles in their corresponding coordinate system $O_i - X_iY_iZ_i$ in the initial state of the 3-RPS parallel mechanism can be obtained as shown in Eq. (3.22):

$$\begin{aligned} A_i &= (R, 0, 0) \\ B_i &= \left(-\frac{R}{2}, -\frac{\sqrt{3}R}{2}, 0\right), \quad i = 1, 2 \\ C_i &= \left(-\frac{R}{2}, \frac{\sqrt{3}R}{2}, 0\right) \end{aligned} \tag{3.22}$$

In the initial state of the 3-RPS parallel mechanism, the moving platform coordinate system $O_2 - X_2Y_2Z_2$ origin O_2 in the coordinate system $O_1 - X_1Y_1Z_1$ position coordinates is obtained as shown in Eq. (3.23):

$$O_2 = (0, 0, -l_0) \tag{3.23}$$

Let the lengths of the three tandem branches A_1A_2, B_1B_2, and C_1C_2 after expansion be l_1, l_2, and l_3, respectively, and the inclination angles to the fixed platform be ϕ_1, ϕ_2, and ϕ_3, respectively, as shown in Figure 3.11.

In Figure 3.11, after the tandem branch drives the moving platform, the projections of the three tandem branches A_1A_2, B_1B_2, and C_1C_2 on the fixed plane slide on the line segments O_1A_1, O_1B_1, and O_1C_1 with the tilt angles of the tandem branches and the fixed platform as ϕ_1, ϕ_2, and ϕ_3 are changed, respectively.

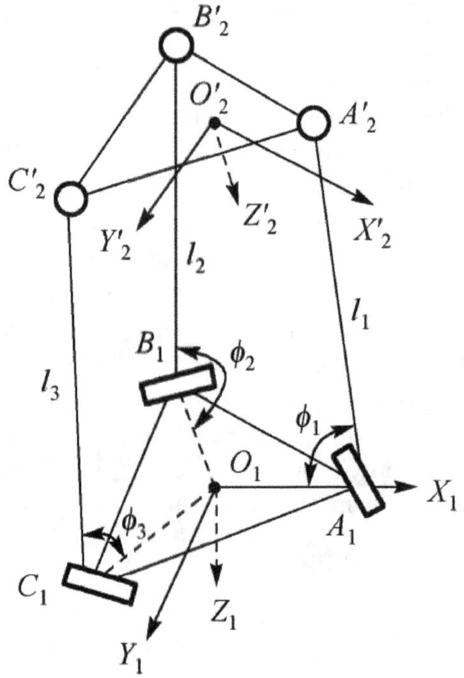

Figure 3.11. Tandem branching structure of the parallel mechanism.

Therefore, after the tandem branch telescoping, combining Eq. (3.22), the three vertices of the moving platform A_2', B_2', and C_2', whose position coordinates in the coordinate system $O_1-X_1Y_1Z_1$ are shown in Eqs. (3.24), (3.25), and (3.26), respectively, are obtained:

$$A_2': \begin{cases} X_{A_2'} = R - l_1 \cos\phi_1 \\ Y_{A_2'} = 0 \\ Z_{A_2'} = -l_1 \sin\phi_1 \end{cases} \tag{3.24}$$

$$B_2': \begin{cases} X_{B_2'} = -\dfrac{R}{2} + l_2 \cos\phi_2 \cos 60^\circ \\ Y_{B_2'} = -\dfrac{\sqrt{3}R}{2} + l_2 \cos\phi_2 \sin 60^\circ \\ Z_{B_2'} = -l_2 \sin\phi_2 \end{cases} \tag{3.25}$$

$$C_2' : \begin{cases} X_{C_2'} = -\dfrac{R}{2} + l_3 \cos \phi_3 \cos 60^\circ \\ Y_{C_2'} = \dfrac{\sqrt{3}R}{2} - l_3 \cos \phi_3 \sin 60^\circ \\ Z_{C_2'} = -l_3 \sin \phi_3 \end{cases} \tag{3.26}$$

It is also known that the length of the side of an equilateral triangle is $\sqrt{3}R$, i.e.,

$$\begin{cases} |A_2'B_2'| = \sqrt{3}R \\ |A_2'C_2'| = \sqrt{3}R \\ |B_2'C_2'| = \sqrt{3}R \end{cases} \tag{3.27}$$

Substituting Eqs. (3.24), (3.25), and (3.26) into Eq. (3.27), the inclination angles of the three tandem branches with respect to the fixed platform can be obtained as ϕ_1, ϕ_2, and ϕ_3. This equation is a transcendental equation with tilt angles ϕ_1, ϕ_2, and ϕ_3 as independent variables, and the solution process is complicated. In this book, only the arithmetic process of converting the parallel mechanism from joint space to actuator space is used, so the detailed solution process of this transcendental equation is not repeated.

In addition, since the inner triangle of the moving platform is an equilateral triangle, the origin of the moving platform coordinate system $O_2 - X_2Y_2Z_2O_2$ is the center of gravity of the positive triangle. When the moving platform moves, its origin is recorded as O_2' and the coordinates of O_2' in the coordinate system $O_1 - X_1Y_1Z_1$ are shown in Eq. (3.28):

$$O_2' : \begin{cases} X_{O_2'} = \dfrac{1}{3}(X_{A_2'} + X_{B_2'} + X_{C_2'}) \\ Y_{O_2'} = \dfrac{1}{3}(Y_{A_2'} + Y_{B_2'} + Y_{C_2'}) \\ Z_{O_2'} = \dfrac{1}{3}(Z_{A_2'} + Z_{B_2'} + Z_{C_2'}) \end{cases} \tag{3.28}$$

Therefore, the directional cosines of the three axes of the moving platform coordinate system X, Y, and Z can be obtained, as shown in Eq. (3.29):

$$\begin{cases} \overrightarrow{u_X} = \dfrac{1}{|O'_2A'_2|}[(X_{A'_2} - X_{O'_2})\overrightarrow{i} + (Y_{A'_2} - Y_{O'_2})\overrightarrow{j} + (Z_{A'_2} - Z_{O'_2})\overrightarrow{k}] \\ \overrightarrow{u_Y} = \dfrac{1}{|B'_2C'_2|}[(X_{C'_2} - X_{B'_2})\overrightarrow{i} + (Y_{C'_2} - Y_{B'_2})\overrightarrow{j} + (Z_{C'_2} - Z_{B'_2})\overrightarrow{k}] \\ \overrightarrow{u_Z} = \overrightarrow{u_X} \times \overrightarrow{u_Y} \end{cases} \tag{3.29}$$

To simplify, Eq. (3.29) is written as

$$\begin{cases} \overrightarrow{u_X} = u_{x1}\overrightarrow{i} + u_{x2}\overrightarrow{j} + u_{x3}\overrightarrow{k} \\ \overrightarrow{u_Y} = u_{y1}\overrightarrow{i} + u_{y2}\overrightarrow{j} + u_{y3}\overrightarrow{k} \\ \overrightarrow{u_Z} = u_{z1}\overrightarrow{i} + u_{z2}\overrightarrow{j} + u_{z3}\overrightarrow{k} \end{cases} \tag{3.30}$$

In summary, according to Eqs. (3.29) and (3.30), the position and orientation matrix of the moving platform in the fixed platform coordinate system $O_1 - X_1Y_1Z_1{}^{O_1}_{O_2}\boldsymbol{T}$ can be obtained, as shown in Eq. (3.31):

$$^{O_1}_{O_2}\boldsymbol{T} = \begin{bmatrix} u_{x1} & u_{y1} & u_{z1} & X_{O'_2} \\ u_{x2} & u_{y2} & u_{z2} & Y_{O'_2} \\ u_{x3} & u_{y3} & u_{z3} & Z_{O'_2} \\ 0 & 0 & 0 & 1 \end{bmatrix} \tag{3.31}$$

3.3 Integrated Practice of Robot Forward Kinematics

3.3.1 *Simulation modeling*

Drawing upon the upper-body model of the robot outlined in the aforementioned section, the kinematics of the robot's two-arm configuration is validated utilizing the MATLAB Robotics toolbox. This specialized toolbox tailored for robotics facilitates the creation and simulation of robot kinematics models, as well as the resolution of forward and inverse kinematics problems and trajectory planning.[17]

The primary operating arm of the robot comprises five rotary joints spanning from the shoulder to the forearm, whereas the auxiliary operating arm consists of six rotary joints extending from the shoulder to the wrist. Both the main robot arm and the auxiliary arm simulation models adopt the robot chest center coordinate system as the reference frame. Leveraging the enhanced DH modeling method, the Link class function of the toolbox is employed to establish their interconnection. Subsequently, the SerialLink class function is utilized to integrate the individual robot links established by the Link class function into a cohesive tandem operating arm system. One use of the Link class function is as follows:

```
L=Link([theta,d,a,alpha,sigma,offset],'modified')
```

where the parameter "theta" denotes the joint angle; the parameter "d" denotes the link offset; the parameter "a" denotes the link length; the parameter "alpha" denotes the link torsion angle; the parameter "sigma" denotes the type of the joints, with the rotational joints being 0 and the moving joints being 1; the parameter "offset" denotes the offset of the joint variables; and "modified" denotes that it is modeled by the improved DH modeling method.

The SerialLink class function connects the joint models created above together as follows:

```
robot=SerialLink([L1,L2,...,L5],'name','right arm')
```

So, the code used to build the dual-arm model of this robot using the MATLAB Robotics toolbox is shown as follows:

Auxiliary operating arm:

```
L1 = Link([0 200 0 pi 0 0 0],'modified'); L1.qlim=[-120,120]*pi/180;
L2 = Link([0 0 0 0 -pi/2 0 -pi/2],'modified'); L2.qlim=[-10,120]*pi/180;
L3 = Link([0 183 0 pi/2 0 -pi/2],'modified'); L3.qlim=[-90,90]*pi/180;
L4 = Link([0 0 0 0 -pi/2 0 0 0],'modified'); L4.qlim=[-120,20]*pi/180;
L5 = Link([0 300 0 pi/2 0 0],'modified'); L5.qlim=[-90,90]*pi/180;
L6 = Link([0 0 0 0 -pi/2 0 0 0],'modified'); L6.qlim=[-15,15]*pi/180;
```

Main operating arm:

```
R1 = Link([0 200 0 0 0 0 0],'modified'); R1.qlim=[-120,120]*pi/180;
R2 = Link([0 0 0 pi/2 0 pi/2],'modified'); R2.qlim=[-120,10]*pi/180;
R3 = Link([0 183 0 -pi/2 0 pi/2],'modified'); R3.qlim=[-90,90]*pi/180;
R4 = Link([0 0 0 pi/2 0 -pi/2],'modified'); R4.qlim=[-120,20]*pi/180;
R5 = Link([0 0 0 pi/2 0 0],'modified'); R5.qlim=[-90,90]*pi/180;
```

In the above-described model, "qlim" denotes the range of the joint variables.

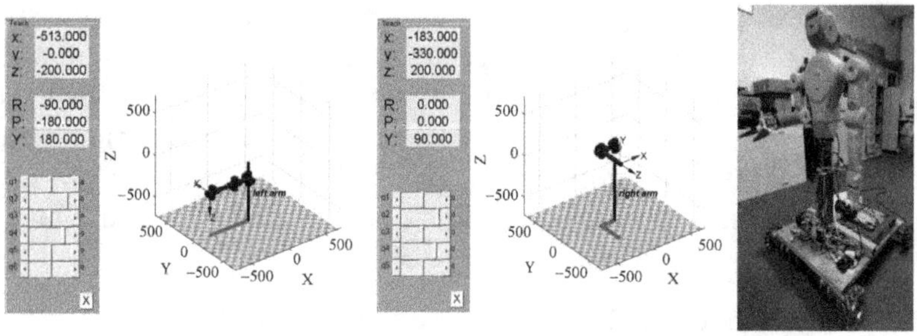

(a) (b) (c)

Figure 3.12. Initial state model of the robot's two arms: (a) auxiliary operating arm, (b) main operating arm, and (c) actual status of the robot.

Finally, the last joint coordinate system of the main and auxiliary operating arms of the robot is transformed to the end-effector, and the output initial state model is shown in Figure 3.12.

According to its initial state, the positive kinematics function is applied to obtain the bit position and orientation of its end-effector $\boldsymbol{T}_{\mathrm{L0}}$, $\boldsymbol{T}_{\mathrm{R0}}$, as shown in Eq. (3.32):

$$\boldsymbol{T}_{\mathrm{L0}} = \begin{bmatrix} 0 & 1 & 0 & -513 \\ 1 & 0 & 0 & 0 \\ 0 & 0 & -1 & -200 \\ 0 & 0 & 0 & 1 \end{bmatrix} \quad \boldsymbol{T}_{\mathrm{R0}} = \begin{bmatrix} 1 & 0 & 0 & -183 \\ 0 & 0 & -1 & -330 \\ 0 & 1 & 0 & 200 \\ 0 & 0 & 0 & 1 \end{bmatrix} \tag{3.32}$$

3.3.2 *Forward kinematics simulation*

When given a set of random angle values for each joint of the robot's dual arms, such as $(-100^\circ, -10^\circ, 30^\circ, -60^\circ, 24^\circ, 5^\circ)$ for the six joints of the auxiliary arm and $(-100^\circ, -10^\circ, 30^\circ, -60^\circ, 24^\circ)$ for the five joints of the main arm, the corresponding states of the output simulation model can be obtained, as shown in Figure 3.13.

Based on the above-mentioned angle values, the position and orientation of the end-effector of the main and auxiliary operating arms are obtained as output $\boldsymbol{T}_{\mathrm{L1}}$ and $\boldsymbol{T}_{\mathrm{R1}}$, as shown in Eq. (3.33):

$$\boldsymbol{T}_{\mathrm{L1}} = \begin{bmatrix} -0.107 & -0.805 & 0.584 & 298.8 \\ -0.872 & 0.358 & 0.334 & -272.4 \\ -0.478 & -0.473 & -0.740 & -0.0416 \\ 0 & 0 & 0 & 1 \end{bmatrix}$$

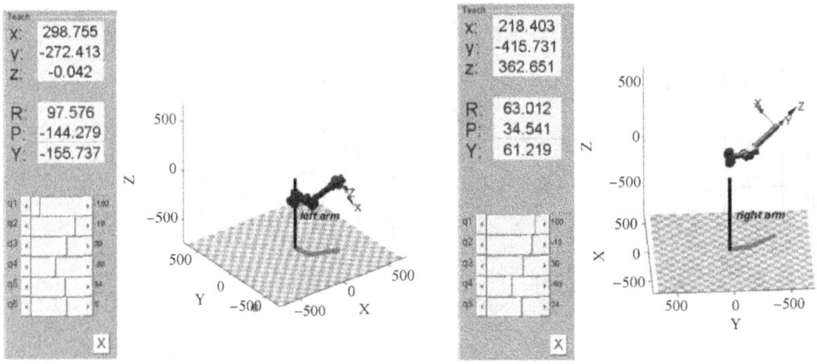

(a) (b)

Figure 3.13. Model states corresponding to random angles: (a) auxiliary operating arm and (b) main operating arm.

$$\boldsymbol{T}_{\mathrm{R1}} = \begin{bmatrix} 0.374 & -0.734 & 0.567 & 218.4 \\ 0.655 & -0.224 & -0.722 & -415.7 \\ 0.657 & 0.641 & 0.397 & 362.7 \\ 0 & 0 & 0 & 1 \end{bmatrix} \tag{3.33}$$

3.3.3 *Forward kinematics practice*

(1) Familiarize yourself with the positive kinematics of the robot, establish the mathematical model of the moving cart, 3-RPS parallel mechanism, and series double arm, and design the positive kinematics algorithm, as shown in Figure 3.14.

(2) Parameter-level robot positive kinematics virtual simulation experiment.

Select "Experiment" → "Robot Control" command to enter the control panel, as shown in Figure 3.15.

In the input box of each joint parameter of the control panel, input the designed kinematic parameters and click the "Confirm Parameters" button; the result is shown in Figure 3.16.

(3) Programming-level robot positive kinematics virtual simulation experiments.

Click the "Online Programming" tab to enter the "Online Programming" interface, write the designed positive kinematics model

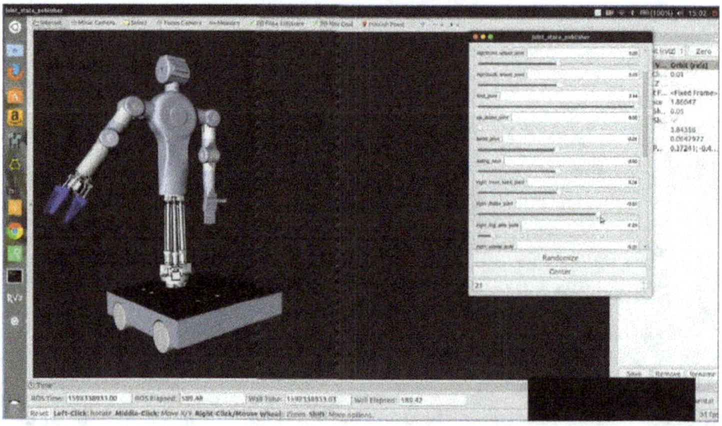

Figure 3.14. Positive kinematics modeling.

Figure 3.15. Parameter-level robot positive kinematics learning.

as a C, Java, or JavaScript program, and click the "Run" button to view the results, as shown in Figure 3.17.

(4) Select "Experiment" → "Robot Control" command, enter the control panel to get the physical experiment interface, as shown in Figure 3.18.

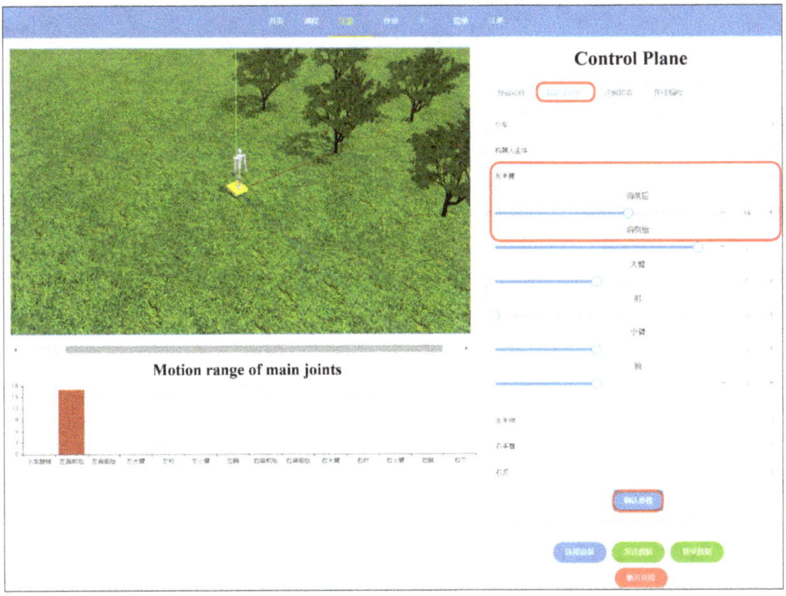

Figure 3.16. Parametric-level robot positive kinematics manipulation.

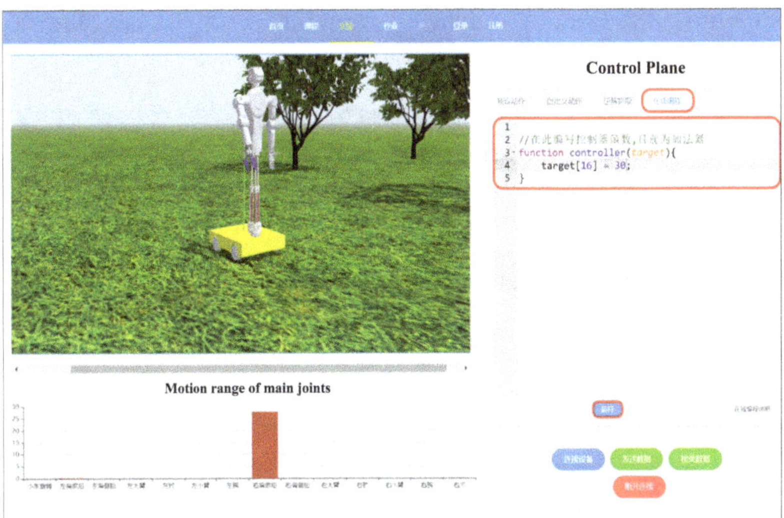

Figure 3.17. Programming level virtual simulation experiment positive kinematics learning.

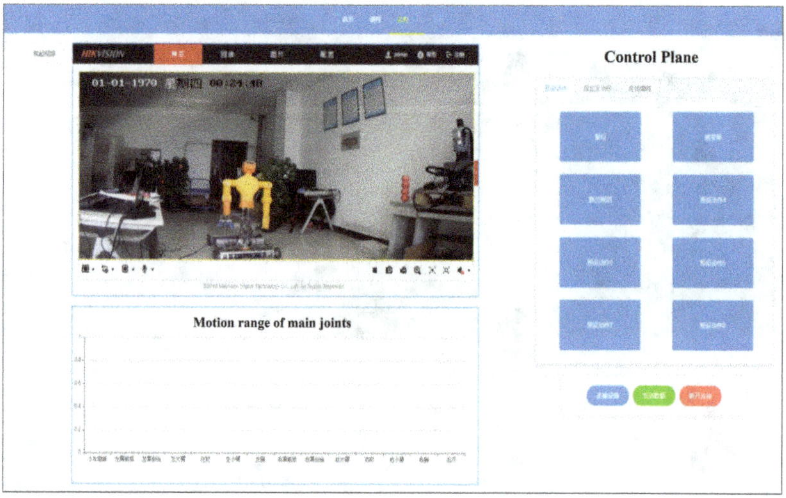

Figure 3.18. Positive kinematics learning for parametric-level robot physical experiments.

In the control panel, according to the design of the positive kinematics parameters, select the positive kinematics, enter the parameters, and click the "Confirm Parameters" button, as shown in Figure 3.19.

Figure 3.19. Parametric-level robot positive kinematics physical experiment.

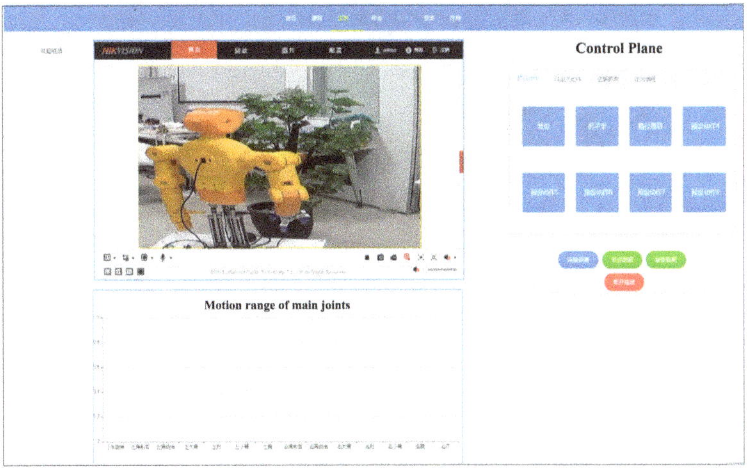

Figure 3.20. Positive kinematics algorithm for programming-level robot physical experimentation.

(5) Programming-level robot positive kinematics physical experiments.

Prepare a flowchart of the designed positive kinematics algorithm, and click the "Online Programming" tab, as shown in Figure 3.20.

Enter the "Online Programming" interface, use the flow chart to write C, Java, or JavaScript language program, click "Connect to the Device" "Send Data" buttons, and then click the "Run" button, as shown in Figure 3.21.

3.4 Summary

This chapter introduces the concept of robot space and the transformation relationship between coordinate systems. It also provides a brief overview of the link description within the robotic system. Additionally, the improved Denavit–Hartenberg modeling method is applied to establish link coordinate systems for the robot perception system, the main operating system, and the auxiliary operating system, based on the actual structure of this robotic platform. Furthermore, the forward kinematics are thoroughly examined and analyzed.

Figure 3.21. Programming level robot physical experiment positive kinematics programming.

3.5 Exercises

3.1 As shown in the accompanying figure, the planar manipulator with two degrees of freedom has joint 1 as a rotational joint and joint 2 as a moving joint, and the joint variables are θ_1 and d_2.

(1) Establish a joint coordinate system and write the equation of motion for the manipulator.

(2) Find the position value of the center of the hand according to the following joint variable parameters.

θ_1	0°	30°	45°	90°
d_2/m	0.30	0.50	0.80	1.00

3.2 Consider again the planar manipulator with two degrees of freedom shown in the figure of Problem 3.1. The values of the coordinates of the center of the hand are known to be X_0 and Y_0. Find the inverse solution θ_1, d_2 of the equations of motion of the manipulator, and verify them in practice.

3.3 A manipulator with three degrees of freedom is shown in the accompanying figure, with the angles of rotation θ_1, θ_2, and θ_3, the link lengths l_1 and l_2, and the distance of the center

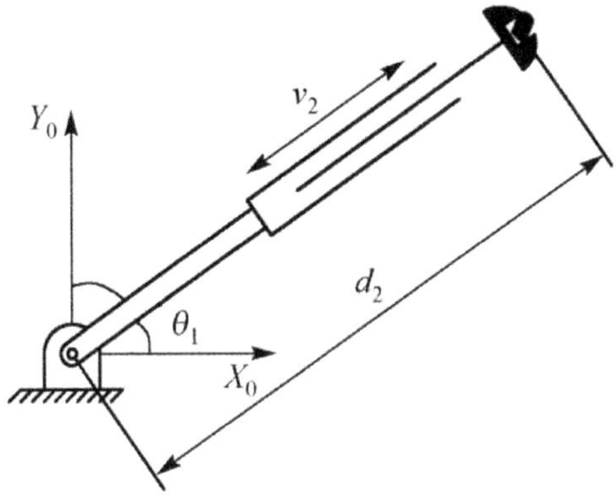

Exercise 3.1 figure

of the hand from the center of the wrist H. Try to establish a link coordinate system, derive the kinematic equations of the manipulator, and verify them in practice.

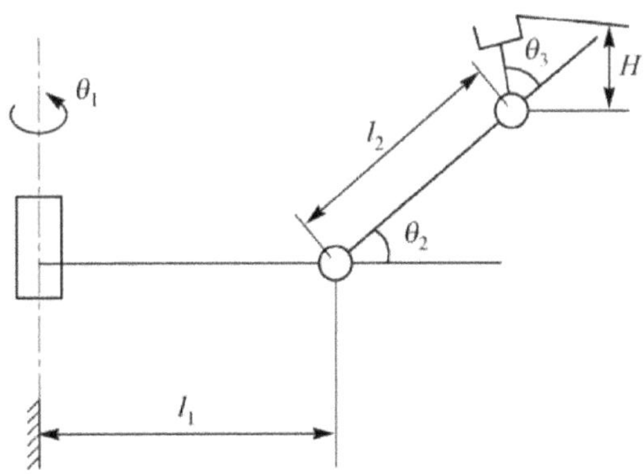

Exercise 3.3 figure

Chapter 4

Design and Practice of Robotic Inverse Kinematics

4.1 Theory of Robot Inverse Kinematics

4.1.1 *Definition of inverse kinematics*

Chapter 3 gave an example of robot positive kinematics, i.e., the robot positive solution problem, and this chapter explores robot inverse kinematics, i.e., the robot inverse solution problem.

For a robot with n degrees of freedom, the kinematic equations can be written as

$$\begin{bmatrix} n_x & o_x & a_x & p_x \\ n_y & o_y & a_y & p_y \\ n_z & o_z & a_z & p_z \\ 0 & 0 & 0 & 1 \end{bmatrix} = \boldsymbol{A}_1 \boldsymbol{A}_2 \boldsymbol{A}_3 \cdots \boldsymbol{A}_{n-1} \boldsymbol{A}_n \tag{4.1}$$

The left-hand side of Eq. (4.1) represents the position and orientation of the end-effector with respect to the base coordinate system. The process of calculating the corresponding joint variables given the end-effector's position and orientation is called robot inverse kinematics, and its inverse operation is called finding the inverse kinematic solution.[18] The inverse operation is called finding the kinematic inverse solution.

4.1.2 *Solution of robot inverse kinematics*

4.1.2.1 *Multiple solutions*

The robot kinematics inverse solution has multiple solutions, as shown in Figure 4.1. For a given end-effector position and orientation, its joint angle variable has two sets of solutions, and both joint angle combinations can be used to realize the target pose.

The inverse solution of robot kinematics produces multiple solutions for the following reasons:

(1) Due to the large number of trigonometric function equations in robot kinematics, solving inverse kinematics inevitably requires solving the inverse trigonometric function equations, and solving the inverse trigonometric functions will produce redundant solutions that the robot cannot realize.
(2) Structurally, there exist multiple ways of combining joint angles (multiple solutions) to achieve the target position and orientation.

While robot inverse kinematics often produces multiple solutions, for a real robot, only one set of solutions is usually optimal, for which judgment must be made to select the appropriate solution.

The general method for eliminating redundant solutions is as follows:

(1) Eliminate redundant solutions for joint angles in solving the inverse trigonometric function based on some parametric requirements (e.g., the link length is not negative).

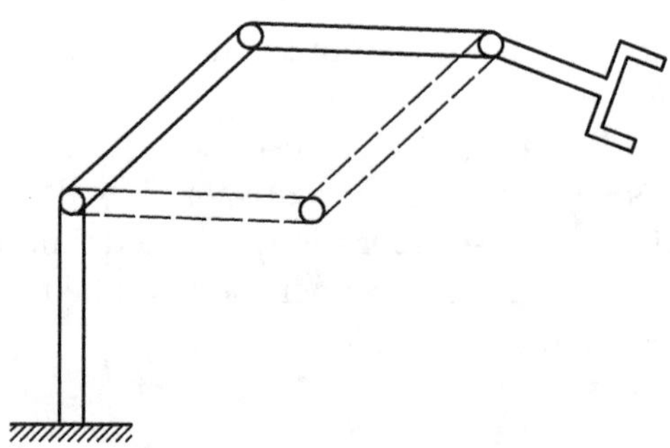

Figure 4.1. Multi-solution nature of the inverse solution of robot kinematics.

(2) Eliminate physically redundant solutions involving unrealizable joint angles based on the motion space of the joints.
(3) Based on the joint motion process, choose a solution that is the closest and easiest to realize.
(4) According to the obstacle avoidance requirements, eliminate the redundant solutions of joint angles limited by obstacles.
(5) Remove redundant solutions level by level.

4.1.2.2 *Solvability*

The solvability of a robot is the ability to find the analytic formula of the inverse solution of the robot's kinematics.

All robot systems with rotating and moving joints are solvable when there are a total of six degrees of freedom (or fewer) in a single tandem chain. For a robot to be analytically solvable, it is important to design it so that the robot's structure is as simple as possible and that it tries to satisfy the special condition that there are a number of intersecting joint axes, or α_i, which equals 0° or ±90°.

For the solution of inverse kinematics, although 12 equations can be obtained through Eq. (4.1), if the 12 equations are to be solved together, their analytical solutions are often difficult to find due to the complexity of the equation expressions, so the method of solving the equations jointly is generally not used.

The solution of inverse kinematics typically involves the inverse $\boldsymbol{A}_i^{-1}$ left-multiplication of a series of transformation matrices, followed by identifying the elements on the right-hand side that are constants or simple expressions. These elements are then equated with their corresponding elements on the left-hand side. This process yields solvable trigonometric equations, from which the corresponding joint angles can be determined. By iterating this procedure, the values of all joint variables are ultimately obtained.

4.2 Integrated Design of Robot Inverse Kinematics

After acquiring the position and orientation information of the external target through the robot's binocular vision and perception systems, it becomes necessary to precisely control the end-effector of both the robot's primary and auxiliary operational systems to accurately reach the desired target point. This requires conducting an inverse kinematics analysis of the robot system.

Specifically, for the specific robot tasks as shown in Figure 2.3, once the target's position and orientation are captured by the sensing system, to enable the primary operational system's end-effector to accurately grasp the target, the inverse kinematics for this system, which has 12 degrees of freedom, must be analyzed. Additionally, if the target is unobstructed during the primary system's operation, the auxiliary system can proceed to grasp other targets. Conversely, if the target is obstructed, the auxiliary arm can be used to remove the obstacle. This necessitates the inverse kinematics analysis of the auxiliary system, which has six degrees of freedom, to determine the desired movement angles. The perception system's role is to convert the position and orientation data acquired by the binocular vision system from the binocular reference base coordinate system into the robot's base coordinate system. As such, only a forward kinematics analysis is required for the perception system. Next, we proceed with the inverse kinematics analysis for both the primary and auxiliary operational systems.

4.2.1 *Inverse kinematics of the host operating system*

According to the above, it is known that the main operating system has 12 degrees of freedom. The general robot kinematics are represented by a homogeneous coordinate transformation matrix, and the matrix has 16 elements, but there are only 12 elements containing the robot's link and joint parameters. For robots with multiple degrees of freedom, 12 nonlinear equations containing multiple parameters are listed when solving the inverse kinematics, and the solution of these equations is very complicated. Therefore, this book refers to the kinematic modeling and solution method of multi-degree-of-freedom robots based on the space robot base suspension theory, and it adopts the model separation method to solve the inverse kinematics of this multi-degree-of-freedom system, i.e., the kinematic model of the main operating system in Figure 3.8 is separated from the waist coordinate system $\{W\}$, and is divided into two parts, namely, the mobile cart to the waist of the robot and the waist of the robot to the end-effector of the main operating arm. They are then solved using inverse kinematics to obtain the inverse kinematic solution. The waist coordinate system $\{W\}$ is attached to both the lower and upper parts of the model.

The overall idea of the "model separation method" is to divide the model into upper and lower parts from one of the separation points according to the structural characteristics of the ground operation multi-degree-of-freedom robots, convert the target point from the target object to the separation point, and then use the position and orientation of the separation point as the target position and orientation of the lower part of the model for inverse kinematics analysis. However, the method must satisfy the condition that the separated upper and lower parts of the model must be seamlessly connected. Therefore, when selecting the separation point, the reachable space of the separation point of the upper half model must be within the reachable space of the end-effector of the lower half model.

Part I: Assume that the end-effector of the upper half of the model has successfully reached the target point, i.e., the upper half of the model is in a "levitated" state at this time. Establish the kinematic model of the end-effector up to the separation point, carry out the positive kinematics calculations, and get the position and orientation information of the separation point in the whole model base coordinate system.

Part 2: The separation point is also used as the end-effector of the second half of the model. The position and orientation information of the separation point are used as the target position and orientation of the end-effector of the lower half of the model, and the inverse kinematic analysis of the lower half of the model is carried out to realize the seamless connection between the lower half of the model and the upper half of the "levitated" system.

In the robot platform, the separation point is selected at the waist, which is denoted as the point W, and the coordinate system is $\{W\}$, i.e., the number of joints in the upper half of the model is five, and the number of joints in the lower half of the model is seven. A schematic diagram of the grasping operation of the robot platform is shown in Figure 4.2.

According to the above, the "model separation method" is summarized in three stages: initial, transition, and execution. In the initial stage, the position and orientation of the waist separation point in the virtual "levitation" state is solved when the upper part successfully grasps the target. In the transition stage, the amount of motion required by each joint is solved when the lower part of the

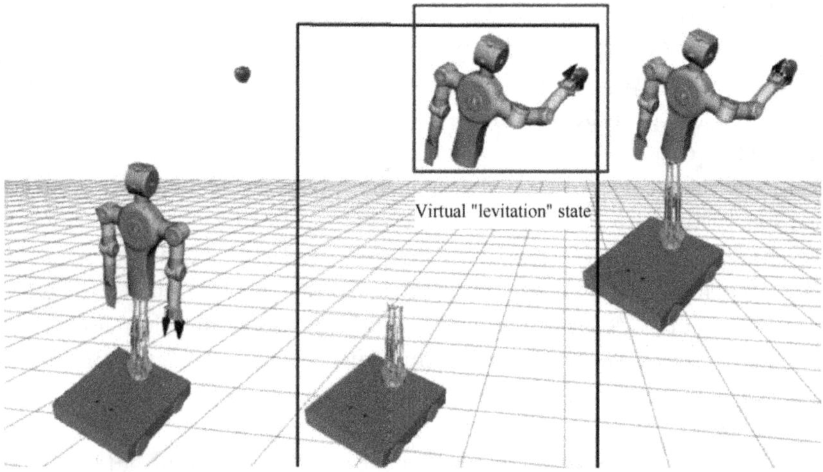

Figure 4.2. Schematic diagram of robotic platform gripping operation: (a) initial status, (b) transition status, and (c) grab status.

robot is docked with the "levitated" upper part. The execution stage involves analyzing the inverse kinematics of the upper part of the system after the waist end-effector of the lower part has reached the specified "levitation" position and orientation and the binocular camera has tracked the target again in order for the upper part of the end-effector to arrive at the target point more accurately.

4.2.1.1 *Initial stage: Solving for the position and orientation of the separation point in the "levitation" state*

Before the robot is ready to work, it is assumed that the upper body end-effector of the main operating system has successfully grasped the target object A, i.e., the coordinate system $\{A\}$ and the end-effector coordinate system $\{12'_{\mathrm{m}}\}$ in Figure 3.8 are completely coincident. At this time, the lower body of the robot has not yet moved, so the upper body is in a virtual "levitation" state. To find the position and orientation of the waist at the separation point in this "levitation" state, the kinematic model from the end-effector to the waist is established. As shown in Figure 4.3, the target point coordinate system {A} is set to be the base, while {W} is set to be the end effector. The links from the base A to the waist W are defined to be $i_{m-} (i_{m-} = 1_{m-}, \ldots, 5_{m-})$, with their coordinate system being $\{i_{m-}\}$.

In the model shown in Figure 4.3, the direction of the rotation axis of each joint coordinate system is considered consistent with the direction of the coordinate system in the model shown in Figure 3.8 for the convenience of calculation. The meanings of the parameters a and b are explained as follows:

a: The distance from the shoulder joint to the center of the chest, which is equal in size to $d_{8_{\mathrm{m}}}$ in the model shown in Figure 3.8.
b: The distance from the center of the chest to the waist, which is equal in size to $a_{7_{\mathrm{m}}}$ in the model shown in Figure 3.8.

Based on the above-described model, the DH parameters of the end-effector P to the waist separation point W are obtained as shown in Table 4.1.

Here, $d_{1_{\mathrm{m}-}}$ denotes the distance from the center of the actuator at the end of the main operating arm to the elbow, and $d_{3_{\mathrm{m}-}}$ denotes the distance from the elbow to the shoulder of the main operating arm. According to the model and Table 4.1, the transformation matrix

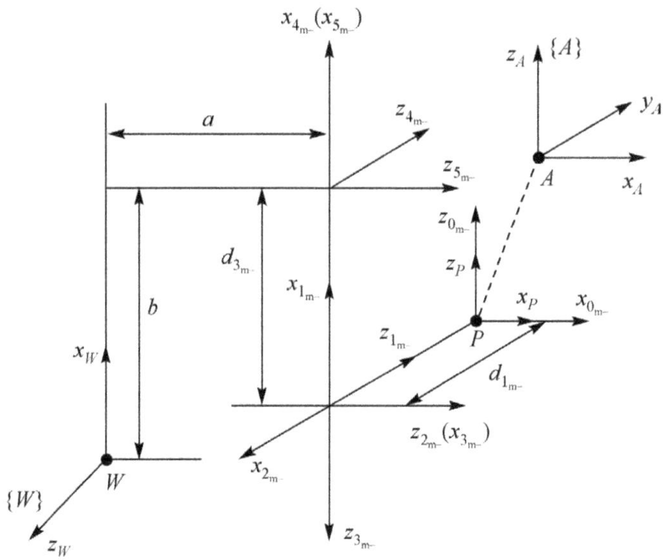

Figure 4.3. Kinematic model from end-effector to waist.

Table 4.1. DH parameters for end-effector P to waist separation point W.

Link i_{m-}	Joint angle $\theta_{i_{m-}}$	Torsion angle $\alpha_{(i-1)_{m-}}$	Link length $a_{(i-1)_{m-}}$	Link offset $d_{i_{m-}}$	$\theta_{i_{m-}}$ angular range
1_{m-}	$\theta_{i_m} - 90°$	$-90°$	0	$-d_{1_{m-}} = -300$ mm	$-90° \sim 90°$
2_{m-}	$\theta_{2_{m-}} + 90°$	$-90°$	0	0	$-20° \sim 120°$
3_{m-}	$\theta_{3_{m-}} - 90°$	$-90°$	0	$-d_{3_{m-}} = -183$ mm	$-90° \sim 90°$
4_{m-}	$\theta_{4_{m-}} - 90°$	$90°$	0	0	$-10° \sim 120°$
5_{m-}	$\theta_{5_{m-}}$	$-90°$	0	0	$-120° \sim 120°$

between each link can be obtained as ${}^{(i-1)_{m-}}_{i_{m-}}\boldsymbol{T}$, as shown in Eq. (4.2):

$$
{}^{0_{m-}}_{1_{m-}}\boldsymbol{T} = \begin{bmatrix} s_{1_{m-}} & c_{1_{m-}} & 0 & 0 \\ 0 & 0 & 1 & -d_{1_{m-}} \\ c_{1_{m-}} & -s_{1_{m-}} & 0 & 0 \\ 0 & 0 & 0 & 1 \end{bmatrix} \quad {}^{1_{m-}}_{2_{m-}}\boldsymbol{T} = \begin{bmatrix} -s_{2_{m-}} & -c_{2_{m-}} & 0 & 0 \\ 0 & 0 & 1 & 0 \\ -c_{2_{m-}} & s_{2_{m-}} & 0 & 0 \\ 0 & 0 & 0 & 1 \end{bmatrix}
$$

$$
{}^{2_{m-}}_{3_{m-}}\boldsymbol{T} = \begin{bmatrix} s_{3_{m-}} & c_{3_{m-}} & 0 & 0 \\ 0 & 0 & 1 & -d_{3_{m-}} \\ c_{3_{m-}} & -s_{3_{m-}} & 0 & 0 \\ 0 & 0 & 0 & 1 \end{bmatrix} \quad {}^{3_{m-}}_{4_{m-}}\boldsymbol{T} = \begin{bmatrix} s_{4_{m-}} & c_{4_{m-}} & 0 & 0 \\ 0 & 0 & -1 & 0 \\ -c_{4_{m-}} & s_{4_{m-}} & 0 & 0 \\ 0 & 0 & 0 & 1 \end{bmatrix}
$$

$$
{}^{4_{m-}}_{5_{m-}}\boldsymbol{T} = \begin{bmatrix} c_{5_{m-}} & -s_{5_{m-}} & 0 & 0 \\ 0 & 0 & 1 & 0 \\ -s_{5_{m-}} & -c_{5_{m-}} & 0 & 0 \\ 0 & 0 & 0 & 1 \end{bmatrix} \tag{4.2}
$$

Therefore, the position and orientation of the shoulder joint in the base coordinate system $\{0_{m-}\}{}^{0_{m-}}_{5_{m-}}\boldsymbol{T}$ is as shown in Eq. (4.3):

$$
{}^{0_{m-}}_{5_{m-}}\boldsymbol{T} = {}^{0_{m-}}_{1_{m-}}\boldsymbol{T}\,{}^{1_{m-}}_{2_{m-}}\boldsymbol{T}\,{}^{2_{m-}}_{3_{m-}}\boldsymbol{T}\,{}^{3_{m-}}_{4_{m-}}\boldsymbol{T}\,{}^{4_{m-}}_{5_{m-}}\boldsymbol{T} \tag{4.3}
$$

Since there are only five joints in the upper body of the robot and the starting joint is the shoulder joint, it is necessary to transform the shoulder joint coordinate system $\{5_{m-}\}$ into the separation point

W via the coordinate transformation relation $\boldsymbol{T}_0$, where $\boldsymbol{T}_0$ is

$$\boldsymbol{T}_0 = \boldsymbol{R}_X(-90^\circ) \times \boldsymbol{D}_X(-b) \times \boldsymbol{D}_Y(a) = \begin{bmatrix} 1 & 0 & 0 & -b \\ 0 & 0 & 1 & 0 \\ 0 & -1 & 0 & -a \\ 0 & 0 & 0 & 1 \end{bmatrix} \quad (4.4)$$

Therefore, by setting the angle values of each joint in the model shown in Figure 4.3, the position and orientation information of the waist separation point W with respect to the target A when the end-effector successfully grasps the target can be obtained as ${}^{A}_{W}\boldsymbol{T}$. In this book, the following method is employed to solve the initial angle values of each joint of the robot upper body, thereby determining the pose and position of the floating separation point.

Assume that the robot end-effector coincides completely with the given target pose, i.e., the end-effector is fixed at the target pose, while the separation point floats arbitrarily in space. Currently, the typical existing inverse kinematics solution methods have to solve the movement of each joint according to the target pose and fixed base coordinate system. Therefore, in this book, the FABRIK method is adopted to determine the position of each link when the robot end-effector reaches the target pose, and thus, the pose and position of the floating separation point could be solved. Specifically, it should satisfy that the pose and position ${}^{G}_{W}\boldsymbol{T}_{\text{float}}$ of the floating separation point obtained using the above-mentioned method be within the reachable working space of the robot's lower body, and in addition, each joint angle should be within the joint limit, as shown by Eq. (4.5):

$${}^{G}_{W}\boldsymbol{T}_{\text{float}} \in \mathbb{R}_L, \quad (\theta_{i_{m-}} \in [\theta_{i_{m-}\min}, \theta_{i_{m-}\max}], i = 1, 2, 3, \ldots) \quad (4.5)$$

Then, according to the floating base pose that meets the conditions, the joint values of the upper body are re-optimized and solved following the third step of the method in the execution stage. During this process, if the obtained joint angles exceed their limits, it is necessary to continue adopting the same method to re-determine the link positions of the robot's upper body, followed by re-solving the floating separation point pose and position that meet the requirements.

In summary, after selecting the angle values of each joint of the upper body, the position and orientation of the waist coordinate system $\{W\}$ with respect to the target object coordinate system $\{A\}$ can be obtained ${}^{A}_{W}\boldsymbol{T}$, as shown in Eq. (4.6):

$$ {}^{A}_{W}\boldsymbol{T} = {}^{0_{\mathrm{m}-}}_{5_{\mathrm{m}-}}\boldsymbol{T}\boldsymbol{T}_0 \tag{4.6} $$

4.2.1.2 *Transition phase: Lower body end-effector docking to designated "levitation" state*

The position and orientation of the target A in the geodetic coordinate system $\{G\}$ is obtained using the binocular camera and the sensing system. At the same time, combining with Eq. (4.6), we can solve the position and orientation of ${}^{G}_{W}\boldsymbol{T}_{\mathrm{float}}$ relative to the geodetic coordinate system $\{G\}$ when the waist is "levitated", as shown in Eq. (4.7):

$$ {}^{G}_{W}\boldsymbol{T}_{\mathrm{float}} = {}^{G}_{A}\boldsymbol{T}{}^{A}_{W}\boldsymbol{T} \tag{4.7} $$

In order to allow the robot's lower body end-effector to complete the docking with its "levitated" upper body start end, inverse kinematics analysis of the lower part model is required, and the target pose is denoted by ${}^{G}_{W}\boldsymbol{T}_{\mathrm{float}}$. The kinematic model of the moving cart up to the waist and its DH parameters are obtained, as shown in Figure 4.4 and Table 4.2, respectively.

The DH parameters of the kinematic model for moving the cart to the waist are shown in Table 4.2.

Here, the transformation matrices of each joint in this model are the first seven matrices in Eq. (4.1). And the last joint coordinate system $\{7_{\mathrm{m}}\}$ in this model is coincident with the waist coordinate system $\{W\}$, and its transformation relationship is the unit matrix $\boldsymbol{I}$, as shown in Eq. (4.8):

$$ {}^{7_{\mathrm{m}}}_{W}\boldsymbol{T} = \boldsymbol{I}_{4\times 4} \tag{4.8} $$

Therefore, the position and orientation ${}^{G}_{W}\boldsymbol{T}$ of the robot's lower body end-effector W with respect to the geodetic coordinate system $\{G\}$ is shown in Eq. (4.9):

$$ {}^{G}_{W}\boldsymbol{T} = {}^{0_{\mathrm{m}}}_{1_{\mathrm{m}}}\boldsymbol{T}{}^{1_{\mathrm{m}}}_{2_{\mathrm{m}}}\boldsymbol{T}{}^{2_{\mathrm{m}}}_{3_{\mathrm{m}}}\boldsymbol{T}{}^{3_{\mathrm{m}}}_{4_{\mathrm{m}}}\boldsymbol{T}{}^{4_{\mathrm{m}}}_{5_{\mathrm{m}}}\boldsymbol{T}{}^{5_{\mathrm{m}}}_{6_{\mathrm{m}}}\boldsymbol{T}{}^{6_{\mathrm{m}}}_{7_{\mathrm{m}}}\boldsymbol{T}{}^{7_{\mathrm{m}}}_{W}\boldsymbol{T} \tag{4.9} $$

In order to find the amount of motion required for moving the cart to each joint of the waist, let the target position and orientation ${}^{G}_{W}\boldsymbol{T}_{\mathrm{float}}$

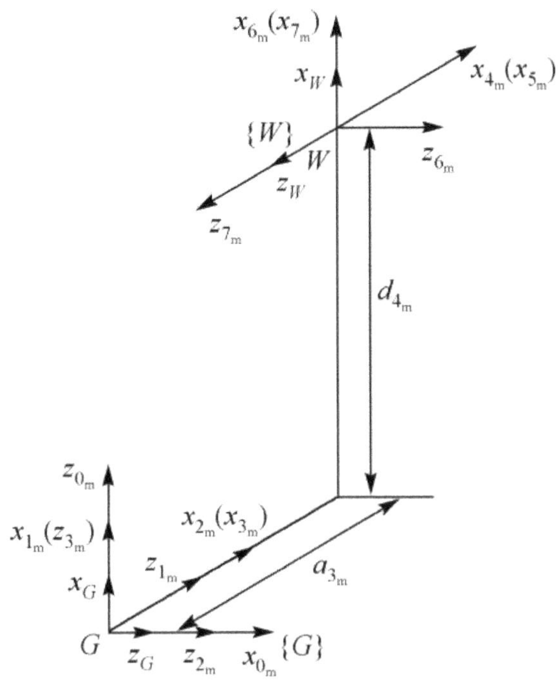

Figure 4.4. Kinematic model of moving the cart to the waist.

Table 4.2. DH parameters of the kinematic model for moving the trolley to the lumbar region.

Link i_m	Joint angle θ_{i_m}	Torsion angle $\alpha_{(i-1)_m}$	Link length $a_{(i-1)_m}$	Link offset d_{i_m}	θ_{i_m} angular range
1_m	$-90°$	$-90°$	0	$d_{1_m} = d_y$ (variables)	A constant (math.)
2_m	$-90°$	$-90°$	0	$d_{2_m} = d_x$ (variables)	A constant (math.)
3_m	θ_{3_m}	$-90°$	0	0	$-180°\sim180°$
4_m	θ_{4_m}	$0°$	$a_{3_m} = 186$ mm	$d_{4_m} =$ 545 mm	$-180°\sim180°$
5_m	$0°$	$0°$	0	$d_{5_m} = h_z$ (variables)	A constant (math.)
6_m	$\theta_{6_m} + 90°$	$90°$	0	0	$-60°\sim60°$
7_m	θ_{7_m}	$-90°$	0	0	$-60°\sim60°$

be known and written as

$$ {}^{G}_{W}\boldsymbol{T}_{\text{float}} = \begin{bmatrix} a_{11} & a_{12} & a_{13} & p_x \\ a_{21} & a_{22} & a_{23} & p_y \\ a_{31} & a_{32} & a_{33} & p_z \\ 0 & 0 & 0 & 1 \end{bmatrix} \tag{4.10} $$

Then, it is obtained from ${}^{G}_{W}\boldsymbol{T} = {}^{G}_{W}\boldsymbol{T}_{\text{float}}$:

$$ \begin{aligned} a_{11} &= c_{7_{\text{m}}} s_{6_{\text{m}}} (c_{3_{\text{m}}} s_{4_{\text{m}}} + c_{4_{\text{m}}} s_{3_{\text{m}}}) - s_{7_{\text{m}}} (c_{3_{\text{m}}} c_{4_{\text{m}}} - s_{3_{\text{m}}} s_{4_{\text{m}}}) \\ a_{21} &= -s_{7_{\text{m}}} (c_{3_{\text{m}}} s_{4_{\text{m}}} + c_{4_{\text{m}}} s_{3_{\text{m}}}) - c_{7_{\text{m}}} s_{6_{\text{m}}} (c_{3_{\text{m}}} c_{4_{\text{m}}} - s_{3_{\text{m}}} s_{4_{\text{m}}}) \\ a_{31} &= c_{6_{\text{m}}} c_{7_{\text{m}}} \\ a_{12} &= -c_{7_{\text{m}}} (c_{3_{\text{m}}} c_{4_{\text{m}}} - s_{3_{\text{m}}} s_{4_{\text{m}}}) - s_{6_{\text{m}}} s_{7_{\text{m}}} (c_{3_{\text{m}}} s_{4_{\text{m}}} + c_{4_{\text{m}}} s_{3_{\text{m}}}) \\ a_{22} &= s_{6_{\text{m}}} s_{7_{\text{m}}} (c_{3_{\text{m}}} c_{4_{\text{m}}} - s_{3_{\text{m}}} s_{4_{\text{m}}}) - c_{7_{\text{m}}} (c_{3_{\text{m}}} s_{4_{\text{m}}} + c_{4_{\text{m}}} s_{3_{\text{m}}}) \\ a_{32} &= -c_{6_{\text{m}}} s_{7_{\text{m}}} \\ a_{13} &= c_{6_{\text{m}}} (c_{3_{\text{m}}} s_{4_{\text{m}}} + c_{4_{\text{m}}} s_{3_{\text{m}}}) \\ a_{23} &= -c_{6_{\text{m}}} (c_{3_{\text{m}}} c_{4_{\text{m}}} - s_{3_{\text{m}}} s_{4_{\text{m}}}) \\ a_{33} &= -s_{6_{\text{m}}} \\ p_x &= d_x - a_{3_{\text{m}}} s_{3_{\text{m}}} \\ p_y &= d_y + a_{3_{\text{m}}} c_{3_{\text{m}}} \\ p_z &= d_{4_{\text{m}}} + h_z \end{aligned} \tag{4.11} $$

After validation, one of the sets of angular values that fit within the limits of the joint is selected, as shown in Eq. (4.12):

$$ \begin{cases} \theta_{7_{\text{m}}} = a\tan 2(-a_{32}, a_{31}), \quad \theta_{6_{\text{m}}} = a\tan 2(-a_{33}, a_{31}/c_{7_{\text{m}}}) \\ \theta_{3_{\text{m}}} + \theta_{4_{\text{m}}} = a\tan 2(a_{13}/c_{6_{\text{m}}}, a_{23}/-c_{6_{\text{m}}}), \quad d_x = p_x + a_{3_{\text{m}}} s_{3_{\text{m}}} \\ d_y = p_y - a_{3_{\text{m}}} c_{3_{\text{m}}}, \quad h_z = p_z - d_{4_{\text{m}}} \end{cases} \tag{4.12} $$

Among them, the angle range for the two rotary joints, $\theta_{6_{\text{m}}}$ and $\theta_{7_{\text{m}}}$, of the robot's waist bending and side waist is $[-60°, 60°]$, so $c_{6_{\text{m}}}$ and $c_{7_{\text{m}}}$ in Eq. (4.12) are not equal to zero.

Meanwhile, in this model, only the sum of the angles θ_3 and θ_4 can be obtained. Therefore, in this book, when the sum of the two angles exceeds the maximum limit of the robot's foot rotation, the mobile cart also needs to rotate by a certain angle; when the sum of the two angles does not exceed the maximum limit of the robot's foot rotation, priority is given to the rotation of the robot's foot, and the mobile cart does not rotate.

4.2.1.3 *Implementation phase: Analysis of the upper body capture process*

(1) The camera tracks the target continuously.

According to the above-mentioned initial and transition phases, the calculated joint angles of the lower body are sent to the underlying drive system through the control system. After the lower body of the robot reaches the designated "levitation" position and orientation, in order to grasp the target more accurately, the binocular camera is used to continuously acquire the position and orientation information of the target A, and inverse kinematics analysis is performed on the upper body. However, in this process, due to the movement of the mobile cart and the joints of the robot from the waist down, the binocular camera will not be able to find the target A, as seen earlier.

Therefore, in this book, it is preset that after the two joints of the robot's neck, the pitch and the left-right rotation, respectively, rotate by δ_1 and δ_2, the binocular camera tracks the previously seen target A, and the tracked target A is in the middle pixel value of the camera's field of view, i.e., the target is in the direction of the x and y axes in the binocular camera's base coordinate system $\{C\}$ with a relatively small deviation value, and the distance in the field of view in the front, i.e., in the direction of z, is restricted to be in the range of 0–τ_d. The position and orientation of the target with respect to the binocular camera base coordinate system $\{C\}$ can be obtained, denoted as (X_C, Y_C, Z_C), and its constraint bars are $X_C \in [-10, 10]$, $Y_C \in [-10, 10]$, and $Z_C \in [0, 600]$, all in millimeters (mm). Here, only the position is considered, while the attitude of the target at this point is set as a unit matrix.

Substituting the angle values of the moving cart to each joint of the waist and the two angle values of the head δ_1 and δ_2 obtained by Eq. (4.12) into Eq. (3.16), the position and orientation of the target in the geodetic coordinate system can be obtained after the waist arrives at the specified "levitation" position and orientation ${}^{G}_{A}\boldsymbol{T}'$. Before the whole platform starts to move, i.e., when the target is recognized by the binoculars, the position and orientation of the target relative to the geodetic coordinate system is ${}^{G}_{A}\boldsymbol{T}$. If we do not consider the error during the whole platform movement, the position of ${}^{G}_{A}\boldsymbol{T}$ and ${}^{G}_{A}\boldsymbol{T}'$ should be exactly the same. But in fact, there is a deviation between the two positions.

According to the practical parameters of the camera considered in this book, the position deviation values in the three directions of ${}^{G}_{A}\boldsymbol{T}$, i.e., x, y, and z, in ${}^{G}_{A}\boldsymbol{T}'$ are limited to $\pm 10\,\text{mm}$, $\pm 10\,\text{mm}$, and $\pm 100\,\text{mm}$, respectively, and used as constraints; in order to allow the target to be at the center of the binocular's field of view, the target function is defined as the minimum of the sum of the absolute values of the target's position in the directions of x and y, i.e., $\min(\text{Abs}(X_C)+\text{Abs}(Y_C))$. The objective function and constraints are as follows:

$$\begin{aligned}&\text{MinFunction} \quad \text{Abs}(X_C) + \text{Abs}(Y_C)\\&\text{s.t.} \quad \begin{cases}\text{Abs}({}^{G}_{A}\boldsymbol{T}(1,4) - {}^{G}_{A}\boldsymbol{T}'(1,4)) \leq 10\\ \text{Abs}({}^{G}_{A}\boldsymbol{T}(2,4) - {}^{G}_{A}\boldsymbol{T}'(2,4)) \leq 10\\ \text{Abs}({}^{G}_{A}\boldsymbol{T}(3,4) - {}^{G}_{A}\boldsymbol{T}'(3,4)) \leq 100\end{cases}\end{aligned} \tag{4.13}$$

By applying the Gauss–Newton method[1]9 and generalized global optimization algorithm,[2]0 we solve the above equations to get the angle values of the rotation of the two rotary joints of the head δ_1 and δ_2, so as to realize the tracking and localization of the target, as shown in Figure 4.5.

(2) Gauss–Newton method and generalized global optimization algorithm.

During the computation of a system of nonlinear equations, if an iterative method is used, a suitable initial value for the iteration must be given. However, the selection of initial values is often difficult. And the most important feature of the generalized global optimization algorithm is that it avoids the difficult problem of setting initial values in using iterative methods in the field of optimization computation.

(a) (b)

Figure 4.5. Tracking of the target by the vision system: (a) before tracking; (b) after tracking.

A least squares problem is as shown in Eq. (4.14):

$$\min_{\boldsymbol{x}} \frac{1}{2} ||f(\boldsymbol{x})||_2^2 \tag{4.14}$$

where the independent variable $\boldsymbol{x} \in \mathbb{R}^n$ and the function f is any nonlinear function with m dimension: $f(\boldsymbol{x}) \in \mathbb{R}^m$. In addition, it is necessary to think about how to find its optimized solution. In general, when the least squares solution cannot be obtained directly, it is necessary to use the gradient descent method to keep searching until the algorithm converges. In order to obtain the descent vector $\Delta\boldsymbol{x}$, which minimizes the value of $||f(\boldsymbol{x}+\Delta\boldsymbol{x})||_2^2$, the Gauss–Newton method is used for the solution.

The Gauss–Newton method is an improved method based on Newton's method, which is an optimization algorithm for nonlinear least squares to solve nonlinear least squares problems. The basic idea of the Gauss-Newton method involves approximating a nonlinear regression model using a Taylor series expansion. Through iterative adjustments, regression coefficients are incrementally refined to converge toward the optimal values for the nonlinear model. Such a process eventually minimizes the residual sum of squares (RSS) of the original model.

Perform a first-order Taylor expansion of $f(\boldsymbol{x})$:

$$f(\boldsymbol{x}+\Delta\boldsymbol{x}) \approx f(\boldsymbol{x}) + \boldsymbol{J}(\boldsymbol{x})\Delta\boldsymbol{x} \tag{4.15}$$

where, $\boldsymbol{J}(\boldsymbol{x})$ is the derivative of $f(\boldsymbol{x})$ with respect to the independent variable $\boldsymbol{x}$ of order $m \times n$, which is called the Jacobi matrix. In order to obtain the descent vector, the following linear least squares equation needs to be solved:

$$\Delta \boldsymbol{x}^* = \arg\min_{\boldsymbol{x}} \frac{1}{2} \left\| f(\boldsymbol{x}) + \boldsymbol{J}(\boldsymbol{x})\Delta \boldsymbol{x} \right\|_2^2 \tag{4.16}$$

According to the extreme value condition, expand Eq. (4.16) and conduct the derivative on $\Delta \boldsymbol{x}$ so that the derivative is equal to zero to obtain the system of equations shown in Eq. (4.17):

$$\boldsymbol{J}(\boldsymbol{x})^{\mathrm{T}} \boldsymbol{J}(\boldsymbol{x})\Delta \boldsymbol{x} = -\boldsymbol{J}(\boldsymbol{x})^{\mathrm{T}} f(\boldsymbol{x}) \tag{4.17}$$

In Eq. (4.17), the variable to be solved is $\Delta \boldsymbol{x}$, so the system of equations is a system of linear equations, which can also be referred to as the Gauss–Newton equations or formal equations. Define $\boldsymbol{J}(\boldsymbol{x})^{\mathrm{T}} \boldsymbol{J}(\boldsymbol{x})$ on the left-hand side of the equation as $\boldsymbol{H}$ and the right-hand side of the equation as $\boldsymbol{g}$, then Eq. (4.17) becomes

$$\boldsymbol{H}\Delta \boldsymbol{x} = \boldsymbol{g} \tag{4.18}$$

During the operation of the Gauss–Newton method, $\boldsymbol{J}^{\mathrm{T}}\boldsymbol{J}$ is approximated as a second-order Hessian matrix, i.e., the computation process of matrix $\boldsymbol{H}$ is reduced. In summary, the steps of the Gauss–Newton method are as follows:

Step 1: Give the initial value $\boldsymbol{x}_0$.
Step 2: Find the Jacobi matrix $\boldsymbol{J}(\boldsymbol{x}_k)$ and the error $f(\boldsymbol{x}_k)$ during the first k iterations.
Step 3: Solve the incremental equation: $\boldsymbol{H}\Delta \boldsymbol{x}_k = \boldsymbol{g}$.
Step 4: If the descent vector $\Delta \boldsymbol{x}_k$ is small enough, stop the operation. Otherwise, set $\boldsymbol{x}_{k+1} = \boldsymbol{x}_k + \Delta \boldsymbol{x}_k$ and return to step 2 to continue the run until the end.

(3) Angle solution for each joint of the robot's upper body.

When the binocular camera tracks the target A, an upper-body kinematic model is built and then kinematically analyzed. This upper-body kinematic model uses the robot chest center coordinate system $\{O\}$ as its base coordinate system, which contains a total of five joints of the main operating arm. The kinematic model of the chest to the right arm end-effector is shown in Figure 4.6.

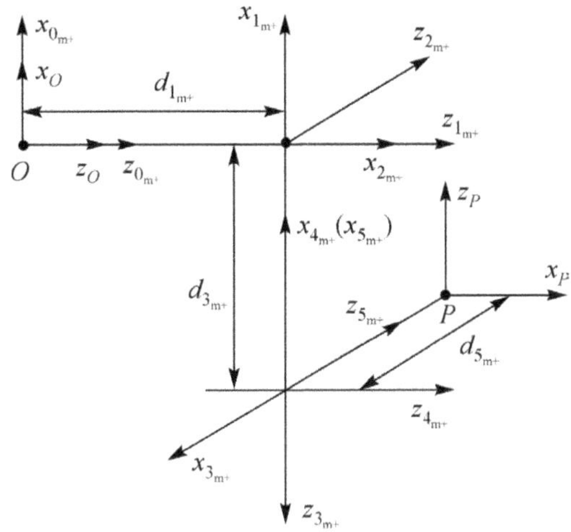

Figure 4.6. Kinematic model of the chest up to the right arm end-effector.

Table 4.3. The Kinematic DH parametric model for the end-effector from its chest to the end of the right arm.

Link i_{m+}	Joint angle $\theta_{i_{m+}}$	Torsion angle $\alpha_{(i-1)_{m+}}$	Link length $a_{(i-1)_{m+}}$	Link offset $d_{i_{m+}}$	$\theta_{i_{m+}}$ angular range
1_{m+}	$\theta_{1_{m+}}$	0°	0	$d_{1_{m+}} =$ 200 mm	−120°∼120°
2_{m+}	$\theta_{2_{m+}} + 90°$	90°	0	0	−120°∼10°
3_{m+}	$\theta_{3_{m+}} + 90°$	−90°	0	$d_{3_{m+}} =$ 183 mm	−90°∼90°
4_{m+}	$\theta_{4_{m+}} - 90°$	90°	0	0	−120°∼20°
5_{m+}	$\theta_{5_{m+}}$	90°	0	0	−90°∼90°

As shown in Figure 4.6, in order to differentiate from the DH parameters and coordinate system in Figure 4.3, the link in the upper-body model is denoted as i_{m+}, and each joint coordinate system is denoted as $\{i_{m+}\}$. At the same time, the coordinate system is established to be completely consistent with the main operating arm joint coordinate system in the model shown in Figure 3.8. According to this model, its DH parameters are obtained, as shown in Table 4.3.

Here, $d_{1_{m+}}$ is the distance from the center of the chest to the shoulder of the main operating arm and $d_{3_{m+}}$ is the distance from the shoulder of the main operating arm to the elbow. According to Table 4.3, the homogeneous coordinate transformation matrix of each joint in this model can be obtained as ${}^{(i-1)_{\mathrm{m+}}}_{i_{\mathrm{m+}}}\boldsymbol{T}$, as shown in Eq. (4.19):

$$
{}^{0_{\mathrm{m+}}}_{1_{\mathrm{m+}}}\boldsymbol{T} = \begin{bmatrix} c_{1_{\mathrm{m+}}} & -s_{1_{\mathrm{m+}}} & 0 & 0 \\ s_{1_{\mathrm{m+}}} & c_{1_{\mathrm{m+}}} & 0 & 0 \\ 0 & 0 & 1 & d_{1_{\mathrm{m+}}} \\ 0 & 0 & 0 & 1 \end{bmatrix} \quad {}^{1_{\mathrm{m+}}}_{2_{\mathrm{m+}}}\boldsymbol{T} = \begin{bmatrix} -s_{2_{\mathrm{m}}} & -c_{2_{\mathrm{m}}} & 0 & 0 \\ 0 & 0 & -1 & 0 \\ c_{2_{\mathrm{m}}} & -s_{2_{\mathrm{m}}} & 0 & 0 \\ 0 & 0 & 0 & 1 \end{bmatrix}
$$

$$
{}^{2_{\mathrm{m+}}}_{3_{\mathrm{m+}}}\boldsymbol{T} = \begin{bmatrix} -s_{3_{\mathrm{m+}}} & -c_{3_{\mathrm{m+}}} & 0 & 0 \\ 0 & 0 & 1 & d_{3_{\mathrm{m+}}} \\ -c_{3_{\mathrm{m+}}} & s_{3_{\mathrm{m+}}} & 0 & 0 \\ 0 & 0 & 0 & 1 \end{bmatrix} \quad {}^{3_{\mathrm{m+}}}_{4_{\mathrm{m+}}}\boldsymbol{T} = \begin{bmatrix} s_{4_{\mathrm{m+}}} & c_{4_{\mathrm{m+}}} & 0 & 0 \\ 0 & 0 & -1 & 0 \\ -c_{4_{\mathrm{m+}}} & s_{4_{\mathrm{m+}}} & 0 & 0 \\ 0 & 0 & 0 & 1 \end{bmatrix}
$$

$$
{}^{4_{\mathrm{m+}}}_{5_{\mathrm{m+}}}\boldsymbol{T} = \begin{bmatrix} c_{5_{\mathrm{m}}} & -s_{5_{\mathrm{m}}} & 0 & 0 \\ 0 & 0 & -1 & 0 \\ s_{5_{\mathrm{m}}} & c_{5_{\mathrm{m}}} & 0 & 0 \\ 0 & 0 & 0 & 1 \end{bmatrix} \tag{4.19}
$$

The model shown in Figure 4.6 is built by lifting the right arm of the robot from the elbow, so in order to transform the last joint in the model into the end-effector center coordinate system $\{P\}$, it is necessary to translate the coordinate system along the z axis direction by $d_{5_{\mathrm{m+}}} = 300\mathrm{mm}$. As a result, the position and orientation ${}^{O}_{P}\boldsymbol{T}$ of the end-effector center of the main working arm in the base coordinate system $\{O\}$ of the model can be obtained, as shown in Eq. (4.20):

$$
{}^{O}_{P}\boldsymbol{T} = {}^{0_{\mathrm{m+}}}_{1_{\mathrm{m+}}}\boldsymbol{T}\,{}^{1_{\mathrm{m+}}}_{2_{\mathrm{m+}}}\boldsymbol{T}\,{}^{2_{\mathrm{m+}}}_{3_{\mathrm{m+}}}\boldsymbol{T}\,{}^{3_{\mathrm{m+}}}_{4_{\mathrm{m+}}}\boldsymbol{T}\,{}^{4_{\mathrm{m+}}}_{5_{\mathrm{m+}}}\boldsymbol{T} \times \boldsymbol{D}_{Z}(d_{5_{\mathrm{m+}}}) \tag{4.20}
$$

After the binocular camera rotates by δ_1 and δ_2 with respect to the two joints of the robot head tilting and rotating left and right, the target is tracked and the position and orientation information of the target in the chest center coordinate system $\{O\}$ can be obtained according to Eq. (3.16), which is written as ${}^{O}_{P}\boldsymbol{T}_{\text{des}}$, as shown in Eq. (4.21):

$$ {}^{O}_{P}\boldsymbol{T}_{\text{des}} = \begin{bmatrix} b_{11} & b_{21} & b_{31} & b_{41} \\ b_{12} & b_{22} & b_{32} & b_{42} \\ b_{13} & b_{23} & b_{33} & b_{43} \\ 0 & 0 & 0 & 1 \end{bmatrix} \tag{4.21} $$

Taking ${}^{O}_{P}\boldsymbol{T}_{\text{des}}$ as the target position and orientation expected to be achieved by the end-effector of the robot's upper-body model, this book solves the inverse kinematics of this upper body system using analytical and numerical methods, respectively.

(1) The analytic method.

The process of inverse kinematics involves finding the value of each joint angle by equating each corresponding element in ${}^{O}_{P}\boldsymbol{T}$ and ${}^{O}_{P}\boldsymbol{T}_{\text{des}}$, and according to ${}^{O}_{P}\boldsymbol{T} = {}^{O}_{P}\boldsymbol{T}_{\text{des}}$, find out the relationship that exists between the corresponding elements, and the final inverse solution is shown in Eq. (4.22):

$$ \begin{aligned} \theta_{2_{\text{m+}}} &= \arctan 2\Bigg((d_{1_{\text{m+}}} - b_{34} + b_{33}d_{5_{\text{m+}}})/d_{3_{\text{m+}}}, \\ &\qquad \pm\sqrt{1 - [(d_{1_{\text{m+}}} - b_{34} + b_{33}d_{5_{\text{m+}}})/d_{3_{\text{m+}}}]^2}\Bigg) \\ \theta_{1_{\text{m+}}} &= \arctan 2((b_{24} - d_{5_{\text{m+}}}b_{23})/(-d_{3_{\text{m+}}}c_{2_{\text{m+}}}), \\ &\qquad (b_{14} - d_{5_{\text{m+}}}b_{13})/(-d_{3_{\text{m+}}}c_{2_{\text{m+}}})) \\ \theta_{4_{\text{m+}}} &= \arctan 2\Bigg(b_{33}s_{2_{\text{m+}}} + b_{13}c_{1_{\text{m+}}}c_{2_{\text{m+}}} + b_{23}c_{2_{\text{m+}}}s_{1_{\text{m+}}}, \\ &\qquad \pm\sqrt{1 - (b_{33}s_{2_{\text{m+}}} + b_{13}c_{1_{\text{m+}}}c_{2_{\text{m+}}} + b_{23}c_{2_{\text{m+}}}s_{1_{\text{m+}}})^2}\Bigg) \\ \theta_{3_{\text{m+}}} &= \arctan 2((c_{2_{\text{m+}}}b_{34} - c_{2_{\text{m+}}}d_{1_{\text{m+}}} - c_{1_{\text{m+}}}b_{14}s_{2_{\text{m+}}} - b_{24}s_{1_{\text{m+}}}s_{2_{\text{m+}}})/ \\ &\qquad (c_{4_{\text{m+}}}d_{5_{\text{m+}}}), (b_{14}s_{1_{\text{m+}}} - c_{1_{\text{m+}}}b_{24})/(c_{4_{\text{m+}}}d_{5_{\text{m+}}})) \\ \theta_{5_{\text{m+}}} &= \arctan 2((-b_{32}s_{2_{\text{m+}}} - b_{12}c_{1_{\text{m+}}}c_{2_{\text{m+}}} - b_{22}c_{2_{\text{m+}}}s_{1_{\text{m+}}})/c_{4_{\text{m+}}}, \\ &\qquad (b_{31}s_{2_{\text{m+}}} + b_{11}c_{1_{\text{m+}}}c_{2_{\text{m+}}} + b_{21}c_{2_{\text{m+}}}s_{1_{\text{m+}}})/c_{4_{\text{m+}}}) \end{aligned} \tag{4.22} $$

In this solution process, the unit the angle values is radians (rad). Also, since trigonometric functions are periodic, there will be multiple sets of solutions, and it is necessary to find the amount of motion that fits within the limits of each joint as the final solution among the sets of solutions obtained.

(2) The numerical method.

According to the equality of each corresponding element in ${}^{O}_{P}\boldsymbol{T}$ and ${}^{O}_{P}\boldsymbol{T}_{\text{des}}$ in Eqs. (4.20) and (4.21), the elements of the matrix containing the link parameters are selected to establish a system of nonlinear equations:

$$\begin{aligned} &\boldsymbol{F}(\boldsymbol{\theta}_{\text{m+}}) = 0, \quad \boldsymbol{F}(\boldsymbol{\theta}_{\text{m+}}) = (f_1, f_2, \ldots, f_{12})^{\text{T}}, \\ &\boldsymbol{\theta}_{\text{m+}} = (\theta_{1_{\text{m+}}}, \theta_{2_{\text{m+}}}, \ldots, \theta_{5_{\text{m+}}})^{\text{T}} \end{aligned} \tag{4.23}$$

During the numerical solution, each iteration yields a different transformation matrix, ${}^{O}_{P}\boldsymbol{T}_{\boldsymbol{\theta}_{\text{m+}}}(\theta^i_{1_{\text{m+}}}, \theta^i_{2_{\text{m+}}}, \theta^i_{3_{\text{m+}}}, \theta^i_{4_{\text{m+}}}, \theta^i_{5_{\text{m+}}})$, where i is the number of iterations, $i = 0, 1, 2, \ldots$. The matrix is obtained by subtracting the 12 corresponding elements of the first three rows of the transformation matrices ${}^{O}_{P}\boldsymbol{T}_{\boldsymbol{\theta}_{\text{m+}}}$ and ${}^{O}_{P}\boldsymbol{T}_{\text{des}}$:

$$\boldsymbol{F}(\boldsymbol{\theta}_{\text{m+}}) = {}^{O}_{P}\boldsymbol{T}_{\theta_{\text{m}-}} - {}^{O}_{P}\boldsymbol{T}_{\text{des}} = 0 \tag{4.24}$$

which are illustrated as below,

$$\begin{cases} f_1(\boldsymbol{\theta}_{\text{m+}}) = {}^{O}_{P}\boldsymbol{T}_{\boldsymbol{\theta}_{\text{m+}}}(1,1) - {}^{O}_{P}\boldsymbol{T}_{\textbf{des}}(1,1) \\ f_2(\boldsymbol{\theta}_{\textbf{m+}}) = {}^{O}_{P}\boldsymbol{T}_{\boldsymbol{\theta}_{\text{m+}}}(1,2) - {}^{O}_{P}\boldsymbol{T}_{\textbf{des}}(1,2) \\ f_3(\boldsymbol{\theta}_{\textbf{m+}}) = {}^{O}_{P}\boldsymbol{T}_{\boldsymbol{\theta}_{\text{m+}}}(1,3) - {}^{O}_{P}\boldsymbol{T}_{\textbf{des}}(1,3) \\ f_4(\boldsymbol{\theta}_{\textbf{m+}}) = {}^{O}_{P}\boldsymbol{T}_{\boldsymbol{\theta}_{\text{m+}}}(1,4) - {}^{O}_{P}\boldsymbol{T}_{\textbf{des}}(1,4) \\ f_5(\boldsymbol{\theta}_{\textbf{m+}}) = {}^{O}_{P}\boldsymbol{T}_{\boldsymbol{\theta}_{\text{m+}}}(2,1) - {}^{O}_{P}\boldsymbol{T}_{\textbf{des}}(2,1) \\ f_6(\boldsymbol{\theta}_{\textbf{m+}}) = {}^{O}_{P}\boldsymbol{T}_{\boldsymbol{\theta}_{\text{m+}}}(2,2) - {}^{O}_{P}\boldsymbol{T}_{\textbf{des}}(2,2) \\ f_7(\boldsymbol{\theta}_{\textbf{m+}}) = {}^{O}_{P}\boldsymbol{T}_{\boldsymbol{\theta}_{\text{m+}}}(2,3) - {}^{O}_{P}\boldsymbol{T}_{\textbf{des}}(2,3) \\ f_8(\boldsymbol{\theta}_{\textbf{m+}}) = {}^{O}_{P}\boldsymbol{T}_{\boldsymbol{\theta}_{\text{m+}}}(2,4) - {}^{O}_{P}\boldsymbol{T}_{\textbf{des}}(2,4) \\ f_9(\boldsymbol{\theta}_{\textbf{m+}}) = {}^{O}_{P}\boldsymbol{T}_{\boldsymbol{\theta}_{\text{m+}}}(3,1) - {}^{O}_{P}\boldsymbol{T}_{\textbf{des}}(3,1) \\ f_{10}(\boldsymbol{\theta}_{\textbf{m+}}) = {}^{O}_{P}\boldsymbol{T}_{\boldsymbol{\theta}_{\text{m+}}}(3,2) - {}^{O}_{P}\boldsymbol{T}_{\textbf{des}}(3,2) \\ f_{11}(\boldsymbol{\theta}_{\textbf{m+}}) = {}^{O}_{P}\boldsymbol{T}_{\boldsymbol{\theta}_{\text{m+}}}(3,3) - {}^{O}_{P}\boldsymbol{T}_{\textbf{des}}(3,3) \\ f_{12}(\boldsymbol{\theta}_{\textbf{m+}}) = {}^{O}_{P}\boldsymbol{T}_{\boldsymbol{\theta}_{\text{m+}}}(3,4) - {}^{O}_{P}\boldsymbol{T}_{\textbf{des}}(3,4) \end{cases} \tag{4.25}$$

where ${}^{O}_{P}\boldsymbol{T}_{\boldsymbol{\theta}_{\mathrm{m+}}}(m,n)$ and ${}^{O}_{P}\boldsymbol{T}_{\mathrm{des}}(m,n)$ are the elements of the mth row and nth column of the corresponding transformation matrices ${}^{O}_{P}\boldsymbol{T}_{\boldsymbol{\theta}_{\mathrm{m+}}}$ and ${}^{O}_{P}\boldsymbol{T}_{\mathrm{des}}$, respectively.

Based on Eq. (4.25), the Jacobi matrix of the system of equations can be determined, as shown in Eq. (4.26):

$$\boldsymbol{J}_i(\boldsymbol{\theta}^i_{\mathrm{m+}}) = \begin{bmatrix} \dfrac{\partial f_1}{\partial \theta^i_{1_{\mathrm{m+}}}} & \dfrac{\partial f_1}{\partial \theta^i_{2_{\mathrm{m+}}}} & \cdots & \dfrac{\partial f_1}{\partial \theta^i_{5_{\mathrm{m+}}}} \\ \dfrac{\partial f_2}{\partial \theta^i_{1_{\mathrm{m+}}}} & \dfrac{\partial f_2}{\partial \theta^i_{2_{\mathrm{m+}}}} & \cdots & \dfrac{\partial f_2}{\partial \theta^i_{5_{\mathrm{m+}}}} \\ \vdots & \vdots & & \vdots \\ \dfrac{\partial f_{12}}{\partial \theta^i_{1_{\mathrm{m+}}}} & \dfrac{\partial f_{12}}{\partial \theta^i_{2_{\mathrm{m+}}}} & \cdots & \dfrac{\partial f_{12}}{\partial \theta^i_{5_{\mathrm{m+}}}} \end{bmatrix} \tag{4.26}$$

Solve the Newton's iterative formula for this system of equations as shown in Eq. (4.27):

$$\boldsymbol{\theta}^{i+1}_{\mathrm{m+}} = \theta^i_{\mathrm{m+}} - \boldsymbol{J}^{-1}\boldsymbol{F}(\boldsymbol{\theta}^i_{\mathbf{m+}}) \tag{4.27}$$

where the Jacobi matrix $\boldsymbol{J}$ is a matrix of order 12×5, and only its pseudo-inverse can be found, so Eq. (4.27) becomes

$$\boldsymbol{\theta}^{i+1}_{\mathrm{m+}} = \theta^i_{\mathrm{m+}} - (\boldsymbol{J}_i^{\boldsymbol{T}}\boldsymbol{J}_i)^{-1}\boldsymbol{J}_i^{\boldsymbol{T}}\boldsymbol{F}(\boldsymbol{\theta}^i_{\mathbf{m+}}) \tag{4.28}$$

where $\boldsymbol{J}^{\mathrm{T}}$ denotes the transpose of the Jacobi matrix $\boldsymbol{J}$. Therefore, Eq. (4.28) can be utilized to solve the inverse problem of the working arm.

Another advantage of applying Newton's iterative method used in this book is that the initial values of the individual joint angle iterations will be chosen more rationally. Although there are small errors in the whole system, the inverse solution obtained for the robot's upper body system will inevitably be in the neighborhood of the initial set of joint angle values given in the previous section, i.e., approximately Eq. (4.5a) or Eq. (4.6b). Therefore, using this as the initial value for iteration will iterate more quickly to find the angle values that meet the joint limits.

It should be noted that the positive and negative signs of the corresponding joint motion values are opposite due to the different

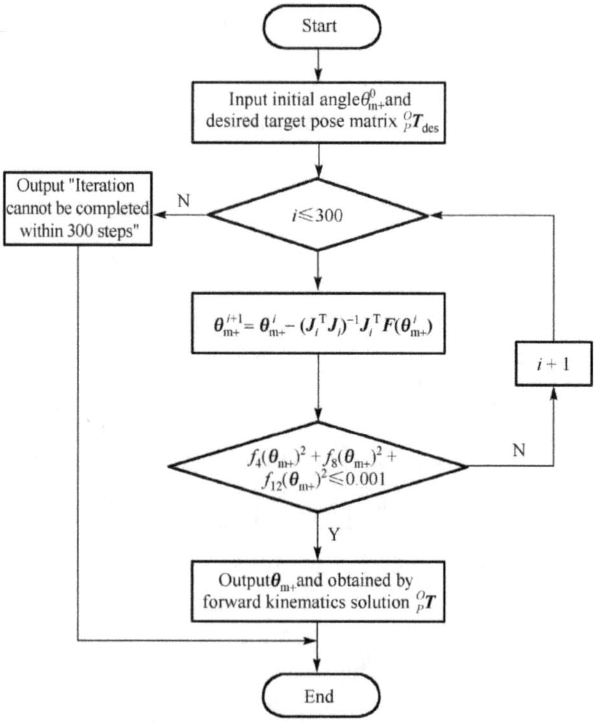

Figure 4.7. Iterative algorithm flowchart.

reference base coordinate systems selected in the two models from end-effector to chest center and from chest center to end-effector in this book. In addition, when grasping the target object, the most influential factor is the position accuracy, so only the position error is considered here, and the iterative algorithm flow is shown in Figure 4.7.

Here, ε denotes the precision value after implementing the iterations. Once the iteration is complete, a set of joint angles that meet the limits of each joint is derived, which is sent down to the bottom layer, and the operating arm can be driven to successfully grasp the target.

So far, the motion values of all the joints of the main operating system have been obtained, and in order to verify the correctness of the algorithm, the 12 joint motion values can be substituted into Eqs. (3.18) and (3.19) for the verification of the positive kinematics.

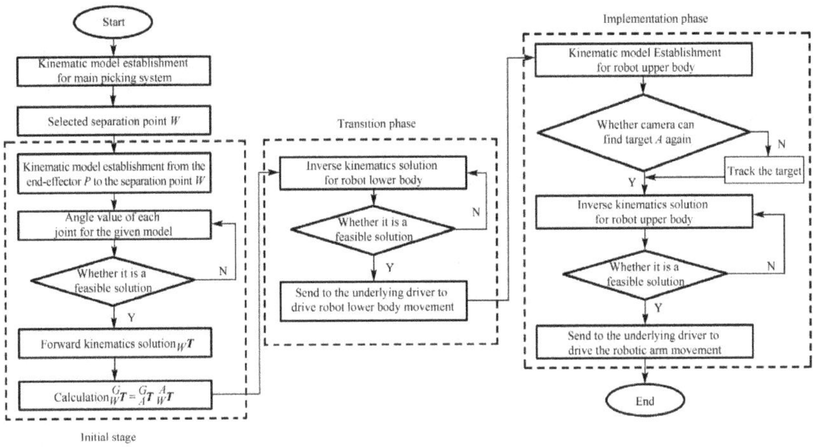

Figure 4.8. Overall flowchart of the inverse kinematic analysis of the main operating system.

Combining the above initial, transition, and execution phases, the overall flowchart of the inverse kinematic analysis of the main operating system is obtained, as shown in Figure 4.8.

4.2.2 *Inverse kinematics of the auxiliary operating system*

The robot platform adopted in this book uses the auxiliary operating arm as an auxiliary operating system to clear away obstacles that obscure the target for the main operating system and facilitate the grasping operation of the main operating arm. Therefore, after acquiring the position and orientation of the obstacle through binocular vision, the motion of each joint of the auxiliary operating system is solved using inverse kinematics analysis. The obstacle to be cleared is denoted as B; the position and orientation of the obstacle acquired by the binocular camera are denoted as ${}^{C}_{B}\boldsymbol{T}$; the base coordinate system of the auxiliary operating system is also selected as the chest center coordinate system $\{O\}$; the point N represents the wrist of the auxiliary operating arm, and the coordinate system $\{N\}$ is defined as the wrist coordinate system, which is completely consistent with the coordinate system $\{6_{\mathrm{a}}\}$ in the model shown in Figure 3.9.

When the motions of the two joints of head pitching and left-right rotation are known, according to Eq. (3.16), the orientation of the obstacle B with respect to the base coordinate system $\{O\}$ can be obtained as ${}^O_B\boldsymbol{T}$:

$$ {}^O_B\boldsymbol{T} = {}^O_C\boldsymbol{T}{}^C_B\boldsymbol{T} \tag{4.29} $$

Robot systems with six rotary joints need to satisfy the Pieper criterion, i.e., the axes of three neighboring joints should intersect at one point or three neighboring joint axes should be parallel to each other when solving the inverse solution. The auxiliary operating arm containing six rotary joints has a more complicated analytical solution calculation process.

In order for the auxiliary operating arm end-effector to reach the target position and orientation, the following condition needs to be met:

$$ {}^O_N\boldsymbol{T} \times \boldsymbol{D}_Y(-d) = {}^O_B\boldsymbol{T} \tag{4.30} $$

Let the end-effector's target position and orientation be known and are denoted as

$$ {}^O_B\boldsymbol{T}({}^{5_a}_{6_a}\mathrm{T} \times \mathrm{D}_Y(-d))^{-1} = \begin{bmatrix} x_{11} & x_{12} & x_{13} & x_{14} \\ x_{21} & x_{22} & x_{23} & x_{24} \\ x_{31} & x_{32} & x_{33} & x_{34} \\ 0 & 0 & 0 & 1 \end{bmatrix} \tag{4.31} $$

The correspondence between each element in the matrix ${}^O_B\boldsymbol{T}$ is obtained according to positive kinematics:

$$ \begin{aligned} {d_{5_a}}^2 + 2d_{3_a}d_{5_a}c_{4_a} + {d_{3_a}}^2 &= (x_{14} + dx_{12})^2 \\ &\quad + (x_{24} + dx_{22})^2 + (x_{34} + dx_{32} + d_{1_a})^2 \end{aligned} \tag{4.32} $$

Then, the relational equation for the sixth joint in the auxiliary arm system is obtained, as shown in Eq. (4.33):

$$ \begin{aligned} {d_{3_a}}^2 &= (x_{14} + x_{12}(d + d_{5_a}c_{6_a}) + x_{11}d_{5_a}s_{6_a})^2 + (x_{24} + x_{22}(d + d_{5_a}c_{6_a}) \\ &\quad + x_{21}d_{5_a}s_{6_a})^2 + (x_{34} + x_{32}(d + d_{5_a}c_{6_a}) + x_{31}d_{5_a}s_{6_a} + d_{1_a})^2 \end{aligned} \tag{4.33} $$

In this book, the bisection method is applied to solve for the sixth joint in Eq. (4.33), which will not be detailed in this book due to space constraints. Similarly, the analytical solutions of the first five joints of the auxiliary arm, θ_{1_a}, θ_{2_a}, θ_{3_a}, θ_{4_a}, and θ_{5_a}, are obtained according to their correspondences with other elements in the matrix and combined with Eq. (4.32), as shown in Eq. (4.34):

$$
\begin{aligned}
\theta_{4_a} &= a\tan 2(\pm\sqrt{1-p_4^2}, p_4),\\
p_4 &= ((x_{14}+dx_{12})^2+(x_{24}+dx_{22})^2+(x_{34}+dx_{32}+d_{1_a})^2\\
&\quad -({d_{5_a}}^2-{d_{3_a}}^2)/(2d_{3_a}d_{5_a})\\
\theta_{2_a} &= a\tan 2(p_2, \pm\sqrt{1-p_2^2}),\\
p_2 &= -(x_{34}+x_{32}(d+d_{5_a}c_{6_a})+x_{31}d_{5_a}s_{6_a}+d_{1_a})/d_{3_a}\\
\theta_{3_a} &= a\tan 2(p_3, \pm\sqrt{1-p_3^2}),\\
p_3 &= (x_{34}+dx_{32}+d_{1_a}+d_{5_a}c_{4_a}s_{2_a}+d_{3_a}s_{2_a})/(-c_{2_a}s_{4_a}d_{5_a})\\
\theta_{5_a} &= a\tan 2(s_{2_a}s_{4_a}-c_{2_a}c_{4_a}s_{3_a}, c_{2_a}c_{3_a})\\
&\quad - a\tan 2\left((c_{6_a}(c_{4_a}s_{2_a}+c_{2_a}s_{3_a}s_{4_a})-x_{32})/s_{6_a},\right.\\
&\quad \left.\pm\sqrt{\begin{aligned}&(c_{2_a}c_{3_a})^2+(s_{2_a}s_{4_a}-c_{2_a}c_{4_a}s_{3_a})^2\\ &\quad -((c_{6_a}(c_{4_a}s_{2_a}+c_{2_a}s_{3_a}s_{4_a})-x_{32})/s_{6_a})^2\end{aligned}}\right)\\
\theta_{1_a} &= a\tan 2((x_{24}+x_{22}(d+d_{5_a}c_{6_a})+x_{21}d_{5_a}s_{6_a})/(d_{3_a}c_{2_a}),\\
&\quad (x_{14}+x_{12}(d+d_{5_a}c_{6_a})+x_{11}d_{5_a}s_{6_a})/(-d_{3_a}c_{2_a}))
\end{aligned}
\tag{4.34}
$$

In summary, the inverse solution of the auxiliary operation arm can be derived, and the derived solution can be substituted into the positive kinematics model for the verification of positive kinematics as a way to verify the correctness of the algorithm. At the same time, one set of solutions within the actual joint limits of the robot is selected and sent down to the bottom layer through the control system to drive the end-effector of the robot's assisted operation arm to reach the desired goal point ${}_B^O\boldsymbol{T}$.

4.2.3 *Inverse kinematics of the 3-RPS parallel mechanism*

In this book, the moving platform of the parallel mechanism is made to turn by an angle β around the X axis and by an angle α around the Y axis by driving the tandem branch, and the translation along the Z axis is z, which needs to be converted into the length of the driving tandem branch motion. In essence, this process is the conversion of the parallel mechanism from its joint space to the driver space. This process is referred to in this book as the inverse kinematics solution process for parallel mechanisms. In addition, the rotation angles β, α, and the translation distance z of this moving platform correspond to θ_6, θ_7, and d_5 in the main robot's operating system and sensing system, respectively.

According to Figure 3.10, at the initial state of the parallel mechanism, the initial position and orientation matrix of the moving platform coordinate system $O_2 - X_2Y_2Z_2$ in the fixed platform coordinate system $O_1 - X_1Y_1Z_1\,\boldsymbol{T}_{\text{init}}$ is as shown in Eq. (4.35):

$$\boldsymbol{T}_{\text{init}} = \begin{bmatrix} 1 & 0 & 0 & 0 \\ 0 & 1 & 0 & 0 \\ 0 & 0 & 1 & -l_0 \\ 0 & 0 & 0 & 1 \end{bmatrix} \tag{4.35}$$

According to the above-mentioned rotational relationship, when the angles β and α as well as $d_5 = z$ are known values, the transformation matrix $\boldsymbol{T}$ of the transformed moving platform coordinate system $O_2' - X_2'Y_2'Z_2'$ with respect to the fixed platform coordinate system $O_1 - X_1Y_1Z_1$ can be obtained:

$$\begin{aligned} \boldsymbol{T} &= \boldsymbol{T}_{\text{init}} \times \boldsymbol{D}_Z(h_0)\boldsymbol{R}_Y(\alpha)\boldsymbol{R}_X(\beta) \\ &= \begin{bmatrix} c\beta & s\alpha s\beta & c\alpha s\beta & 0 \\ 0 & c\alpha & -s\alpha & 0 \\ -s\beta & s\alpha c\beta & c\alpha c\beta & -l_0 - d_5 \\ 0 & 0 & 0 & 1 \end{bmatrix} \end{aligned} \tag{4.36}$$

When the movable platform only rotates, the center distance between the upper and lower platforms does not change, and the center distance between the two changes only when the movable platform has a lifting motion $|d_5|$, and its value is the sum of the initial distance and the lifting distance of the movable platform $|l_0 + d_5|$.

The value of d_5 can be positive or negative. According to Figure 3.10, when d_5 is negative, it means that the moving platform descends in the positive direction along the Z_1 axis, and when d_5 is positive, it means that the moving platform rises in the negative direction along the Z_1 axis.

For the above-defined transformation matrix $\boldsymbol{T}$, its orientation is denoted as a matrix of order 3×3, $\boldsymbol{T}_\mathrm{R}$, and its position is denoted as a matrix of order 3×1, $\boldsymbol{P}$, as shown in Eq. (4.37):

$$\boldsymbol{T}_\mathrm{R} = \begin{bmatrix} c\beta & s\alpha s\beta & c\alpha s\beta \\ 0 & c\alpha & -s\alpha \\ -s\beta & s\alpha c\beta & c\alpha c\beta \end{bmatrix}; \quad \boldsymbol{P} = \begin{bmatrix} 0 \\ 0 \\ -l_0 - d_5 \end{bmatrix} \tag{4.37}$$

The coordinates of the position of the vertices of the square triangles on the moving platform, $\boldsymbol{A}_2'$, $\boldsymbol{B}_2'$, and $\boldsymbol{C}_2'$, in the fixed coordinate system $O_1 - X_1Y_1Z_1$ are written, respectively, as

$$\begin{aligned} \boldsymbol{A}_2' &= (X_{\boldsymbol{A}_2'}, Y_{\boldsymbol{A}_2'}, Z_{\boldsymbol{A}_2'}) \\ \boldsymbol{B}_2' &= (X_{\boldsymbol{B}_2'}, Y_{\boldsymbol{B}_2'}, Z_{\boldsymbol{B}_2'}) \\ \boldsymbol{C}_2' &= (X_{\boldsymbol{C}_2'}, Y_{\boldsymbol{C}_2'}, Z_{\boldsymbol{C}_2'}) \end{aligned} \tag{4.38}$$

Therefore, according to Eqs. (3.22) and (4.38), the coordinate transformation relationship between the position coordinates $\boldsymbol{A}_2'$, $\boldsymbol{B}_2'$, and $\boldsymbol{C}_2'$ after the transformation of the moving platform and the position of the vertices of the positive triangles $\boldsymbol{A}_2$, $\boldsymbol{B}_2$, and $\boldsymbol{C}_2$ in the initial state can be obtained, as shown in Eq. (4.39):

$$\begin{aligned} (\boldsymbol{A}_2')^\mathrm{T} &= \boldsymbol{T}_\mathrm{R} \times \boldsymbol{A}_2^\mathrm{T} + \boldsymbol{P} \\ (\boldsymbol{B}_2')^\mathrm{T} &= \boldsymbol{T}_\mathrm{R} \times \boldsymbol{B}_2^\mathrm{T} + \boldsymbol{P} \\ (\boldsymbol{C}_2')^\mathrm{T} &= \boldsymbol{T}_\mathrm{R} \times \boldsymbol{C}_2^\mathrm{T} + \boldsymbol{P} \end{aligned} \tag{4.39}$$

the above equations could also be written as below,

$$\begin{bmatrix} X_{A_2'} \\ Y_{A_2'} \\ Z_{A_2'} \end{bmatrix} = \boldsymbol{T}_{\mathrm{R}} \times \begin{bmatrix} R \\ 0 \\ 0 \end{bmatrix} + \boldsymbol{P}; \quad \begin{bmatrix} X_{B_2'} \\ Y_{B_2'} \\ Z_{B_2'} \end{bmatrix} = \boldsymbol{T}_{\mathrm{R}} \times \begin{bmatrix} -\dfrac{R}{2} \\ -\dfrac{\sqrt{3}R}{2} \\ 0 \end{bmatrix} + \boldsymbol{P};$$

$$\begin{bmatrix} X_{C_2'} \\ Y_{C_2'} \\ Z_{C_2'} \end{bmatrix} = \boldsymbol{T}_{\mathrm{R}} \times \begin{bmatrix} -\dfrac{R}{2} \\ \dfrac{\sqrt{3}R}{2} \\ 0 \end{bmatrix} + \boldsymbol{P} \tag{4.40}$$

From this, the expansion and contraction of the drive link l_1, l_2, and l_3 can be obtained, as shown in Eq. (4.41):

$$\begin{aligned} l_1 &= |\boldsymbol{A}_2\boldsymbol{A}_2'| - l_0 \\ l_2 &= |\boldsymbol{B}_2\boldsymbol{B}_2'| - l_0 \\ l_3 &= |\boldsymbol{C}_2\boldsymbol{C}_2'| - l_0 \end{aligned} \tag{4.41}$$

In the formula, $|\boldsymbol{A}_2\boldsymbol{A}_2'|$, $|\boldsymbol{B}_2\boldsymbol{B}_2'|$, and $|\boldsymbol{C}_2\boldsymbol{C}_2'|$ denote the distances between the vertices of equilateral triangles on the upper and lower platforms, and l_0 is the initial distance between the upper and lower platforms.

When β, α, and z correspond to the joint space θ_6, θ_7, and d_5, respectively, the expansion and contraction of the driving link, l_1, l_2, and l_3, can be obtained, as shown in Eq. (4.42):

$$\begin{aligned} l_1 &= \sqrt{(c_6R - R)^2 + (-d_5 - l_0 - s_6R)^2} - l_0 \\ l_2 &= \sqrt{\begin{aligned} &(0.5R - 0.5Rc_6 - 0.5\sqrt{3}Rs_6s_7)^2 + (0.5\sqrt{3}R - 0.5\sqrt{3}Rc_7)^2 \\ &+(-d_5 - l + 0.5Rs_6 - 0.5\sqrt{3}Rc_6s_7)^2 - l_0 \end{aligned}} \\ l_3 &= \sqrt{\begin{aligned} &(0.5R + 0.5\sqrt{3}Rs_6s_7 - 0.5Rc_6)^2 + (0.5\sqrt{3}Rc_7 - 0.5\sqrt{3}R)^2 \\ &+(-d_5 - l + 0.5Rs_6 + 0.5\sqrt{3}Rc_6s_7)^2 - l_0 \end{aligned}} \end{aligned} \tag{4.42}$$

In summary, the motion of the robot's waist can be converted from its joint space to the drive space of the parallel mechanism.

4.3 Integrated Practice of Robotic Inverse Kinematics

(1) Familiarize yourself with robot inverse kinematics.
(2) Parameter-level robot inverse kinematics virtual simulation experiment.

Select "Experiment" → "Robot Control" command to enter the control panel, as shown in Figure 4.9.

According to the designed inverse kinematics parameters in the control panel, select "Grab Apple" in the "Preset Action" to view the whole process of grabbing the target, as shown in Figure 4.10.
(3) Programming-level robot inverse kinematics virtual simulation experiments.

Click the "Online Programming" tab, enter the "Online Programming" interface, write a C, Java, or JavaScript language programs according to the designed algorithms, and click the "Run" button to view the results, as shown in Figure 4.11.

(4) Physical experiments on inverse kinematics of parameter-level robots.

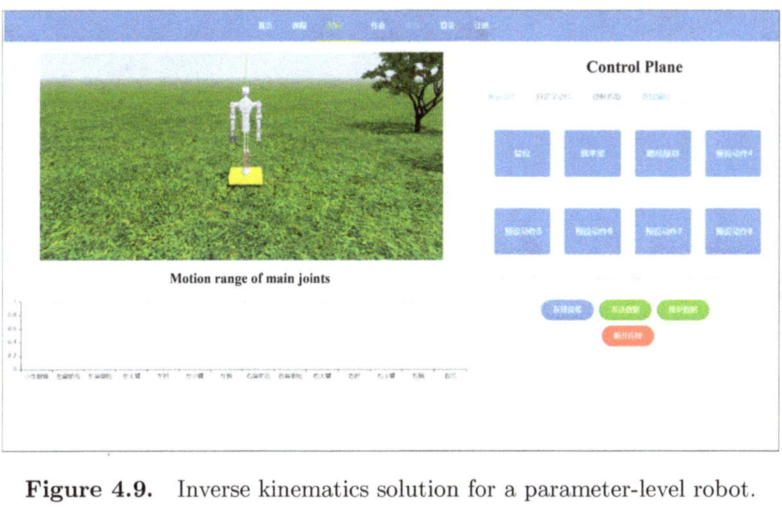

Figure 4.9. Inverse kinematics solution for a parameter-level robot.

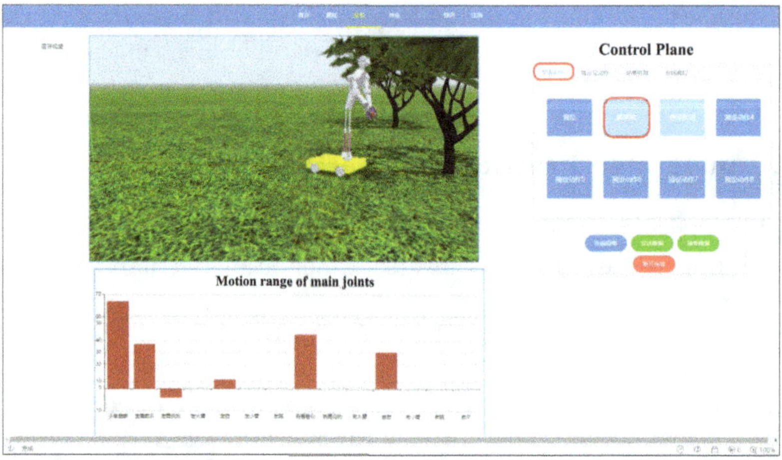

Figure 4.10. Inverse kinematic parameters.

Figure 4.11. Programming-level inverse kinematics.

Select "Experiment" → "Robot Control" command to enter the control panel to obtain the physical experiment interface, as shown in Figure 4.12.

In the control panel, according to the designed parameters, select the inverse kinematics, enter the parameters, and click the "Confirm Parameters" button, as shown in Figure 4.13.

(5) Programming-level robot inverse kinematics physical experiment.

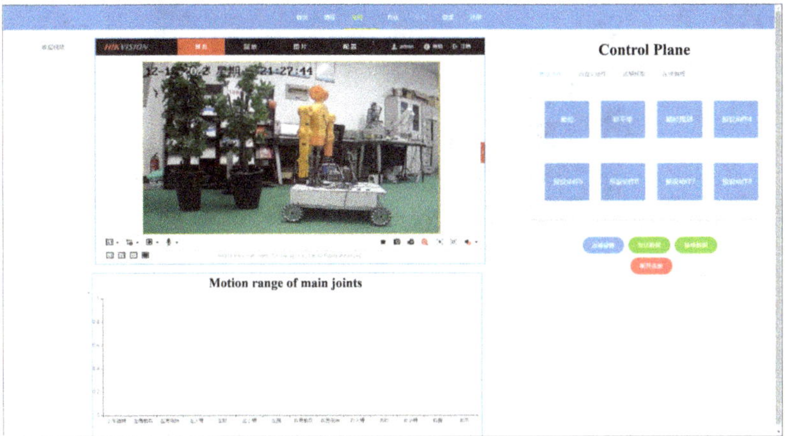

Figure 4.12. Parameter-level physical inverse kinematics experiment.

Figure 4.13. Parameter-level physical inverse kinematics operation.

Prepare a flowchart of the designed inverse kinematics algorithm, and click the "Online Programming" tab, as shown in Figure 4.14.

After entering the "Online Programming" interface, write a C, Java, or JavaScript language program, click the "Connect Device" "Send Data" button, and then click the "Run" button, as shown in Figure 4.15.

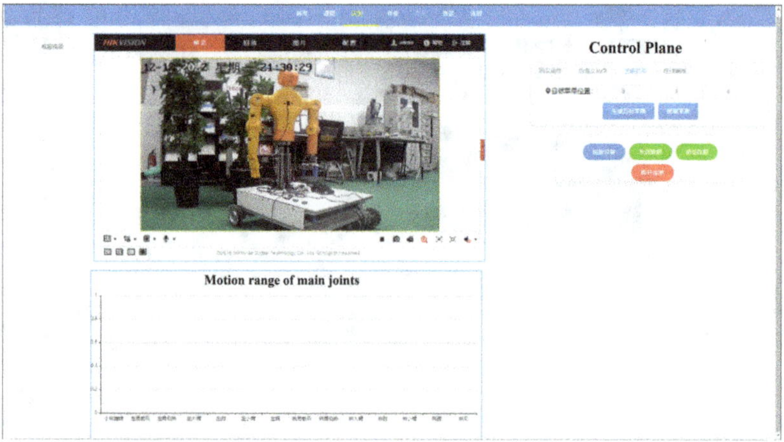

Figure 4.14. Programming-level physical inverse kinematics experiment.

Figure 4.15. Programming-level physical inverse kinematics programming experiment.

4.4 Summary

In the inverse kinematics solution process, the complexity escalates when the robot system has an increased number of degrees of freedom. This is because the trigonometric polynomials in the kinematics equations expand geometrically with more degrees of freedom, leading to equations characterized by high-order nonlinearity and strong coupling. These factors make solving the inverse kinematics highly complicated. This chapter introduces and examines specific inverse kinematics solution methods for a main operating system with 12 degrees of freedom and an auxiliary operating system with six degrees of freedom, which does not meet the Pieper criterion. For both systems, it addresses how to calculate the required motions for each rotating or moving joint to position and orient the end-effector at the desired target location. Additionally, it explores the transformation of motion in the 3-RPS parallel mechanism between joint space and actuator space.

4.5 Exercises

4.1 The accompanying figure shows a manipulator with two degrees of freedom; the lengths of the two links are 0.5 m. Try to establish the coordinate system of each link, find the homogeneous coordinate transformation matrix $\boldsymbol{A}_1$ and $\boldsymbol{A}_2$ and the kinematic inverse of the manipulator, and complete the practical verification.

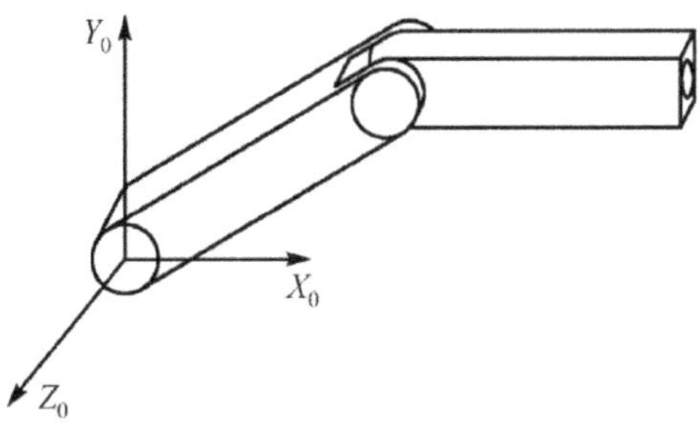

Exercise 4.1 figure

4.2 The accompanying figure shows a manipulator with three degrees of freedom. Try to establish the DH coordinate system of each link, find the homogeneous coordinate transformation matrices $\boldsymbol{A}_1$, $\boldsymbol{A}_2$, and $\boldsymbol{A}_3$, and complete the practical verification.

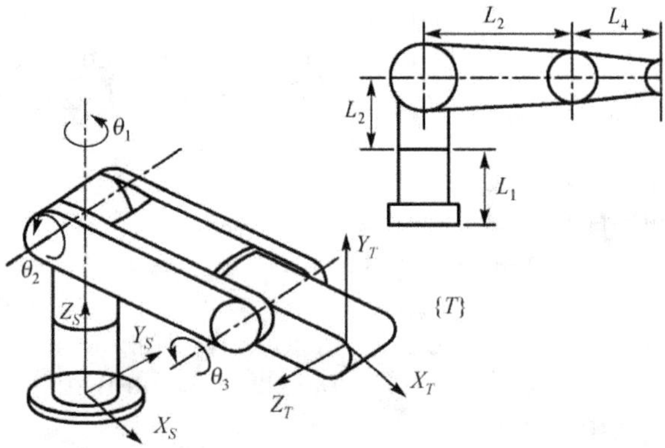

Exercise 4.2 figure

4.3 Design and practice the inverse kinematics algorithms for 3-RPS parallel robots.

4.4 Explain which of the main problems of robotics are addressed by inverse kinematics.

Chapter 5

Robot Visual Perceptual and SLAM

5.1 Introduction

For humanoid mobile robots to complete operational tasks, the first step is to understand the environment. This involves using sensors to obtain real-time, effective, and reliable information about the external scene. Subsequently, it's essential to build an accurate localization system, followed by adopting a suitable navigation strategy for motion planning. This chapter delves into the study of humanoid mobile robots and aims to establish an accurate and reliable indoor global environment by designing an RGB-D SLAM system that boasts high accuracy and real-time performance. Initially, we explore the common camera imaging model and various matrices related to position and orientation transformations under the camera coordinate system. This provides theoretical support for the subsequent recovery of 3D scene information through the vision system. Then, focusing on the five common components of the RGB-D SLAM framework — front-end visual odometry, motion estimation, back-end optimization, loop closure detection, and environment construction — we compare multiple methods for each component. By selecting the most appropriate algorithm for each, we aim to realize the overall framework of the RGB-D SLAM. The flowchart of the RGB-D SLAM framework is illustrated in Figure 5.1.

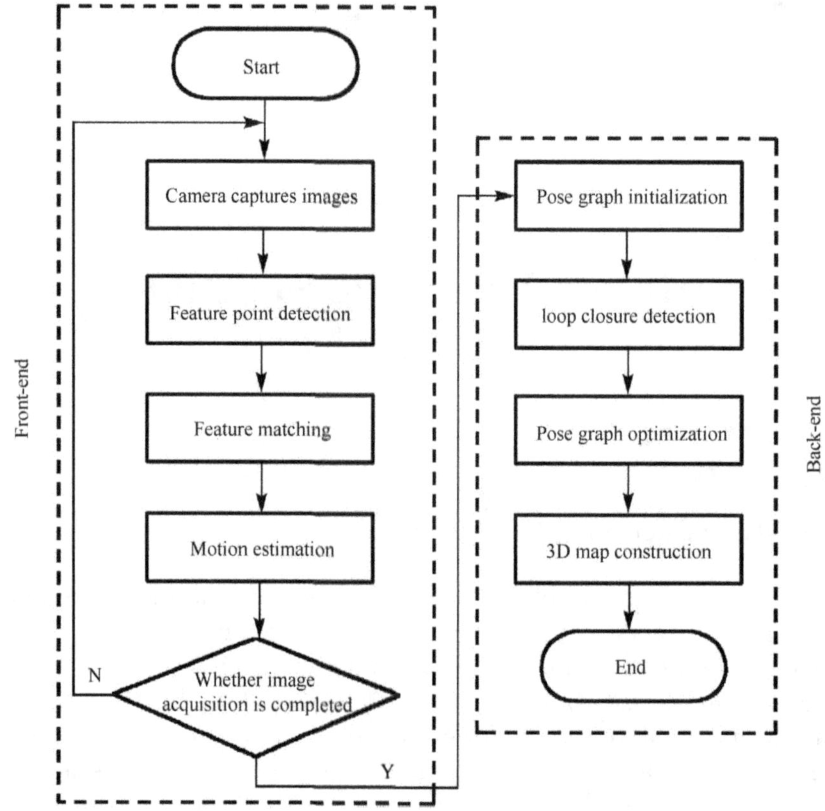

Figure 5.1. RGB_D SLAM framework flowchart.

5.2 Theory of Robot Visual Perceptual System

5.2.1 *Camera imaging model*

One of the most common camera imaging models available is the pinhole model,[22] as shown in Figure 5.2.

In Figure 5.2, $O - xyz$ is the camera coordinate system, the z axis of this coordinate system is perpendicular to the camera plane, x axis direction is to the right, y axis direction is downward, and O is the pinhole model of the hole and also the optical center of the camera. In the world space, there is a point $\boldsymbol{P}$, whose coordinates are $(X, Y, Z)^{\mathrm{T}}$, its physical imaging point in the camera after the

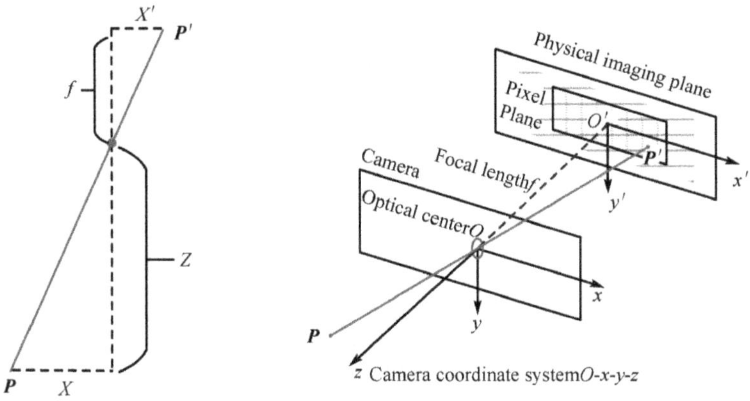

(a) (b)

Figure 5.2. Pinhole model: (a) two-dimensional diagram and (b) three-dimensional stereogram.

projection through the hole O is $\boldsymbol{P}'$, and the coordinates of $\boldsymbol{P}'$ are $(X', Y', Z')^{\mathrm{T}}$, assuming that the physical focal length is f. Based on the principle of similar triangles, we can get formula (5.1):

$$\frac{Z}{f} = -\frac{X}{X'} = -\frac{Y}{Y'} \tag{5.1}$$

In this equation, the negative sign represents the inverted and reversed image under the pinhole model. In order to simplify the model, the resulting image is usually symmetrically inverted to the front of the camera of equal magnitude. Then, we can obtain formula (5.2):

$$\frac{Z}{f} = \frac{X}{X'} = \frac{Y}{Y'} \tag{5.2}$$

This is organized to obtain Eq. (5.3):

$$\begin{aligned} X' &= f\frac{X}{Z} \\ Y' &= f\frac{Y}{Z} \end{aligned} \tag{5.3}$$

Equation (5.3) describes the mapping relationship from the 3D space to the 2D plane as a mathematical expression of the pinhole model.

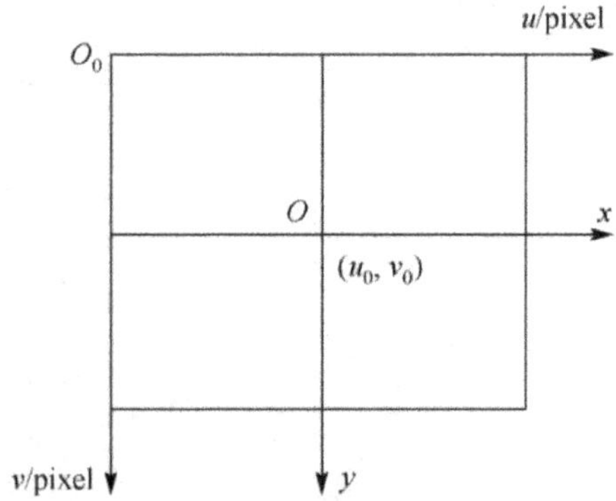

Figure 5.3. Pixel plane.

5.2.2 *Transformation between coordinate systems*

5.2.2.1 *Pixel coordinate system*

In computer vision, a pixel plane $o - u - v$ is often assumed to conveniently describe the conversion of light into image pixels by a sensor, which is shown in Figure 5.3. The coordinates on the plane are called pixel coordinates (u, v), and there exist $M \times N$ pixel points on the pixel plane, stored in a matrix of equal size, with a number on each pixel point representing the gray value of the pixel. The value at each pixel point is the gray value of the image pixel. The origin of the pixel plane, O_0, is at the upper-leftmost corner of the imaging plane, with the u axis oriented horizontally to the right and the v axis oriented vertically down.

5.2.2.2 *Image coordinate system*

In order to display the image, we also need an image coordinate system, whose coordinates are defined in the following way: x axis and u axis directions parallel to the numerical difference of α times; y axis and v axis directions parallel to the numerical difference of β times; The origin o, defined as the intersection point between the imaging plane and the camera's optical axis, is translated by $(u_0, v_0)^T$ with respect to the pixel coordinate system. Assuming that there exists a point $\boldsymbol{P}'$ whose coordinates on the image plane are $(X', Y')^{\mathrm{T}}$ and the corresponding coordinates of this point under the pixel coordinate system are $(u_x, v_y)^{\mathrm{T}}$, then the transformation relationship can be obtained according to the definition of image coordinates, as shown in Eq. (5.4):

$$\begin{aligned} u_x &= \alpha X' + u_0 \\ v_y &= \beta Y' + v_0 \end{aligned} \tag{5.4}$$

Substituting Eq. (5.3) into Eq. (5.4) yields Eq. (5.5):

$$\begin{cases} u = f_x \dfrac{X}{Z} + u_0 \\ v = f_y \dfrac{Y}{Z} + v_0 \end{cases} \tag{5.5}$$

where $f_x = \alpha f$, α is the scaling factor on the u axis; $f_y = \beta f$, β is the scaling factor on the v axis; $u = u_x$; and $v = v_y$.

Converting Eq. (5.5) into the chi-square coordinate form yields Eq. (5.6):

$$\begin{bmatrix} u \\ v \\ 1 \end{bmatrix} = \frac{1}{Z} \begin{bmatrix} f_x & 0 & u_0 \\ 0 & f_y & v_0 \\ 0 & 0 & 1 \end{bmatrix} \begin{bmatrix} X \\ Y \\ Z \end{bmatrix} \Leftrightarrow \frac{1}{Z} \boldsymbol{KP} \tag{5.6}$$

Simplify Eq. (5.6) to obtain Eq. (5.7):

$$Z \begin{bmatrix} u \\ v \\ 1 \end{bmatrix} = \begin{bmatrix} f_x & 0 & u_0 \\ 0 & f_y & v_0 \\ 0 & 0 & 1 \end{bmatrix} \begin{bmatrix} X \\ Y \\ Z \end{bmatrix} \Leftrightarrow \boldsymbol{KP} \tag{5.7}$$

where $\boldsymbol{K}$ is the internal parameter matrix of the camera, which is determined by the manufacturer.

5.2.2.3 *Camera coordinate system*

A camera coordinate system is mainly used to describe the position and orientation of a point in space relative to the camera. It is defined as follows: Taking the optical center of the camera as the origin, establish a 3D right-angle coordinate system, $O_c - X_c Y_c Z_c$, whose Z_c axis is perpendicular to the image coordinate system, X_c axis direction is parallel to the x axis direction of the image coordinate system, and Y_c axis direction is parallel to the y axis direction of the image coordinate system.

5.2.2.4 *World coordinate system*

In the objective world, there also exists a world coordinate system for describing the position and orientation of objects, $O_{\mathrm{w}} - X_{\mathrm{w}}Y_{\mathrm{w}}Z_{\mathrm{w}}$. In fact, when we usually represent the position of a point in space, it is described under the world coordinate system. Assuming that there is a point $\boldsymbol{P}_{\mathrm{w}}$ in the space, its coordinates under the world coordinate system are denoted as $(x_{\mathrm{w}}, y_{\mathrm{w}}, z_{\mathrm{w}})^{\mathrm{T}}$, and the current coordinates are transformed into $(x_{\mathrm{c}}, y_{\mathrm{c}}, z_{\mathrm{c}})^{\mathrm{T}}$ under the camera coordinate system, the relationship between the two can be expressed as Eq. (5.8):

$$\begin{bmatrix} x_{\mathrm{c}} \\ y_{\mathrm{c}} \\ z_{\mathrm{c}} \end{bmatrix} = \boldsymbol{R} \begin{bmatrix} x_{\mathrm{w}} \\ y_{\mathrm{w}} \\ z_{\mathrm{w}} \end{bmatrix} + \boldsymbol{t} \tag{5.8}$$

where $\boldsymbol{R}$ is the rotation matrix of the camera position and orientation while $\boldsymbol{t}$ is the translation vector of the camera.

Convert Eq. (5.8) into a homogeneous coordinate transformation matrix. as shown in Eq. (5.9):

$$\begin{bmatrix} x_{\mathrm{c}} \\ y_{\mathrm{c}} \\ z_{\mathrm{c}} \\ 1 \end{bmatrix} = \begin{bmatrix} \boldsymbol{R} & \boldsymbol{t} \\ \boldsymbol{0}^{\mathrm{T}} & 1 \end{bmatrix} \begin{bmatrix} x_w \\ y_w \\ z_w \\ 1 \end{bmatrix} \tag{5.9}$$

By substituting Eq. (5.9) into Eq. (5.7), the correspondence between the pixel coordinates and the world coordinates can be obtained, as shown in Eq. (5.10):

$$Z\boldsymbol{P}_{uv} = Z \begin{bmatrix} u \\ v \\ 1 \end{bmatrix} = \boldsymbol{K}(\boldsymbol{R}\boldsymbol{P}_{\mathrm{w}} + \boldsymbol{t}) = \boldsymbol{K}^{\mathrm{T}}\boldsymbol{P}_{\mathrm{w}} \tag{5.10}$$

where $\boldsymbol{R}$ and $\boldsymbol{t}$ are the outer parameter matrices of the camera and $\boldsymbol{K}$ is the inner parameter matrix of the camera.

The inner parameter matrix can be obtained from the camera manual or through calibration, which represents the imaging relationship of the camera. And the outer parameter matrix represents

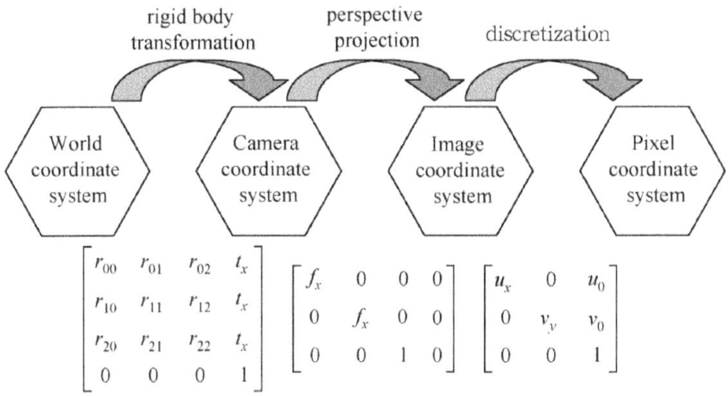

Figure 5.4. Conversion relationship between coordinate systems.

the motion trajectory of the camera, which needs to be estimated through computation and is the core problem to be solved in SLAM systems.

The conversion relationship of the above four coordinate systems is shown in Figure 5.4.

5.3 Integrated Design of Robot Visual Perceptual System

5.3.1 *Camera selection*

In this book, an RGB_D camera developed by Intel, Realsense D435, is selected. The camera has the characteristics of high real-time, high accuracy, and good imaging performance, small size, USB power supply form, etc., and carries four main modules.

In Figure 2.4, from left to right, the left IR camera, IR spot projector, right IR camera, and RGB camera are shown. The Realsense D435 camera adopts the structured light principle and has a very complete optical depth solution, making it better suited for the requirements of the experiments in this book. The parameters of the Realsense D435 camera are shown in Table 5.1.

Using the IR camera to get the coordinates of the point $\boldsymbol{P}$ in the space is $[x_{\mathrm{ir}}, y_{\mathrm{ir}}, z_{\mathrm{ir}}]^{\mathrm{T}}$, combined with the transformation relationship between the coordinates, we can find out the coordinates of the point

Table 5.1. Parameters of the Realsense D435 camera.

Norm	Performances
Operating environment	Indoor/Outdoor
Depth stream output resolution/px	Up to 1280 × 720
Depth stream output rate/fps	Up to 90
Depth flow output distance/m	0.2–10
RGB output resolution/px	Up to 1920 × 1080
RGB output rate/FPS	Up to 30

in the RGB camera coordinate system and the transformation relationship for formula (5.11):

$$Z \begin{bmatrix} x_{\text{rgb}} \\ y_{\text{rgb}} \\ z_{\text{rgb}} \end{bmatrix} = \boldsymbol{R}_{\text{ir}}^{\text{rgb}} \begin{bmatrix} x_{\text{ir}} \\ y_{\text{ir}} \\ z_{\text{ir}} \end{bmatrix} + \boldsymbol{t}_{\text{ir}}^{\text{rgb}} \tag{5.11}$$

where $\boldsymbol{R}_{\text{ir}}^{\text{rgb}}$ is the position and orientation rotation matrix from the IR camera to the RGB camera, $\boldsymbol{t}_{\text{ir}}^{\text{rgb}}$ is the translation vector, and Z is the corresponding depth value of the pixel point. Through Eq. (5.11), the color image and depth image can be aligned, which provides the theoretical basis for object recognition later and for finding the relative spatial position and orientation.

5.3.2 *Camera calibration*

Typically, aberrations are caused by the presence of machining errors in the camera lens. The more influential ones are radial and tangential aberrations. Radial aberrations are usually corrected using a Taylor series expansion:

$$\begin{cases} x_{\text{corrected}} = x(1 + k_1 r^2 + k_2 r^4 + k_3 r^6 + \cdots) \\ y_{\text{corrected}} = y(1 + k_1 r^2 + k_2 r^4 + k_3 r^6 + \cdots) \end{cases} \tag{5.12}$$

where (x, y) is the original position of the aberration point in the image, r is the distance from the aberration point to the center of the imaging plane, $(x_{\text{corrected}}, y_{\text{corrected}})$ denote the corrected position, and k_1, k_2, k_3 are the coefficients of the Taylor series.

Tangential aberration can be corrected using Eq. (5.13):

$$\begin{cases} x_{\text{corrected}} = x + [2p_1xy + p_2(r^2 + 2x^2)] \\ y_{\text{corrected}} = y + [2p_2xy + p_1(r^2 + 2y^2)] \end{cases} \tag{5.13}$$

where (x, y) denote the original position of the aberration point in the image, r is the distance from the aberration point to the center of the imaging plane, and $(x_{\text{corrected}}, y_{\text{corrected}})$ denote the corrected position.

This book mainly utilizes the first two low-order terms of the radial and tangential aberrations for correction. By combining Eqs. (5.12) and (5.13), the camera's distortion model is finally obtained, as shown in Eq. (5.14):

$$\begin{cases} x_{\text{corrected}} = x(1 + k_1r^2 + k_2r^4) + [2p_1xy + p_2(r^2 + 2x^2)] \\ y_{\text{corrected}} = y(1 + k_1r^2 + k_2r^4) + [2p_2xy + p_1(r^2 + 2y^2)] \end{cases} \tag{5.14}$$

After calibrating the camera, its internal and external parameters and the aberration parameters can be obtained, from which the image information of the point to be measured can be accurately determined. The IR camera is relatively stable, so this book only calibrates the RGB camera, and the specific parameters obtained are shown in Table 5.2.

The internal and external parameters of the camera and the aberration parameters (K_1 and K_2 are the radial aberration parameters

Table 5.2. RGB camera correction parameters.

Parameters	Numerical value
f_x mm	563.3862
f_y mm	559.1434
u_o px	329.6513
v_o px	231.1877
k_1 a.u.	0.02592
k_2 a.u.	−0.09263
p_1 a.u.	0.00174
p_2 a.u.	0.00612

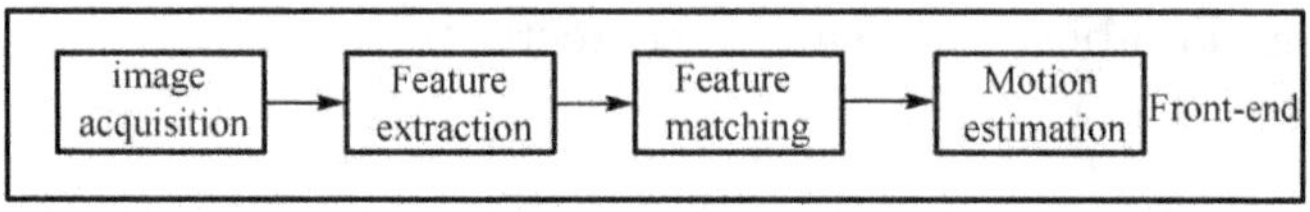

Figure 5.5. RGB_D SLAM front-end operation.

in the x and y directions, respectively, and p_1 and p_2 are the tangential aberration parameters) are obtained, from which the image information of the point to be measured are accurately determined.

5.3.3 *Visual odometer design*

In RGB-D SLAM systems, the front end, also known as visual odometry,[23] primarily addresses the challenge of using image information to preliminarily estimate the camera's motion trajectory. The process begins with the extraction of feature point information from the acquired image. Subsequently, these feature points are precisely matched. Finally, the camera's motion trajectory is roughly estimated based on the mathematical relationships between the matched pairs. The front-end operation provides essential data with robust performance for subsequent back-end optimization. Within the front end, the effectiveness of image feature point detection and matching performance is crucial, as they directly impact the overall quality of the SLAM system. The operation of the RGB-D SLAM front end is illustrated in Figure 5.5.

5.3.4 *Feature point detection algorithm selection*

Target recognition can be effectively achieved by leveraging salient regional features to characterize objects. When utilizing computers to analyze a target object, it's crucial to consider both the differences and the unique attributes of the target. In the field of computer vision, methods based on feature points for target detection have emerged as a significant area of research.

A feature point represents a unique location within an image that remains relatively stable amidst camera movements and is unaffected by external factors. This stability is particularly evident across multiple image frames, where feature points consistently withstand affine transformations such as translation, rotation, and scaling. Feature points are characterized by their repeatability, distinctiveness, locality, and efficiency, and they comprise two main components: the keypoint and the descriptor. Currently, the three most classic and widely implemented algorithms are the Scale Invariant Feature Transform (SIFT), the Speeded Up Robust Features (SURF), and the Oriented FAST and Rotated BRIEF (ORB) algorithms.

5.3.4.1 *SIFT algorithm*

The SIFT[24] algorithm is one of the most commonly used algorithms for key-point detection and characterization. This algorithm was proposed by Lowe. The method of SIFT feature extraction is based on the local information of the image, which has the following advantages:

(1) With translation invariance, rotation invariance, scale invariance, viewpoint invariance and luminance invariance, it has great advantages for the expression of target feature information.
(2) SIFT feature points can robustly adjust the parameters, and the number of feature points can be better adjusted according to the scene when performing feature description.

The SIFT algorithm for extracting local feature points of an image can be divided into the following four steps:

Step 1: Putative Keypoint Detection. SIFT establishes a scale space relative to the image and detects keypoints by computing the Difference of Gaussians (DoG) operator, an approximation of the Laplacian of Gaussian (LoG). This approach effectively characterizes both the scale and orientation of keypoints.

Step 2: Remove pseudo key-points. Calculate the curvature of the feature point, when the M value corresponding to the main curvature of the detection point meets the threshold requirement.

Step 3: Gradient and direction assignment of key-points.

Step 4: Feature vector generation. Each SIFT feature generates feature vectors of 128 dimensions, which are all rotation and scale invariant.

SIFT feature points contain the scale and direction of the key-points, so it is necessary to construct the scale space of the image, and its

(a) (b)

Figure 5.6. SIFT feature point detection: (a) original scene image and (b) SIFT feature point detection map.

scale space $D(x, y, \sigma)$ satisfies Eq. (5.15):

$$\begin{aligned} D(x, y, \sigma) &= f(x, y) \times G(x, y, \sigma) \\ G(x, y, \sigma) &= \frac{1}{2\pi\sigma^2} e^{-(x^2+y^2)/2\sigma^2} \end{aligned} \tag{5.15}$$

where σ is the variance factor, also known as scale factor; and x and y are the horizontal and vertical coordinates of the pixels in the image, respectively.

When calculating the feature points, the Gaussian difference kernel at different scales is utilized for the selection of key-points, which is shown in Eq. (5.16):

$$D(x, y, k, \sigma) = L(x, y, k\sigma) - L(x, y, \sigma) \tag{5.16}$$

where σ is the variance factor, also known as the scale factor; k is the scale amplification factor; and L is the Laplace operator.

A scene from a laboratory is selected for SIFT feature point detection, and the results obtained are shown in Figure 5.6.

5.3.4.2 *SURF algorithm*

The SURF[25] feature point descriptor was proposed by Bay *et al.* This algorithm is a great improvement over the SIFT algorithm. When the SIFT algorithm is not accelerated by an image processor or related hardware, there is a problem of untimely processing, while the SURF algorithm solves this problem better. The SURF algorithm has the following advantages:

(a) (b)

Figure 5.7. SURF feature point detection: (a) original scene image and (b) SURF feature point detection map.

(1) It is robust and faster to detect.
(2) It takes into account the rotation invariance and scale invariance of the SIFT algorithm.
(3) It has strong robustness to changes in illumination, imitation, and transmission.

The key-points of the SURF algorithm are determined by calculating the Hessian matrix and finding the extreme points in the scale space. The Hessian matrix is shown in Eq. (5.17):

$$\boldsymbol{H}(x,\sigma) = \begin{bmatrix} L_{xx}(x,\sigma) & L_{xy}(x,\sigma) \\ L_{xy}(x,\sigma) & L_{yy}(x,\sigma) \end{bmatrix} \tag{5.17}$$

where $L_{xx}(x,\sigma)$ is the second-order partial derivative of the Gaussian function, calculated as $\frac{\partial^2 G(x,y,\sigma)}{\partial x^2}$. The rest of the derivatives are similar.

SURF feature point detection is illustrated in Figure 5.7.

5.3.4.3 *ORB algorithm*

ORB[26] is an improved detection and description technique for feature points based on FAST corner points, which was proposed by Rublee. The ORB feature description has the following main advantages:

(1) It has rotation invariance and scale invariance.
(2) It has strong robustness to noise, perspective imitation, etc.
(3) It provides fast operation.

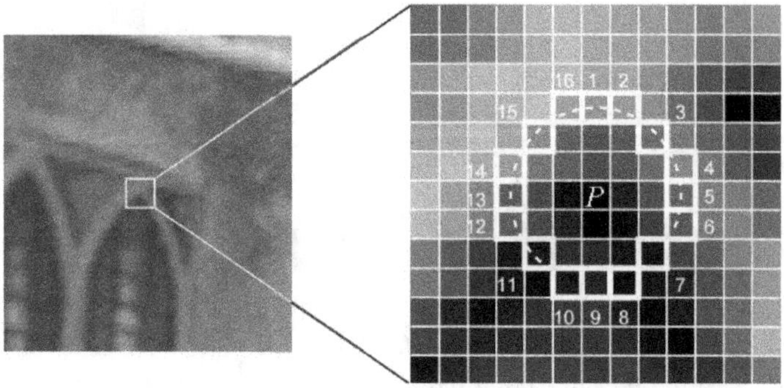

Figure 5.8. Schematic diagram of FAST corner points.

The ORB feature point detection is divided into the following two main steps:

Step 1: Direction FAST feature point detection.

FAST corner detection is the core of ORB feature point detection. FAST corner detection is based on machine learning. FAST key-points have directionality. The detection involves judging 16 pixel points around the point of interest in turn. When judging a point of interest under too bright or too dark conditions, it can be assumed that the point is located at a corner. The FAST corner schematic diagram is shown in Figure 5.8.

The calculation process of FAST key-points is simpler, and the detection effect is significantly improved. Its specific detection steps are as follows:

(1) Select the pixel of interest in the image P, assuming its luminance value as I_P.
(2) Set the brightness threshold T.
(3) With the point P as the central point, select a circle with a radius of 3 pixels and find 16 pixel points on the circle to be compared.
(4) Select N consecutive points on the circle, and compare the brightness of each point I_i with the threshold. If the brightness values of all N points are greater than $I_P + T$ or less than $I_P - T$, then the point P is a corner point. Usually N is taken as 12 or 9.
(5) Traverse the image and repeat the steps 1–4 in a loop to judge the selection of each pixel point.

The FAST corner points are characterized at multiple scales by building an image pyramid, and then the intensity centroid method is introduced to determine the orientation of the feature points. The center intensity of the feature point can be determined using Eq. (5.18):

$$C = \left(\frac{m_{10}}{m_{00}} - \frac{m_{01}}{m_{00}}\right) \tag{5.18}$$

where m_{pq} is the $p+q$ order distance of the neighborhood of the feature point, which is defined as $m_{pq} = \sum_{x,y} x^p y^q f(x,y)$.

The main direction of the feature point is defined as shown in Eq. (5.19):

$$\theta = \arctan(m_{10}/m_{01}) \tag{5.19}$$

Step 2: BRIEF characterization.

The BRIEF descriptor is mainly used to form the region of interest by randomly selecting a number of points in the region around the point of interest. The grayscale of the region of interest is binarized and parsed into a binary coding string, which is the descriptor of the feature point. This binary string is composed of 2^n points around the feature point. These points can be formed into a matrix as shown in Eq. (5.20):

$$\boldsymbol{S} = \begin{bmatrix} x_1 & x_2 & \cdots & x_n \\ y_1 & y_2 & \cdots & y_n \end{bmatrix} \tag{5.20}$$

The main direction of the image θ is formed by the neighborhood around the key-point, combined with the corresponding rotation matrix $\boldsymbol{R}_\theta$. $\boldsymbol{S}$ is linearly transformed to obtain the linear expression matrix $\boldsymbol{S}_\theta$, as shown in Eq. (5.21):

$$\boldsymbol{S}_\theta = \boldsymbol{R}_\theta \boldsymbol{S} = \begin{bmatrix} \cos\theta & \sin\theta \\ -\sin\theta & \cos\theta \end{bmatrix} \boldsymbol{S} \tag{5.21}$$

In calculating $\boldsymbol{S}_\theta$, the direction angle quantization is first performed by splitting it into 12 sub-blocks. Then, $\boldsymbol{S}_\theta$ is calculated for each sub-block, and finally the BRIEF descriptor is calculated by constructing a mapping table.

ORB feature point detection is illustrated in Figure 5.9.

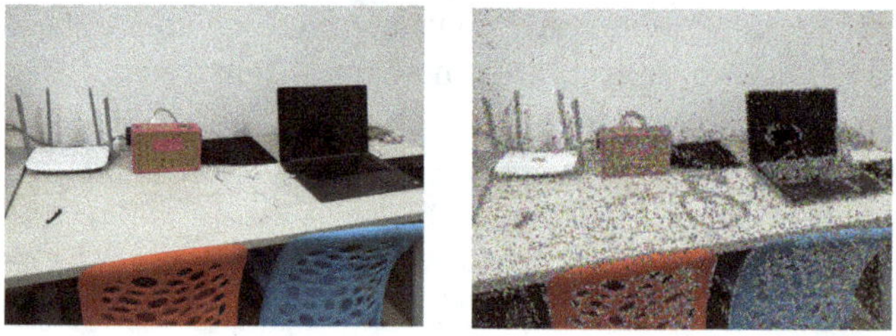

(a) (b)

Figure 5.9. ORB feature point detection: (a) original scene image and (b) ORB feature point detection map.

5.3.5 *Camera motion estimation*

In Section 5.3.4, the feature points between two neighboring frames can be acquired and combined with the matching algorithm to obtain the matching pairs between the two frames, then the relative motion relationship of the camera between the two neighboring frames can be calculated based on the matching relationship. In kinematics, the rotation matrix $\boldsymbol{R}$ and the translation vector $\boldsymbol{t}$ are used to describe the motion relationship between two points. In RGB_D SLAM, the matrix $\boldsymbol{R}$ is used to describe the rotational change in the camera, which is a 3×3 dimensional matrix, while the vector $\boldsymbol{t}$ is used to describe the translation of the camera, which is a 3×1 dimensional vector. The camera's position and orientation coordinate at time t is $\boldsymbol{P}_t$, and the camera's position and orientation coordinate at time $t+1$ is $\boldsymbol{P}_{t+1}$, which can be obtained as Eq. (5.22) according to the coordinate transformation relationship in kinematics:

$$\boldsymbol{P}_{t+1} = \boldsymbol{R}\boldsymbol{P}_t + \boldsymbol{t} \tag{5.22}$$

There are three methods for camera motion estimation based on feature point matching, namely 2D-2D, 3D-3D, and 3D-2D.

The 2D-2D method is often used in SLAM systems with monocular cameras and is also referred to as the epipolar geometry method.[27] It is based on the already completed feature point matching according to the pixel position corresponding to the matching points in the two images where the matching has been completed, the fundamental matrix $\boldsymbol{F}$ and the essential matrix $\boldsymbol{E}$ are derived, and then the two matrices are decomposed using the epipolar constraints to derive $\boldsymbol{R}$ and $\boldsymbol{t}$. However, at least five points are required to solve the matrices, and there are problems with initialization, pure rotation, and scaling.

The 3D-2D method is often used in SLAM systems with RGB_D or binocular cameras and is also known as the Perspective-n-Point.[28] (PnP) method. When the 3D coordinates of a point and its corresponding projected position are known, it can be solved by using the Perspective-three-Point (P3P) method, the Direct Linear Transformation (DLT) method, the Efficient PnP (EPnP) method, the Bundle Adjustment (BA) method, etc. The 3D-2D method has many solution methods, and it can ensure obtaining better motion estimation information when there are fewer matched pairs. However, there are some problems, such as when the 3D information or 2D matching points are affected by noise or there is a mismatch, which have a greater impact on position and orientation estimation.

The 3D-3D method is often used in SLAM systems with RGB_D or binocular cameras. When a set of matched 3D points is already available, the matching relationship between the two point sets is directly utilized without considering the camera's bit position and orientation to find the rotation matrix $\boldsymbol{R}$ and the translation vector $\boldsymbol{t}$. The problem is generally solved using the Iterative Closest Point (ICP) algorithm.[29] Solving the running estimate using the 3D-3D method can simplify the solution process.

The Realsense D435 depth camera is considered in this book, and accurate depth information can already be obtained after calibrating the camera, so the ICP algorithm is chosen in this book to estimate the motion of the camera. There are two ways to solve the ICP algorithm: One is to use linear algebra (represented by the SVD algorithm); the other is to use nonlinear optimization (similar to BA). In this book, the SVD solution method is selected, and its specific procedure is as follows.

Assuming that the sets of matched pairs of points in two neighboring images of two frames are $\boldsymbol{P} = \{\boldsymbol{p}_1, \boldsymbol{p}_2, \ldots, \boldsymbol{p}_n\}$ and $\boldsymbol{Q} = \{\boldsymbol{q}_1, \boldsymbol{q}_2, \ldots, \boldsymbol{q}_n\}$, each point in the point set $\boldsymbol{P}$ is matched with the point in $\boldsymbol{Q}$ correspondingly, and there is only a difference of a rotation matrix $\boldsymbol{R}$ and a translation vector $\boldsymbol{t}$ between the two, then there exists a relationship for the points in any point sets $\boldsymbol{P}$ and $\boldsymbol{Q}$, as shown in Eq. (5.23):

$$\forall i, \quad \boldsymbol{p}_i = \boldsymbol{R}\boldsymbol{q}_i + \boldsymbol{t} \tag{5.23}$$

First, define the error term for each set to represent the error that exists between matched pairs, as shown in Eq. (5.24):

$$\boldsymbol{e}_i = \boldsymbol{p}_i - (\boldsymbol{R}\boldsymbol{q}_i + \boldsymbol{t}) \tag{5.24}$$

Next, a least squares model is constructed to minimize the overall squared error through an iterative function, which in turn leads to the minima of $\boldsymbol{R}$ and $\boldsymbol{t}$. The least squares variance is shown in Eq. (5.25):

$$\min_{\boldsymbol{R},\boldsymbol{t}} E = \frac{1}{2}\sum_{i=1}^{n} \|\boldsymbol{p}_i - (\boldsymbol{R}\boldsymbol{q}_i + \boldsymbol{t})\|_2^2 \tag{5.25}$$

Solving the above least squares model is usually performed using linear algebra in the following steps:

Step 1: Define the center of mass of the two point sets:

$$\boldsymbol{p} = \frac{1}{n}\sum_{i=1}^{n}(\boldsymbol{p}_i), \quad \boldsymbol{q} = \frac{1}{n}\sum_{i}^{n}(\boldsymbol{q}_i) \tag{5.26}$$

Step 2: Do the following in the error function:

$$\begin{aligned}
\min_{\boldsymbol{R},\boldsymbol{t}} E &= \frac{1}{2}\sum_{i=1}^{n} \|\boldsymbol{p}_i - (\boldsymbol{R}\boldsymbol{q}_i + \boldsymbol{t})\|_2^2 \\
&= \frac{1}{2}\sum_{i=1}^{n} \|\boldsymbol{p}_i - \boldsymbol{R}\boldsymbol{q}_i - \boldsymbol{t} - \boldsymbol{p} + \boldsymbol{R}\boldsymbol{q} + \boldsymbol{p} - \boldsymbol{R}\boldsymbol{q}\|_2^2 \\
&= \frac{1}{2}\sum_{i=1}^{n} \|(\boldsymbol{p}_i - \boldsymbol{p} - \boldsymbol{R}(\boldsymbol{q}_i - \boldsymbol{q})) + (\boldsymbol{p} - \boldsymbol{R}\boldsymbol{q} - \boldsymbol{t})\|_2^2 \\
&= \frac{1}{2}\sum_{i=1}^{n} (\|\boldsymbol{p}_i - \boldsymbol{p} - \boldsymbol{R}(\boldsymbol{q}_i - \boldsymbol{q})\|^2 + \|\boldsymbol{p} - \boldsymbol{R}\boldsymbol{q} - \boldsymbol{t}\|_2^2 \\
&\quad + 2(\boldsymbol{p}_i - \boldsymbol{p} - \boldsymbol{R}(\boldsymbol{q}_i - \boldsymbol{q}))^{\mathrm{T}}(\boldsymbol{p} - \boldsymbol{R}\boldsymbol{q} - \boldsymbol{t}))
\end{aligned} \tag{5.27}$$

In the cross term part, $(\boldsymbol{p}_i - \boldsymbol{p} - \boldsymbol{R}(\boldsymbol{q}_i - \boldsymbol{q}))$ is equal to 0 after summation, so the objective optimization function can be reduced to

$$\min_{\boldsymbol{R},\boldsymbol{t}} E = \frac{1}{2}\sum_{i=1}^{n} (\|\boldsymbol{p}_i - \boldsymbol{p} - \boldsymbol{R}(\boldsymbol{q}_i - \boldsymbol{q})\|_2^2 + \|\boldsymbol{p} - \boldsymbol{R}\boldsymbol{q} - \boldsymbol{t}\|_2^2) \tag{5.28}$$

In this case, $\|\boldsymbol{p}_i - \boldsymbol{p} - \boldsymbol{R}(\boldsymbol{q}_i - \boldsymbol{q})\|_2^2$ contains only the rotation matrix $\boldsymbol{R}$, and $\|\boldsymbol{p} - \boldsymbol{R}\boldsymbol{q} - \boldsymbol{t}\|_2^2$ contains $\boldsymbol{R}$ and $\boldsymbol{t}$, but only with respect to the center of mass. Therefore, by obtaining $\boldsymbol{R}$, we can make $\|\boldsymbol{p}-\boldsymbol{R}\boldsymbol{q}-\boldsymbol{t}\|_2^2$ equal to 0, and we can find $\boldsymbol{t}$.

Step 3: Calculate the de-meaned coordinates of the two point sets such that

$$\boldsymbol{M} = \boldsymbol{p}_i - \boldsymbol{p}, \quad \boldsymbol{N} = \boldsymbol{q}_i - \boldsymbol{q} \tag{5.29}$$

Then, the optimization objective function can be written as

$$\begin{aligned} \boldsymbol{R}^* &= \arg\min_{\boldsymbol{R}} \frac{1}{2}\sum_{i=1}^{n} \|\boldsymbol{M} - \boldsymbol{R}\boldsymbol{N}\|_2^2 \\ \boldsymbol{t}^* &= \boldsymbol{p} - \boldsymbol{R}\boldsymbol{q} \end{aligned} \tag{5.30}$$

Equation (5.30) is expanded to obtain the error term with respect to $\boldsymbol{R}$:

$$\frac{1}{2}\sum_{i=1}^{n} \|\boldsymbol{M} - \boldsymbol{R}\boldsymbol{N}\|_2^2 = \frac{1}{2}\sum_{i=1}^{n}(\boldsymbol{M}^{\mathrm{T}}\boldsymbol{M} + \boldsymbol{N}^{\mathrm{T}}\boldsymbol{R}^{\mathrm{T}}\boldsymbol{R}\boldsymbol{N} - 2\boldsymbol{M}^{\mathrm{T}}\boldsymbol{R}\boldsymbol{N}) \tag{5.31}$$

In Eq. (5.31), there is no $\boldsymbol{R}$ term in the first term within the parentheses. The second term within the parentheses, $\boldsymbol{R}^{\mathrm{T}}\boldsymbol{R} = \boldsymbol{I}$, is also independent of $\boldsymbol{R}$. In fact, the objective optimization function can then be simplified to Eq. (5.32):

$$\sum_{i=1}^{n} -\boldsymbol{M}^{\mathrm{T}}\boldsymbol{R}\boldsymbol{N} = \sum_{i=1}^{n} -\mathrm{tr}(\boldsymbol{R}\boldsymbol{N}\boldsymbol{M}^{\mathrm{T}}) = -\mathrm{tr}\left(\boldsymbol{R}\sum_{i=1}^{n}\boldsymbol{N}\boldsymbol{M}^{\mathrm{T}}\right) \tag{5.32}$$

where tr denotes the trace of the matrix. The matrix $\boldsymbol{W}$ is defined by solving Eq. (5.32) using SVD:

$$\boldsymbol{W} = \sum_{i=1}^{n} \boldsymbol{M}\boldsymbol{N}^{\mathrm{T}} \tag{5.33}$$

The SVD decomposition and orientation of $\boldsymbol{W}$ can then be obtained:

$$\boldsymbol{W} = \boldsymbol{U}\sum \boldsymbol{V}^{\mathrm{T}} \tag{5.34}$$

where $\sum$ is a diagonal matrix consisting of singular values and $\boldsymbol{U}$ and $\boldsymbol{V}$ are orthogonal diagonal matrices. When $\boldsymbol{W}$ is a non-singular matrix, it can be found that

$$\boldsymbol{R} = \boldsymbol{U}\boldsymbol{V}^{\mathrm{T}} \tag{5.35}$$

After solving for $\boldsymbol{R}$, it can be used to solve for $\boldsymbol{t}$ by substituting it into Eq. (5.28).

5.3.6 *Closed-loop detection and keyframe selection strategy*

5.3.6.1 *Keyframe selection*

With the increase in motion time and distance, the camera will acquire a large number of image frames. If all the acquired image frames are subjected to feature point extraction and matching, on the one hand, it will drastically increase the size of the subsequent pose graph model and increase the difficulty of back-end optimization, while, on the other hand, it will drastically increase the computational cost of the whole SLAM system, making the system highly redundant. Therefore, the obtained image frames need to be processed.

Keyframes refer to images that are more obviously different from the previous frame when the camera rotates or pans to a certain viewpoint. The selection of keyframes has always been an important part of back-end optimization. By selecting keyframes, the number of image frames can be effectively reduced without losing important feature information. In this book, the Realsense D435 camera acquires images at a frequency of 30 fps, and the image acquisition is fast, so a reasonable number of image frames need to be selected as keyframes. The number of keyframes should neither be too many nor too few. The consequences of too many were previously mentioned, while too few will make the motion between two neighboring image frames too large, making the process of feature point matching very difficult and prone to dropped frames or mismatching of images.

Currently, there are three more commonly used methods for selecting keyframes: selecting keyframes based on time, selecting keyframes based on space, and selecting keyframes based on the similarity between images. This book adopts the third method, i.e., to develop a keyframe selection strategy based on the similarity between images.

Assuming that frame image A and frame image B can be matched, A is a known keyframe, and the rotation matrix $\boldsymbol{R}$ and the translation vector $\boldsymbol{t}$ are known, set a standard threshold D and metric relation E, as shown in Eq. (5.36):

$$E = \xi_1 \|\Delta \boldsymbol{t}\|_2 + \xi_2 \|\Delta \boldsymbol{n}\|_2 \tag{5.36}$$

where E is the metric relationship, indicating the motion connection between frame image B and frame image A; $\Delta \boldsymbol{t}$ is the translation vector, expressed as $[x, y, z]^{\mathrm{T}}$; $\Delta \boldsymbol{n}$ is the rotation vector, expressed as a quaternion $[q_0, q_1, q_2, q_3]$; and ξ_1 and ξ_2 are the weight parameters of translation and rotation.

In practical experiments, the Realsense D435 camera was found to be more sensitive to rotation than to translation. For example, when the robot is in a state of uniform linear motion, the amount of change in the image is small, while when the robot is in a state of uniform rotation, the change in the viewing angle is significantly larger. So, a larger value should be assigned to ξ_2. Based on extensive experimental verification and reference to related literature, we finally determine $\xi_1 = 0.4$, $\xi_2 = 1.0$, $D = 0.3$, i.e., when the calculated E value exceeds D, we consider the frame image B to be the keyframe, and save it in the keyframe library for subsequent loop closure detection and global optimization; otherwise, we discard the frame image B, and continue to obtain new image frames to repeat the calculation.

5.3.6.2 *Closed-loop detection*

Closed-loop detection, also known as loop closure detection, is an important part of the SLAM system, which mainly solves the problem of global consistency.[30] The effect of closed-loop detection is shown in Figure 5.10. In the process of robot movement, there are various computational errors, and each frame of the image corresponding to the camera position and orientation information is based on the previous frame of the image, so there will be cumulative errors, ultimately producing the problem of localization drift, so that the estimated camera trajectory is inaccurate. Closed-loop detection is a method to eliminate the error. The main elements of closed-loop detection are as follows: firstly, combining the similarity between the two image frames to determine whether the robot is in the historical position; then establishing a link between the current image frame and the historical image frame, and if it returns to the historical position, then using the historical image frame to make a new motion estimation. In SLAM systems, historical data are always more accurate than current data, and based on this *a priori* knowledge, the

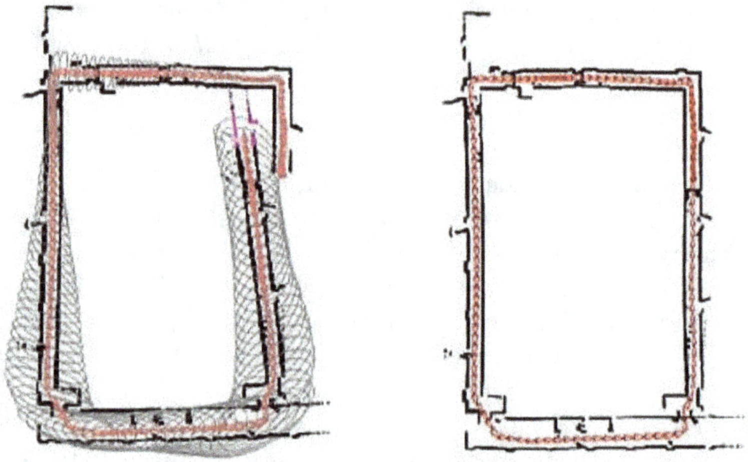

(a) (b)

Figure 5.10. Closed-loop detection effect (historical position found by closed-loop detection): (a) without closed-loop detection and (b) with closed-loop detection.

cumulative error can be dispersed into each small closed loop, and the error of the whole system can be reduced by utilizing historical data.

Currently, there are three judgment methods to construct closed-loop detection: matching based on the environment, matching based on images, and matching based on both the environment and images. Since the constructed environment itself has cumulative errors, the use of environment-based methods will inevitably cause inaccuracies in closed-loop detection, and the one that is used most commonly is matching based on images. Utilizing the similarity between images, a scoring mechanism is used to show the degree of similarity, and when the score reaches a certain value, the two frames can be considered to be approximately the same. In this book, the bag-of-words (BoW) model based on the degree of image similarity is utilized for closed-loop detection.

5.3.6.3 *Bag-of-words model*

The BoW model was originally used an unordered set of words to describe a document or express a text. The BoW model uses a set of description vectors to describe the text. The description vectors only count the frequency of occurrence of each word and do not store sequential information, which is expressed in the form of vectors.

(1) Image-based BoW model construction

In the process of dictionary construction, clustering is usually used to generate dictionaries, and the commonly used method is the K-means algorithm.[31] The K-means algorithm is an iterative approach to classify data samples according to their different characteristics, continuously dividing the samples into a number of small piles until a certain number of iterations is reached. However, the K-means algorithm has the disadvantage of randomly selecting centers, resulting in different clustering results, so this book uses the better K-means++ algorithm. A flowchart depicting the construction of the BoW model is shown in Figure 5.11. In the BoW model, K-fork trees are commonly used to express the dictionary content.

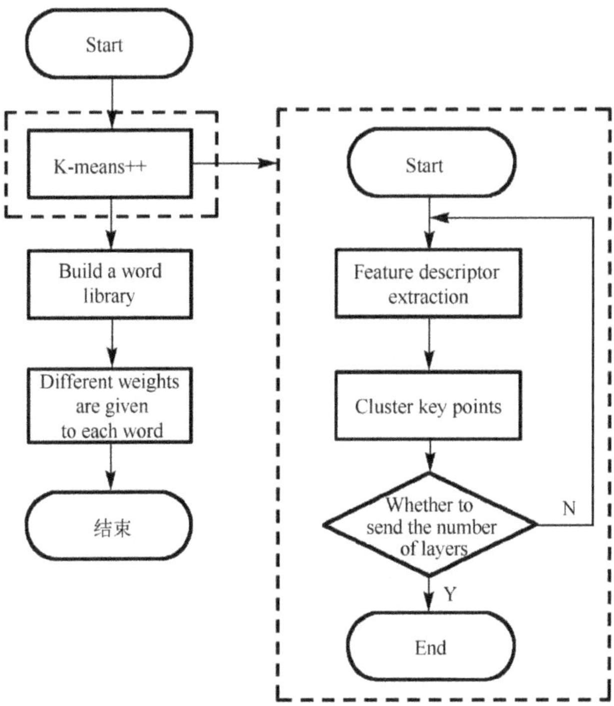

Figure 5.11. Flowchart for constructing a bag-of-words model.

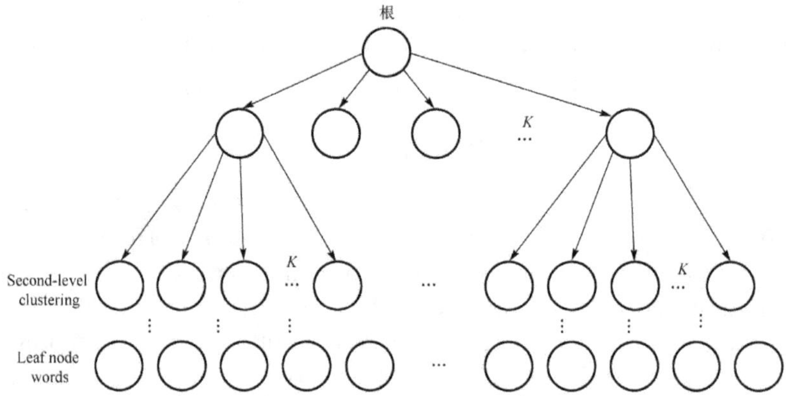

Figure 5.12. Schematic diagram of a K-fork tree.

The so-called K-fork tree is a tree in which each root node can fork into K child nodes. The first layer on the root node contains the K classes obtained by clustering using the K-means++ algorithm, and then each class continues to be clustered and classified until the next layer, and so on. The schematic diagram of a K-fork tree is shown in Figure 5.12.

(2) Similarity calculation

After obtaining the dictionary that describes the image, there is a problem to be considered that each word has a different importance for each frame of the image. So, there is a need to further differentiate the importance of each word. Different weights are given to each word, and the representativeness of each word is measured based on the weights. The most commonly used statistic in text detection is TF-IDF, which is used to measure the importance of a word to the text.

TF is defined as shown in Eq. (5.37):

$$\mathrm{TF}_i = \frac{n_i}{n} \tag{5.37}$$

where n is the total number of occurrences of all words in a single frame and n_i is the number of occurrences of a feature word w_i.

Therefore, TF works as the feature word, where a higher number of occurrences in a single frame expresses that frame better.

The IDF is defined as shown in Eq. (5.38):

$$\mathrm{IDF}_i = \lg \frac{n}{n_i} \tag{5.38}$$

where n is the number of all features and n_i is the number of features contained in the word w_i.

Therefore, IDF is useful for those feature words that occur less frequently to better differentiate the image.

From Eqs. (5.37) and (5.38), the weight η_i corresponding to the word w_i can be defined as shown in Eq. (5.39):

$$\eta_i = \mathrm{TF}_i \times \mathrm{IDF}_i \tag{5.39}$$

At this point, a visual vector $\boldsymbol{v}$ can be constructed to describe the image P:

$$P = \{(w_1, \eta_1), (w_2, \eta_2), \ldots, (w_n, \eta_n)\} \Leftrightarrow \boldsymbol{v}_P \tag{5.40}$$

For images A and B, the similarity can be calculated using the L_1 norm of the vector, as shown in Eq. (5.41):

$$s(\boldsymbol{v}_A - \boldsymbol{v}_B) = 2\sum_{i=1}^{N} |\boldsymbol{v}_{Ai}| + |\boldsymbol{v}_{Bi}| - |\boldsymbol{v}_{Ai} - \boldsymbol{v}_{Bi}| \tag{5.41}$$

5.3.7 *Pose graph optimization*

After the closed-loop detection, an optimization process is required for the whole SLAM system, which serves to reduce the cumulative errors brought about by the long mapping duration, and this process is called back-end optimization. The process of optimizing the camera's position and orientation and the graph of points in space is called BA optimization.[32] It can better solve the problem of localization and mapping. But the computational efficiency of BA optimization decreases significantly with the increase in time and the extension of the trajectory to be optimized. So, a new idea is needed to solve this problem.

Pose graph is a graph that only considers the relationship between camera poses, while ignoring the positions of landmarks, and retains only the trajectories of keyframes. This can effectively reduce the

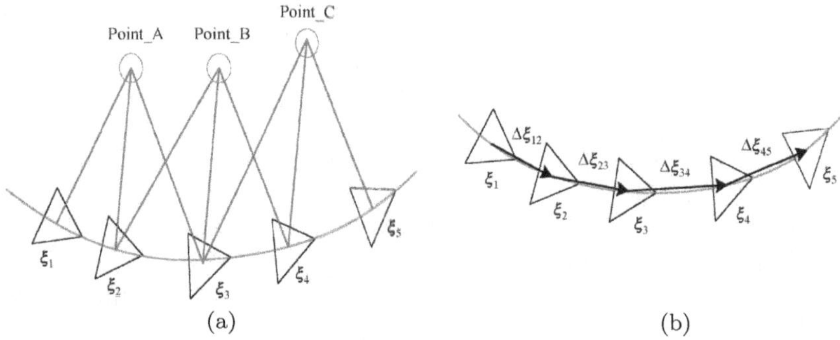

Figure 5.13. Schematic diagram of back-end optimization. (a) BA optimization graph. (b) Pose diagram.

amount of computation. Pose graph is a kind of graph composed of nodes and edges. In this book, its nodes are the camera's bit position and orientations $\boldsymbol{\xi}_i$, while the edges refer to the motion estimations between neighboring keyframes $\Delta\boldsymbol{\xi}$. If the closed-loop detection is successful, then a link can be added between two non-adjacent frames, and a constraint is added to the bit-pose graph, as shown in Figure 5.13.

It can be seen that utilizing pose diagram for the description of the camera motion process can significantly simplify the graphical representation. In Figure 5.13(b), the node is denoted by $\boldsymbol{\xi}_1, \boldsymbol{\xi}_2, \ldots, \boldsymbol{\xi}_n$, and the relative motion between the node $\boldsymbol{\xi}_i$ and the node $\boldsymbol{\xi}_j$ is denoted by $\Delta\boldsymbol{\xi}_{ij}$ with the expression

$$\Delta\boldsymbol{\xi}_{ij} = \boldsymbol{\xi}_i^{-1} \circ \boldsymbol{\xi}_j = \ln(\exp((-\boldsymbol{\xi}_i)^\wedge)\exp(\boldsymbol{\xi}_j^\wedge))^\vee \tag{5.42}$$

where $\wedge$ is the operator that converts a vector into a skew-symmetric matrix and $\vee$ is the operator that converts a skew-symmetric matrix into a vector.

Write Eq. (5.42) according to Liqun's expression:

$$\Delta\boldsymbol{T}_{ij} = \boldsymbol{T}_i^{-1}\boldsymbol{T}_j \tag{5.43}$$

In practice, Eq. (5.43) holds only approximately, so it is necessary to build a least squares problem to solve for the error and the derivative of the error with respect to the optimized variable. The expression for its error function is

$$\begin{aligned} \boldsymbol{e}_{ij} &= \ln(\Delta \boldsymbol{T}_{ij}^{-1} \boldsymbol{T}_i^{-1} \boldsymbol{T}_j)^{\vee} \\ &= \ln(\exp((-\boldsymbol{\xi}_{ij})^{\wedge}) \exp((-\boldsymbol{\xi}_i)^{\wedge}) \exp(\boldsymbol{\xi}_j^{\wedge}))^{\vee} \end{aligned} \tag{5.44}$$

In Eq. (5.44), the variables to be optimized are $\boldsymbol{\xi}_i$ and $\boldsymbol{\xi}_j$. Therefore, we need to solve for the derivatives of $\boldsymbol{e}_{ij}$ with respect to $\boldsymbol{\xi}_i$ and $\boldsymbol{\xi}_j$. Following the solution method of Lie algebra, add a left perturbation to $\boldsymbol{\xi}_i$ and $\boldsymbol{\xi}_j$: $\delta\boldsymbol{\xi}_i$ and $\delta\boldsymbol{\xi}_j$, respectively. Then,

$$\boldsymbol{e}_{ij} = \ln(\boldsymbol{T}_{ij}^{-1} \boldsymbol{T}_i^{-1} \exp((-\delta\boldsymbol{\xi}_i)^{\wedge}) \exp(\delta\boldsymbol{\xi}_j^{\wedge}) \boldsymbol{T}_j)^{\vee} \tag{5.45}$$

Using the BSH approximation, moving the perturbation term to both sides of the equation, and then introducing a new concomitant term, exchanging $\boldsymbol{T}$ between the left and right sides of the perturbation term, derive the Jacobi matrix in right-multiplied form:

$$\begin{aligned} \boldsymbol{e}_{ij} &= \ln(\boldsymbol{T}_{ij}^{-1} \boldsymbol{T}_i^{-1} \exp((-\delta\boldsymbol{\xi}_i)^{\wedge}) \exp(\delta\boldsymbol{\xi}_j^{\wedge}) \boldsymbol{T}_j)^{\vee} \\ &= \ln(\boldsymbol{T}_{ij}^{-1} \boldsymbol{T}_i^{-1} \boldsymbol{T}_j \exp((-\mathrm{Ad}(\boldsymbol{T}_j^{-1})\delta\boldsymbol{\xi}_i)^{\wedge}) \exp((\mathrm{Ad}(\boldsymbol{T}_j^{-1})\delta\boldsymbol{\xi}_j)^{\wedge})^{\vee} \\ &\approx \ln(\boldsymbol{T}_{ij}^{-1} \boldsymbol{T}_i^{-1} \boldsymbol{T}_j [I - (\mathrm{Ad}(\boldsymbol{T}_j^{-1})\delta\boldsymbol{\xi}_i)^{\wedge} + (\mathrm{Ad}(\boldsymbol{T}_j^{-1})\delta\boldsymbol{\xi}_j)^{\wedge}])^{\vee} \\ &\approx \boldsymbol{e}_{ij} + \frac{\partial \boldsymbol{e}_{ij}}{\partial \delta\boldsymbol{\xi}_i}\delta\boldsymbol{\xi}_i + \frac{\partial \boldsymbol{e}_{ij}}{\partial \delta\boldsymbol{\xi}_j}\delta\boldsymbol{\xi}_j \end{aligned} \tag{5.46}$$

Following the Lie algebra derivation rule, the Jacobi matrix of the error with respect to the two bit-position and orientations can be found, where the derivatives with respect to T_i are

$$\frac{\partial \boldsymbol{e}_{ij}}{\partial \delta\boldsymbol{\xi}_i} = -\boldsymbol{J}_r^{-1}(\boldsymbol{e}_{ij}) \mathrm{Ad}(\boldsymbol{T}_j^{-1}) \tag{5.47}$$

The derivative with respect to T_j is

$$\frac{\partial \boldsymbol{e}_{ij}}{\partial \delta\boldsymbol{\xi}_j} = \boldsymbol{J}_r^{-1}(\boldsymbol{e}_{ij}) \mathrm{Ad}(\boldsymbol{T}_j^{-1}) \tag{5.48}$$

where

$$\mathrm{Ad}(\boldsymbol{T}) = \begin{bmatrix} \boldsymbol{R} & \boldsymbol{t}^{\wedge}\boldsymbol{R} \\ 0 & \boldsymbol{R} \end{bmatrix}$$

Considering the complexity of the expression $\boldsymbol{J}_r^{-1}$, an approximation is usually taken. If the error is close to 0, then $\boldsymbol{J}_r^{-1}$ can be approximated as

$$\boldsymbol{J}_r^{-1}(\boldsymbol{e}_{ij}) \approx \boldsymbol{I} + \frac{1}{2}\begin{bmatrix} \boldsymbol{\varphi}_e^{\wedge} & \boldsymbol{\rho}_e^{\wedge} \\ 0 & \boldsymbol{\varphi}_e^{\wedge} \end{bmatrix} \tag{5.49}$$

At this point, a least squares problem can be constructed, denoting ε as the set of all edges. Then, the final objective function is

$$F = \min_{\varepsilon} \frac{1}{2} \sum_{i,j\in\varepsilon} \boldsymbol{e}_{ij}^{\mathrm{T}} \sum_{ij}^{-1} \boldsymbol{e}_{ij} \tag{5.50}$$

This least squares problem can be solved using the general graph optimization (G2O) library, and after obtaining the optimized bit position and orientation, an accurate global map of the environment can be constructed.

5.4 Integrated Practice of Robotic Visual Perceptual System

5.4.1 *Feature point matching comparative analysis*

Matching operation on feature points can effectively solve the data association problem in vision SLAM, provide solution parameters for the sensor's position and orientation estimation, provide initial data for the back end, and play a decisive role in the whole SLAM system.[33]

Currently, there are two types of feature matching methods based on OpenCV library functions: Brute Force (BF) Matching and Fast Library for Approximate Nearest Neighbors (FLANN).[34] The main principle of BF is to extract all the feature points in two frames and calculate the "distance" between each pair of feature points in turn. FLANN refers to the nearest neighbor search on a large amount of data combined with high-dimensional feature vectors.

To determine whether there is a match between two feature points $\boldsymbol{x}_i$ and $\boldsymbol{y}_i$ in an image, the "distance" is usually compared with a set threshold. There are several ways to represent the distance, as follows:

(1) Euclidean distance

$$d_{ij} = \sqrt{(\boldsymbol{D}_{xi} - \boldsymbol{D}_{yi})^{\mathrm{T}}(\boldsymbol{D}_{xi} - \boldsymbol{D}_{yi})} \tag{5.51}$$

(2) Mahalanobis distance

$$d_{ij} = \sqrt{(\boldsymbol{D}_{xi} - \boldsymbol{D}_{yi})^{\mathrm{T}}\boldsymbol{S}^{-1}(\boldsymbol{D}_{xi} - \boldsymbol{D}_{yi})} \tag{5.52}$$

(3) Heininger distance

$$d_{ij} = \frac{1}{\sqrt{2}}\sqrt{\left(\sqrt{\boldsymbol{D}_{xi}} - \sqrt{\boldsymbol{D}_{yi}}\right)^{\mathrm{T}}\left(\sqrt{\boldsymbol{D}_{xi}} - \sqrt{\boldsymbol{D}_{yi}}\right)} \tag{5.53}$$

(4) Hamming distance

$$d_{ij} = \sum \boldsymbol{x}_i \oplus \boldsymbol{y}_i \tag{5.54}$$

Typically, the Hamming distance is used to measure binary descriptors, and the Euclidean distance is used to describe descriptors of floating point type. The SIFT and SURF descriptors mentioned in Section 5.3.4 have a float type data type and can be used with either BF or FLANN matching. As for the ORB algorithm, the data type of its descriptors is binary number, and it can only use the BF matching method. In the following, five combinations of the above three feature point descriptors and two matching algorithms are studied, and the related experimental data are recorded and analyzed for comparison.

(1) Comparative Experiment 1: Detection of speed.

The detection time is first compared for the feature point descriptors, and the specific results are shown in Tables 5.3–Table 5.5, where n denotes the number of feature points.

Table 5.3. SIFT feature point detection time.

Detection algorithm SIFT$(n)_1$	Three groups of experimental feature point detection time/s	Number of feature points/each
SIFT(100)	0.687396, 0.680885, 0.671914	100
SIFT(300)	0.666022, 0.675797, 0.703832	300
SIFT(400)	0.704548, 0.712716, 0.690659	400
SIFT(500)	0.673704, 0.675325, 0.665886	500
SIFT(700)	0.679966, 0.683346, 0.663919	700
SIFT(900)	0.683441, 0.788594, 0.665335	900
SIFT(1000)	0.702454, 0.668677, 0.666914	1000

Code 1: SIFT feature point detection.

```
import cv2
import numpy as np
import time

img = cv2.imread('left.jpg')

gray = cv2.cvtColor(img, cv2.COLOR_BGR2GRAY)
sift = cv2.xfeatures2d.SIFT_create()
start = time.clock()
kp = sift.detect(gray, None) # find the key-points
kp_time = (time.clock() - start)
print("kp_time:", '% 4f' % (kp_time)) # ms
print(len(kp))

img = cv2.drawkey-points(img, kp, img) # draw key-points
cv2.namedWindow('sift_point', 0)
cv2.imshow('sift_point', img)

cv2.waitKey(0)
cv2.destroyAllWindows()
```

Table 5.4. SURF feature point detection time.

Detection algorithm $SURF(n)_2$	Three groups of experimental feature point detection time/s	Number of feature points/each
SURF(100)	0.990093, 1.003616, 0.993129	3790
SURF(300)	0.676772, 0.690055, 0.717059	1427
SURF(500)	0.602716, 0.601148, 0.593609	895
SURF(700)	0.583628, 0.545550, 0.532169	662
SURF(900)	0.509631, 0.502946, 0.514535	537
SURF(1000)	0.498344, 0.507757, 0.492688	490
SURF(1500)	0.460943, 0.491957, 0.462416	333

Code 2: SURF feature point detection.

```
import cv2
import numpy as np
import time
img = cv2.imread('left.jpg')

# Gray-scale processed images
gray = cv2.cvtColor(img, cv2.COLOR_BGR2GRAY)

# parameter is the threshold of the Hessian matrix
surf = cv2.xfeatures2d.SURF_create()
start = time.clock()
# Find key-points and descriptors
key_query= surf.detect(gray, None)
surf_time = (time.clock() - start)
print("surf_time:", '% 4f' % (surf_time)) # ms
# print(key_query, desc_query)
print(len(key_query))
# Mark feature points onto the image
img=cv2.drawkey-points(img, key_query, img)
cv2.namedWindow('testSURF', 0)
cv2.imshow('testSURF', img)
cv2.waitKey(0)
cv2.destroyAllWindows()# clear all windows
```

Table 5.5. ORB feature point detection data.

Detection algorithm $FAST(n)_3$	Three groups of experimental feature point detection time/s	Number of feature points/each
FAST(0)	0.019902, 0.019757, 0.020130	23437
FAST(5)	0.002199, 0.002261, 0.002216	4338
FAST(10)	0.001356, 0.001411, 0.001351	1661
FAST(20)	0.000876, 0.000881, 0.000887	673
FAST(50)	0.000553, 0.000591, 0.000595	107
FAST(100)	0.000514, 0.000555, 0.000523	9

Code 3: ORB feature point detection.

```
import numpy as np
import cv2
from matplotlib import pyplot as plt
import time
imgname1 = 'left.jpg'
imgname2 = 'right.jpg'
orb = cv2.ORB_create()
img1 = cv2.imread(imgname1)
gray1 = cv2.cvtColor(img1, cv2.COLOR_BGR2GRAY) # grayscale the image
img2 = cv2.imread(imgname2)
gray2 = cv2.cvtColor(img2, cv2.COLOR_BGR2GRAY)
start = time.clock()
kp1, des1 = orb.detectAndCompute(img1,None)# des is descriptor
kp2, des2 = orb.detectAndCompute(img2,None)
kp_time = (time.clock() - start)
print("kp_time:", '% 4f' % (kp_time)) # ms
img3 = cv2.drawkey-points(img1,kp1,img1,color=None)
img4 = cv2.drawkey-points(img2,kp2,img2,color=None)
hmerge = np.hstack((img3, img4)) # horizontal stitching
cv2.namedWindow('point', 0)
cv2.imshow("point", hmerge) # splice show as gray
```

Comparing Tables 5.3–5.5 in terms of the detection speed of the feature point descriptors, all of them take about 650 feature points as an example. The detection time of SIFT is in the range of 0.66–0.68 s; the detection time of SURF is in the range of 0.53–0.58 s; and the detection time of FAST is only approximately 0.00088 s. It can be said that there is an order-of-magnitude improvement in the detection speed of FAST. Comparing the groups of experiments as a whole, the speed of FAST feature point detection algorithm is also much higher than that of SIFT and SURF feature point detection algorithms; at the same time, the number of feature points detected by the SURF detection algorithm is much higher than that detected by the SIFT detection algorithm, which reflects that the performance of SURF is superior to that of SIFT from the side.

(2) Comparative Experiment 2: Comparison of matching effects.

In order to compare the experimental effect of matching, all the feature points with about 800 are matched, and the obtained matching effect is as follows.

From Figures 5.14–5.18, it can be seen that for the three aforementioned feature point descriptors, there are a large number of obvious wrong matching pairs whether using the BF matching method or the FLANN matching method, which requires further optimization processing.

Figure 5.14. SIFT+BF feature point matching method (897 feature points).

Code 4: SIFT+BF feature point matching.

```
import numpy as np
import cv2
from matplotlib import pyplot as plt

imgname1 = 'left.jpg'
imgname2 = 'right.jpg'

sift=cv2.xfeatures2d.SIFT_create()
img1 = cv2.imread(imgname1)
gray1 = cv2.cvtColor(img1, cv2.COLOR_BGR2GRAY) # grayscale the image
kp1, des1 = sift.detectAndCompute(img1,None) # des is descriptor
img2 = cv2.imread(imgname2)

gray2 = cv2.cvtColor(img2, cv2.COLOR_BGR2GRAY)# grayscale the image
kp2, des2 = sift.detectAndCompute(img2,None) # des is descriptor
print(des1) hmerge = np.hstack((gray1, gray2)) # horizontal stitching
```

```
img3 = cv2.drawkey-points(img1,kp1,img1,color=(None)) # draw feature points
img4 = cv2.drawkey-points(img2,kp2,img2,color=(None)) # draw feature points
hmerge = np.hstack((img3, img4)) # horizontal stitching

# BFMatcher resolves matches
bf = cv2.BFMatcher()
matches = bf.knnMatch(des1,des2, k=2)

img5 = cv2.drawMatchesKnn(img1,kp1,img2,kp2,matches,None,flags=2)
cv2.namedWindow('BFmatch', 0)
cv2.imshow("BFmatch", img5)
cv2.waitKey(0)
cv2.destroyAllWindows()
```

Figure 5.15. SURF+BF feature point matching (895 feature points).

Code 5: SURF+BF feature point matching.

```
import cv2
import numpy as np

# 1) Open the image
img1 = cv2.imread("left.jpg")
img2 = cv2.imread("right.jpg")

# 2) Image grayscaling
img1_gray = cv2.cvtColor(img1, cv2.IMREAD_GRAYSCALE)
img2_gray = cv2.cvtColor(img2, cv2.IMREAD_GRAYSCALE)

# 3) SURF feature calculation
surf = cv2.xfeatures2d.SURF_create()
kp1, des1 = surf.detectAndCompute(img1, None)
kp2, des2 = surf.detectAndCompute(img2, None)
```

```
# 4) BFmatcher with default parms
bf = cv2.BFMatcher(cv2.NORM_L2)
matches1 = bf.knnMatch(des1, des2, k=2)

bf_img_out = cv2.drawMatchesKnn(img1, kp1, img2, kp2, matches1[:], None,
flags=2)
print(len(matches))

cv2.namedWindow('bfmatch', 0)
cv2.imshow('bfmatch', bf_img_out)# show image
cv2.waitKey(0)# wait for the key to be pressed
cv2.destroyAllWindows()# clear all windows
```

Figure 5.16. ORB+BF feature point matching (752 feature points).

Code 6: ORB+BF feature point matching.

```
import numpy as np
import cv2
from matplotlib import pyplot as plt

imgname1 = 'left.jpg'
imgname2 = 'right.jpg'

orb = cv2.ORB_create()
img1 = cv2.imread(imgname1)
gray1 = cv2.cvtColor(img1, cv2.COLOR_BGR2GRAY) # grayscale the image
kp1, des1 = orb.detectAndCompute(img1,None)# des is descriptor

img2 = cv2.imread(imgname2)
gray2 = cv2.cvtColor(img2, cv2.COLOR_BGR2GRAY)
kp2, des2 = orb.detectAndCompute(img2,None)
hmerge = np.hstack((gray1, gray2)) # horizontal stitching
```

```
img3 = cv2.drawkey-points(img1,kp1,img1,color=(255,0,255))
img4 = cv2.drawkey-points(img2,kp2,img2,color=(255,0,255))
hmerge = np.hstack((img3, img4)) # horizontal stitching
cv2.namedWindow('point', 0)
cv2.imshow("point", hmerge) # splice show as gray

# BFMatcher resolves matches
bf = cv2.BFMatcher()
matches = bf.knnMatch(des1,des2, k=2)
img5 = cv2.drawMatchesKnn(img1,kp1,img2,kp2,matches,None,flags=2)
cv2.namedWindow('ORB', 0)
cv2.imshow("ORB", img5)
cv2.waitKey(0)
cv2.destroyAllWindows()
```

Figure 5.17. SIFT+FLANN feature point matching (897 feature points).

Code 7: SIFT+FLANN feature point matching.

```
import numpy as np
import cv2
from matplotlib import pyplot as plt

imgname1 = 'left.jpg'
imgname2 = 'right.jpg'

sift = cv2.xfeatures2d.SIFT_create()

# FLANN parameterization
FLANN_INDEX_KDTREE = 0
index_params = dict(algorithm = FLANN_INDEX_KDTREE, trees = 3)
```

```
search_params = dict(checks=100)
flann = cv2.FlannBasedMatcher(index_params,search_params)

img1 = cv2.imread(imgname1)
gray1 = cv2.cvtColor(img1, cv2.COLOR_BGR2GRAY) # grayscale the image
kp1, des1 = sift.detectAndCompute(img1,None)# des is descriptor

img2 = cv2.imread(imgname2)
gray2 = cv2.cvtColor(img2, cv2.COLOR_BGR2GRAY)
kp2, des2 = sift.detectAndCompute(img2,None)

hmerge = np.hstack((gray1, gray2)) # horizontal stitching
cv2.namedWindow('gray', 0)
cv2.imshow("gray", hmerge) # splice display as gray
cv2.waitKey(0)

img3 = cv2.drawkey-points(img1,kp1,img1,color=(255,0,255))
img4 = cv2.drawkey-points(img2,kp2,img2,color=(255,0,255))

hmerge = np.hstack((img3, img4)) # horizontal stitching
cv2.namedWindow('point', 0)
cv2.imshow("point", hmerge) # splice show as gray
# cv2.imwrite('point024.jpg', hmerge)
cv2.waitKey(0)
matches = flann.knnMatch(des1,des2,k=2)
matchesMask = [[0,0] for i in range(len(matches))]

img5 = cv2.drawMatchesKnn(img1,kp1,img2,kp2,matches,None,flags=2)
cv2.namedWindow('FLANN', 0)
cv2.imshow("FLANN", img5)
cv2.waitKey(0)
cv2.destroyAllWindows()
```

Figure 5.18. SURF+FLANN feature point matching (895 feature points).

Code 8: SURF+FLANN feature point matching.

```
import cv2
import numpy as np

# 1) Open the image
img1 = cv2.imread("left.jpg")
img2 = cv2.imread("right.jpg")

# 2) Image grayscaling
img1_gray = cv2.cvtColor(img1, cv2.IMREAD_GRAYSCALE)
img2_gray = cv2.cvtColor(img2, cv2.IMREAD_GRAYSCALE)

# 3) SURF feature calculation
surf = cv2.xfeatures2d.SURF_create()

kp1, des1 = surf.detectAndCompute(img1, None)
kp2, des2 = surf.detectAndCompute(img2, None)

# 4) FLANN feature matching
flann_index_kdtree = 1
index_params = dict(algorithm=FLANN_INDEX_KDTREE, trees=5)
search_params = dict(checks=400)
flann = cv2.FlannBasedMatcher(index_params, search_params)
matches = flann.knnMatch(des1, des2, k=2)

flann_img_out = cv2.drawMatchesKnn(img1, kp1, img2, kp2, matches1[:], None,
flags=2)

print(len(matches))
cv2.namedWindow('flannmatch', 0)
cv2.imshow('flannmatch', flann_img_out)# show the image
cv2.waitKey(0)# wait for key to be pressed
cv2.destroyAllWindows()# clear all windows
```

(3) Comparative Experiment 3: Total time consumed.

Record the feature point matching and total time consumed in Figures 5.14–5.18, and the results obtained are shown in Table 5.6.

Table 5.6. Feature point matching and total consumption time.

Matching method	Match points/each	Total consumption time/s
SIFT+BF	897	0.815216
SURF+BF	895	0.631137
SIFT+FLANN	897	0.98234
SURF+FLANN	895	0.753199
ORB+BF	752	0.030334

As can be seen from Table 5.6, the total consumption time of the matching method using ORB+BF is one order of magnitude faster than all other methods. Only this approach can guarantee the requirement of real-time matching during camera movement. Therefore, in summary, the matching method using ORB+BF is selected as the feature point matching method in this book.

5.4.2 *Mismatched feature point elimination*

When feature point matching is performed, a large number of wrongly matched feature point pairs will appear. The quality of the matching will have a significant impact on the subsequent experiments, so it is necessary to carry out culling operation on the mismatched feature point pairs. In this book, the K-nearest neighbor (KNN) algorithm and Random Sampling Consensus (RANSAC) algorithm are combined to eliminate the mismatched feature pairs.

KNN algorithm[35]: Assuming that there are point sets $M = \{m_1, m_1, \ldots, m_n\}$ and $N = \{n_1, n_2, \ldots, n_n\}$, a feature point m_i is selected from the point set M, and the K nearest matches to the feature point m_i in the point set N *are* found to be $n_1, n_2, \ldots, n_K$, which is usually $K = 2$. At this point, we find the closest point n_{i1} and the second-closest point n_{i2}. If n_{i1} is much smaller than n_{i2}, then we can consider m_i and n_{i1} as the best matching points, otherwise they are not. In the actual algorithm, define ration $= n_{i1}/n_{i2}$, if the ration is less than a certain threshold, then it is considered to be the correct matching pair. In this book, after many trials, we finally choose ration $= 0.8$.

RANSAC algorithm[36]: The overall idea is to select a set of data point samples containing "outer points" from the entire point set, and obtain a better mathematical model to fit the samples through multiple iterations of the solution. The entire point set contains the "inner points" and "outer points" of the data. An inner point is the point that conforms to the mathematical model of the data, while an outer point is the point generated by noise, calculation error, measurement error, assumption error, etc. Outer points need to be eliminated in the feature point matching. The basic flowchart of the RANSAC algorithm is shown in Figure 5.19.

Preliminary mismatch point elimination of ORB+BF matching results using the KNN algorithm is shown in Figure 5.20.

It can be seen that the matching pairs in Figure 5.21 can eliminate most of the mismatches after being eliminated by the KNN algorithm. However, in some regions, there are still false matches. As shown in the box in Figure 5.22, there are more obvious mismatched pairs, so another filtering is needed.

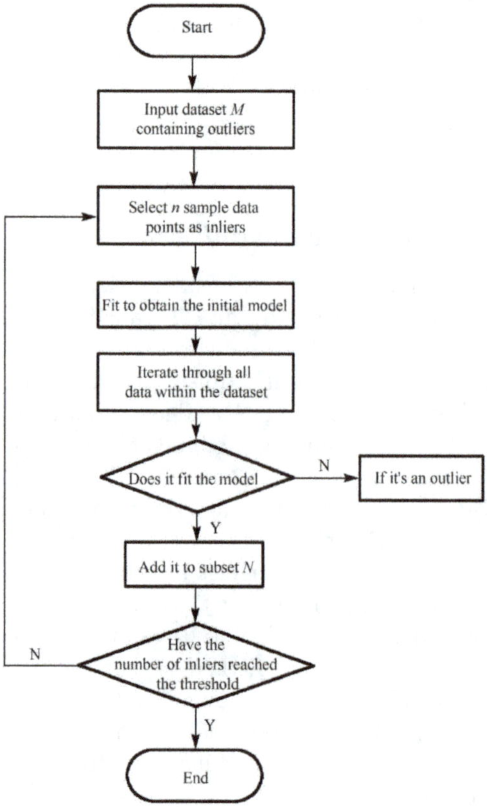

Figure 5.19. Basic flowchart of the RANSAC algorithm.

Figure 5.20. ORB+BF matching original (500 feature points).

Figure 5.21. KNN algorithm filtered match result.

Figure 5.22. KNN + RANSAC algorithm filtering matching map.

Code 9: RANSAC feature point filtering.

```
import numpy as np
import cv2
from matplotlib import pyplot as plt
import time

img01 = cv2.imread('left.jpg')
img02 = cv2.imread('right.jpg')
min_match_count = 10

img1 = cv2.imread('left.jpg',0)
img2 = cv2.imread('right.jpg',0)
ORB = cv2.ORB_create()
kp1, des1 = ORB.detectAndCompute(img1, None)
kp2, des2 = ORB.detectAndCompute(img2, None)
bf = cv2.BFMatcher(cv2.NORM_L2)
matches = bf.knnMatch(des1, des2, k=2)
good = []
# Match optimization
```

```
strat = time.clock()
for m, n in matches:
     if m.distance < 0.8 * n.distance.
          good.append(m)
good1 = np.expand_dims(good, 1)
if len(good)>min_match_count.
     src_pmts = np.float32([kp1[m.queryIdx].pt for m in good]).reshape (-1,1,2)
     dst_pts = np.float32([kp2[m.trainIdx].pt for m in good]).reshape (-1,1,2)
     m, mask = cv2.findHomography(src_pmts, dst_pts, cv2.RANSAC, 5.0)
     matchesmask1= mask.ravel().tolist() # Get the height and width of the
original image

ransac_time = time.clock() - strat

img3 = img1
img3 = cv2.drawMatches(img01, kp1, img02, kp2, good, img3, None, None,
matchesmask1, 2)

cv2.namedWindow('ransac', 0)
cv2.imshow('ransac', img3)
print("ransac_time:", '% 4f' % (ransac_time))

cv2.waitKey(0)# wait for key to be pressed
cv2.destroyAllWindows()# clear all windows
```

The result after filtering by the KNN + RANSAC algorithm is shown in Figure 5.22.

At this point, there are 32 matching pairs in Figure 5.22, all of which are accurately matched, and there are no incorrect matches, so that the accurate matching information can satisfy the requirements for use.

It can be seen from the above experimental results that the use of ORB + BF + KNN + RANSAC algorithm can ensure that the matching work is carried out accurately in the shortest possible time, providing more accurate information for subsequent SLAM links.

5.4.3 *Requirements and classification of mapping establishment*

5.4.3.1 *Requirements and categorization of mapping*

Establishing a well-performing environment is important for the robot's motion. This is because the strategy used for mapping affects the representation of the robot's current position. In navigation problems, the accuracy of the environment can limit the accuracy of the robot's position representation. When choosing a representation to depict the environment, the following three fundamental relationships must be considered:

(1) The accuracy of the environment is matched with the accuracy of the robot's movement to the target point.
(2) The method of characterizing the environmental representation is consistent with the type of data returned by the sensor.
(3) The complexity of the robot's navigation, localization, and reasoning processes is influenced by the complexity of the environment representation.

Therefore, the mapping approach significantly influences the system architecture of robot localization and navigation. In practice, a balance is often achieved through context-specific implementations of simultaneous localization and mapping (SLAM). Common map representations include occupancy grid maps, point cloud maps, topological maps, and feature maps. A comparison of the advantages and disadvantages of different maps is shown in Table 5.7.

5.4.3.2 *Octree maps*

Comparing the four types of maps in Table 5.7, it can be seen that the choice of occupancy grid maps for navigation planning is a much more reasonable one. Octree map (see Figure 5.23) is a more widely used form of occupancy grid map. It has good expandability and updatability, small volume occupation, easily adjustable resolution, and open-source code, while this map can be directly mapped down to a two-bit occupancy grid map, which can be applied to subsequent navigation and path planning.

Octree, as the name suggests, is a tree used to extend storage content. As shown in Figure 5.23, each square is evenly divided into eight smaller squares of equal size, and this step is repeated until the final smallest square meets the actual needs of the project. In this "tree", each small square is called a child node of the parent node on the upper level, itself representing the content of the probability of being occupied. In Figure 5.23, black square means occupied with a probability of 1, white square means not occupied with a probability of 0, and gray square means doubtful, which needs to be further divided for observation. After each observation, the map needs to be updated, and if a square is observed to be occupied, its probability value is increased, otherwise its probability value is decreased. Also, in order to adjust the probability to lie in the range [0, 1], the logarithm of probability is used to describe the occupation, as shown in Eq. (5.55):

$$\begin{aligned} X &= \lg\mathrm{it}(p) = \lg\left(\frac{p}{1-p}\right) \\ p &= \lg\mathrm{it}^{-1}(X) = \frac{\exp(X)}{1+\exp(X)} \end{aligned} \tag{5.55}$$

Table 5.7. Comparison of the advantages and disadvantages of different maps.

Map representation	Advantages	Drawbacks
Point cloud map	Provides an intuitive and complete picture of the environment in three dimensions without the need for excessive processing of sensor information	Extremely demanding on the storage performance of the device, it is not possible to directly express the occupied/idle/unknown situation of a point in space, and requires a secondary understanding before it **can be applied to the navigation of the robot, and at the same time, it is not possible to update the map**
Occupancy grid map	Environmental information in multi-dimensional space can be described in more detail, neatly arranged, and the map storage information can be updated with adjustable resolution, making it easy for direct robot localization and path planning	As the surrounding scene gets larger, the higher the map accuracy, the higher the complexity of the calculations, and the higher the memory requirements for the computer
Feature maps	Small storage space, compact structure, and expression more similar to human perception of the environment	Inability to accurately represent the real information of the environment, which is not conducive to the robot's navigation planning
Topological map	Can be expanded and updated, and is capable of rapid route planning	Robots are unable to achieve accurate localization because the information expressed is too abstracted

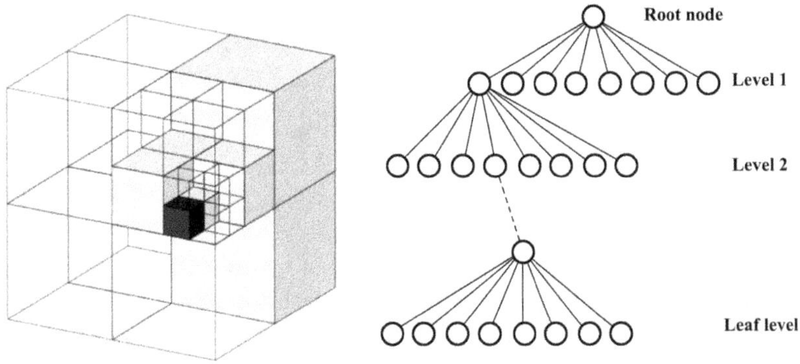

Figure 5.23. Octree map.

where p denotes the observed probability value of a particular node at the current moment and X denotes the occupation of a particular node.

Taking node n as an example, if the logarithmic value of the probability observed at the first t moments is $L(n|z_{1:t})$, then the logarithmic value of the probability at $t+1$ moments is

$$L(n|z_{1:t+1}) = L(n|z_{1:t-1}) + L(n|z_t) \tag{5.56}$$

With logarithmic probability, the map can be updated and processed based on the RGB_D data. Meanwhile, the resolution of the octree map is adjustable, with larger squares corresponding to a lower resolution and smaller squares corresponding to a higher resolution. By querying the node information of the octree, it is possible to determine whether the node information is empty or not, thus providing data information for navigation.

5.5 Summary

In this chapter, we design the comprehensive RGB-D SLAM framework aimed at exploring the mapping capabilities of humanoid mobile robots in a laboratory setting. At the SLAM system's front end, the ORB+BF algorithm is utilized for feature point matching. To enhance matching accuracy, two mismatch elimination methods are employed, and the SVD algorithm is applied for camera motion estimation. On the back end, we propose a selection strategy for keyframe identification, employ the bag-of-words model to establish a closed-loop detection mechanism, and use the G2O optimization library to refine the pose graph. Ultimately, based on the specific requirements of this study, the octree mapping approach is adopted. Overall, this chapter outlines the development of a more robust and complete RGB-D SLAM system.

5.6 Exercises

5.1 Briefly describe the characteristic indicators of a sensor.
5.2 What are the internal sensors of a robot?
5.3 What are the external sensors of a robot?
5.4 Briefly categorize computer vision algorithms.
5.5 Briefly describe the steps in computer vision processing.
5.6 Design and practice computational stereo vision.
5.7 Design and practice binocular vision systems.

Chapter 6

Integrated Design and Practice of Robot Object Recognition

6.1 Introduction

Chapter 5 addressed the issue of robot scene construction, while this chapter focuses on the robot's understanding of its environment, specifically target recognition. For robots tasked with specific operations, the fundamental capability is distinguishing between the target object and the surrounding environment. Additionally, the presence of randomly positioned obstacles in the target environment increases the risk of collisions between the robot's operating arm and these obstacles, potentially damaging the robot and its components. Therefore, it is crucial to perform 3D spatial mapping of environmental obstacles to equip the robot with precise 3D spatial data regarding these obstacles.

This chapter concentrates on the challenges of identifying target objects and environmental obstacles, as well as reconstructing these obstacles in three dimensions. Initially, it suggests utilizing a neural network algorithm for recognizing targets and environmental obstacles. Following obstacle recognition, the chapter proposes grouping environmental obstacles using bidirectional constraints. Finally, it employs polynomial fitting to fill in areas of unrecognized environmental obstacles, thereby achieving the three-dimensional reconstruction of branches and trunks.

6.2 Theory of Robot Object Recognition

6.2.1 *Traditional object recognition*

Object recognition has entered the era of deep learning, but it remains necessary to understand traditional methods. The ideas behind deep learning methods also originated from traditional methods. The literature on traditional methods is extensive, but we only need to understand the following three, which can be considered landmark methods: VJ detector, histogram of oriented gradient (HOG), and deformable part model (DPM). These three methods are briefly described in the following.

6.2.1.1 *VJ detector*

In 2001, P. Viola and M. Jones achieved the first real-time detection of faces without any constraints (e.g., skin color segmentation). Running on a 700 MHz Pentium III CPU, the detector was tens, or even hundreds, of times faster than other algorithms with the same level of detection accuracy. The detection algorithm became known as the "Viola–Jones (VJ) detector", named after the authors in honor of their significant contributions.

(1) Principle

The VJ detector uses the most straightforward detection method, which is to slide a window across all possible positions and proportions in the image to see if any of the windows contain the object to be recognized. Although this seems to be a very simple process, the calculations behind it are far beyond the capabilities of computers at that time. The VJ detector combines three important techniques: "Integral Image", "Feature Selection", and "Detection Cascade", which greatly increase the speed of detection.

Integral image: Integral image is a computational method that accelerates the box filtering or convolution process. As with other target detection algorithms at the time, Haar wavelets are used to represent image features in the VJ detector. The integral image makes the computational complexity of each window in the VJ detector independent of its window size.

Feature selection: The AdaBoost algorithm is used to select a small set of features from a huge pool of random features (about 18,000 dimensions) that are most helpful for target detection.

Detection cascade: A multi-stage detection paradigm (also known as a "detection cascade") is introduced in the VJ detector to reduce computational overhead by reducing the amount of background window computation and increasing target detection computation.

(2) Advantages and disadvantages analysis

Advantage: The combination of the three key techniques, "integral image", "feature selection", and "detection cascade", greatly improves the detection speed of the VJ detector.

Disadvantages: (1) Haar-like features are relatively simple features with low stability. (2) The weak classifier uses a simple decision tree, which is prone to overfitting. As a result, the algorithm is good for frontal faces but not so good for special and complex cases such as occlusion, gestures, and facial expressions. (3) In the classifier design based on the VJ cascade, after entering the next stage, the previous information is discarded, and the classifier evaluates a sample that will not be based on its performance in the previous stage, which makes the classifier less robust.

6.2.1.2 *Histogram of Oriented Gradient*

A HOG detector is an orientation gradient-based histogram detection algorithm with the following principles, advantages, and disadvantages.

(1) Principle

The HOG feature is a type of feature description used in computer vision and image processing for object detection. It constitutes a feature by computing and counting the histogram of orientation gradients in localized regions of an image. (1) Main idea: In an image, the appearance and shape of a local target can be well described by the directional density distribution of gradients or edges. (2) Specific implementation method: Firstly, the image is divided into small connected regions, which we call cellular units, and then the directional histograms of gradients or edges of the pixel points in the cellular units are captured; finally, the feature descriptor can be composed by combining these histograms. (3) Improvement of performance: These local histograms are normalized for contrast over a larger range of the image (we call it an interval or block) by first calculating the density of each histogram within the interval (block), and then normalizing each cell in the interval according to this density. By normalizing, better results can be obtained for light changes and shadows.

(2) Advantages and disadvantages analysis

Pros: (1) HOG represents edge (gradient) structural features, so it can describe local shape information. (2) The quantization of position and orientation space can, to some extent, suppress the effects brought by translation and rotation. (3) Adopting normalized histograms in local areas can partially offset the effects of illumination changes. (4) Since the effect of illumination color on the image is ignored to a certain extent, the dimension of the characterization data required for the image is reduced. (5) The relationship among local pixel points in the image can be well characterized due to the block-and-cell processing method.
Disadvantages: (1) The descriptor generation process is lengthy, resulting in slow speed and poor real-time performance. (2) It can be difficult to deal with occlusion. (3) The descriptor is quite sensitive to noise due to its gradient nature.

6.2.1.3 *Deformable part model*

The DPM is a component-based detection algorithm. The algorithm was proposed by Felzenszwalb in 2008 and received the PASCAL VOC "Lifetime Achievement Award" in 2010. The principle of DPM algorithm and its advantages and disadvantages are as follows.

(1) Principle

The general idea of DPM is consistent with that of the HOG method. First, calculate the HOG and then use the support vector machine (SVM) training to get the gradient model of the object (Model). Such a template can be used directly for classification, making it simple to understand the model and for target matching. DPM has primarily focused on implementing numerous enhancements to the model.

The DPM algorithm adopts the improved HOG features, SVM classifier, and sliding windows detection concept, as well as the component strategy for the problem of multiple viewpoints of the target and the pictorial structure-based component model strategy for the problem of deformation of the target itself. In addition, the model category to which the sample belongs and the location of the component model are used as latent variables, and multiple-instance learning is used to determine them automatically.

(2) Advantages and disadvantages analysis

Advantages: Since the DPM algorithm itself is a component-based detection algorithm, it is very good at detecting distorted, gendered, multi-pose, multi-angle targets (including those that usually do not have large deformation and can be approximated as rigid bodies, and DPM-based algorithms can handle the target detection problem quite well).

Disadvantages: Because the algorithm is too complex, the calculation becomes complicated when judging, and it becomes difficult to meet the real-time requirements. A series of improvements have been subsequently made to the process, such as adding cascade classifiers and the use of integral graph method, but none have reached the efficiency of the VJ detector, so they are seldom used in projects. Instead, the algorithms of the AdaBoost framework are generally used.

6.2.2 *Deep learning-based target recognition algorithms*

In recent years, target detection algorithms have made great breakthroughs. The more popular algorithms can be categorized into two groups: one is the candidate box-based R-CNN family of algorithms (R-CNN,[37] Fast R-CNN,[38] and Faster R-CNN[39]), which require heuristics (selective search) or CNN networks (RPN) to generate candidate boxes first, followed by classification and regression on those candidate boxes. And the other category includes YOLO.[40] and SSD.[41] One-stage algorithms, such as YOLO and SSD, use only one CNN to directly predict the categories and locations of different targets. The first class of algorithms is somewhat more accurate but slower, while the second class of algorithms is faster but less accurate. This can be seen in Figure 6.1.

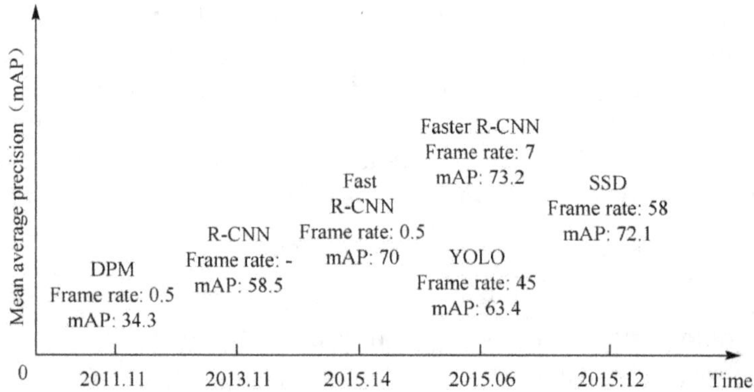

Figure 6.1. Progress and comparison of target detection algorithms.

6.2.2.1 *Deep learning target detection algorithm based on candidate boxes*

Convolutional Neural Networks (CNNs) are a central component of the Region Propoal algorithm. CNNs were first proposed by Prof. Yann LeCun, and in the early days, CNNs were mainly used as classifiers, e.g., for image recognition. However, CNNs have the following three structural properties: local receptive fields, weight sharing, and spatial or temporal subsampling. These properties make CNNs somewhat translation invariant, scaling invariant, and warping invariant. In 2006, Hinton proposed the use of deep neural networks to automatically learn high-level features from large amounts of data. Candidate boxes build on this foundation to solve the two main problems of traditional target detection. The more commonly used candidate box methods are Selective Search and Edge Boxes. Since then, CNNs have been rapidly developed, and the Top-5 Error of Microsoft's latest ResNet and Google's Inception V4 model have dropped to less than 4%, so that after the target detection is obtained from the candidate box, the accuracy and detection speed of using CNNs to classify the images using it are improved.

6.2.2.1.1 R-CNN R-CNN, whose full name is Region-CNN, is the first algorithm to successfully apply deep learning to target detection. Fast R-CNN, Faster R-CNN, and mask-CNN, to be discussed later, are all built on R-CNN.

Most of the traditional target detection algorithms are based on image recognition. Generally, you can use the exhaustive method or sliding window method on the image to select all the objects that may appear in the region frame, extract features for these region frames and use the image recognition classification method, get all the classification success of the region, and then output the results through the non-maximum value suppression.

(1) Principle

R-CNN follows the traditional idea of target detection while also adopting the method of extracting boxes. It uses the three steps of feature extraction, image classification, and non-maximal value suppression for each frame to carry out target detection, except that it has been partially improved, which is specifically manifested in the following: (1) Classical target detection algorithms use the sliding-window method to judge all the possible regions in turn. Here, a series of candidate boxes that may be objects are extracted in advance, and after that, features are extracted only on these candidate boxes for judgment, which greatly reduces the amount of computation. (2) The traditional features (e.g., SIFT and HOG features) are exchanged for features extracted by deep convolutional networks.

(2) Advantages and disadvantages analysis

Advantages: Although the recognition framework of R-CNN is not very different from that of traditional methods, thanks to the excellent feature extraction ability of CNNs, the results of R-CNN are still much better than those obtained using traditional methods. For example, on the VOC2007 dataset, the highest mAP of traditional methods is around 40%, while the mAP of R-CNN reaches 58.5%.
Disadvantages: The disadvantage of R-CNN is the large amount of computation required. The R-CNN process is more involved, including the selection of candidate boxes, training the CNN, training the SVM, and training the regression volume, meaning that the training time is very long (84 h) and it occupies a large amount of disk space. In the process of training CNN for each candidate box to calculate the convolution, which repeats too many unnecessary calculations, imagine an image that can result in more than 2,000 candidate boxes, most of which overlap. Then, the candidate box-based convolution

requires excessive computation, which is exactly the problem that Fast R-CNN mainly solved later.

6.2.2.1.2 SPP-Net Prior to this, all neural networks are required to input fixed size images, such as 224×224 (ImageNet), 32×32 (LenNet), and 96×96. Thus, when we wish to detect various sizes of images, we need to go through a series of operations, such as cropping or deformation scaling, which all lead to the loss and deformation of the image information to a certain extent, limiting the recognition accuracy. Moreover, from a physiological point of view, when the human eye sees a picture, the brain will first think that it is a whole and will not perform cropping or deformation scaling operations.

(1) Principle

SPP-Net improves on the shortcomings present in these networks. The basic idea is to input the whole image, extract the feature map of the whole image, and then use spatial relations to extract the features of each candidate box from the feature map of the whole image in the Spatial Pyramid Pooling (SPP) layer. A normal deep network consists of two parts: the convolutional part and the fully connected part, and the reason for requiring the input image to be of fixed size is not the convolutional part but the fully connected part. So, the SPP layer acts after the last layer of convolution, and the output of the SPP layer has a fixed size. SPP-Net allows different sizes of images to be input not only during testing but also during training, and overfitting can be prevented by having different scales of images. Compared to R-CNN, which extracts 2,000 candidate boxes, SPP-Net only needs to throw the whole image in to get the features, which increases the speed of operation by about 100 times.

(2) Advantages and disadvantages

Advantages: SPP-Net solves the problem of repeated extraction of candidate region features from R-CNN regions by proposing the SPP layer, which makes the input candidate box either large or small. R-CNN needs to calculate the convolution for each region, while SPP-Net only needs to calculate it once, so the efficiency of SPP-Net is much higher than that of R-CNN.

Disadvantages: SPP-Net for target detection is actually an improvement on R-CNN, which improves the recognition speed but not the recognition accuracy.

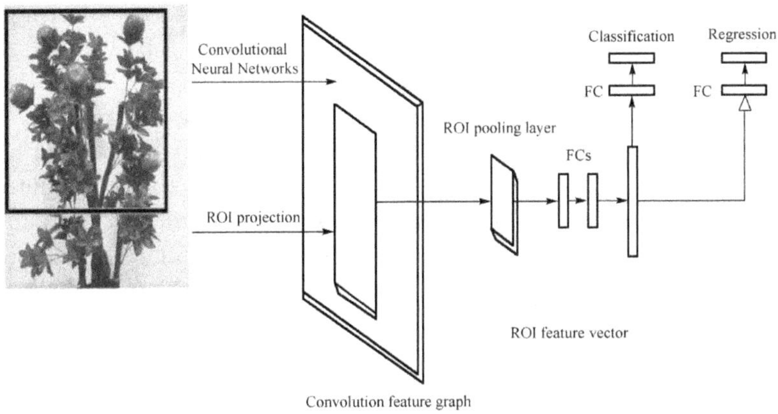

Figure 6.2. Fast R-CNN flowchart.

6.2.2.1.3 Fast R-CNN Fast R-CNN is an improvement of the previous two algorithms and has the following principles, advantages, and disadvantages.

(1) Principle

The Fast R-CNN flowchart is shown in Figure 6.2. The inputs to this network are the original image and the candidate box, and the outputs are the classification categories and the Bbox regression values. For the candidate box region in the original image, which is the same as in SPP-Net, it is mapped to the region corresponding to the convolutional features (region of interest, or ROI, projection in Figure 6.2), and then input to the ROI pooling layer, which can get a fixed-size feature map. This feature map is then passed through two fully connected layers to get the ROI features, then the features are passed through the fully connected layers. The softmax function is used to get the classification, and regression is used to get the edge regression. The main structure of the CNN can be taken from AlexNet or VGGNet.

(2) Advantages and disadvantages analysis

Advantages: Fast R-CNN is equivalent to a comprehensive improvement of the original two algorithms (SPP-Net algorithm and R-CNN algorithm), which not only reduces the number of training steps but also eliminates the need to save the features on disk. The Fast R-CNN algorithm based on VGG-16 is nearly nine times faster than R-CNN and roughly three times faster than SPP-Net in terms of training speed; it is also 213 times faster than R-CNN and 10 times faster than SPP-Net in terms of testing speed. The mAP on VOC2012 is around 66%.

Disadvantages: Fast R-CNN cannot perform end-to-end training either, so its training still consists of some tedious steps. There is also an awkward problem in Fast R-CNN: Sometimes, to detect an image, most of the time is not spent on calculating the classification of the neural network but rather on Selective Search, which is a slower method, to extract candidate boxes.

6.2.2.1.4 Faster R-CNN Faster R-CNN is based on the improvement of Fast R-CNN; its principle, advantages, and disadvantages are as follows.

(1) Principle

From R-CNN to Fast R-CNN to Faster R-CNN, the four basic steps of target detection (candidate box generation, feature extraction, classification, and bounding box regression) are finally unified within a single deep network framework, as shown in Figure 6.3. All computations are not duplicated and are done entirely in the GPU, which greatly improves the running speed.

Faster R-CNN can be simply regarded as a system of "Regional Generation Network (RPN) + Fast R-CNN", which replaces the Selective Search method in Fast R-CNN with Regional Generation Network, and its network structure is shown in Figure 6.4.

The steps of Faster R-CNN algorithm are as follows.

Step 1: Input any size ($M \times N$) image to the CNN network (ZF or VGG-16).

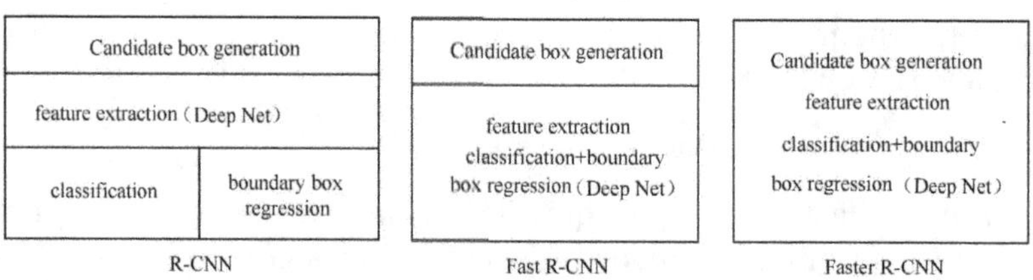

Figure 6.3. Four-step evolutionary normalization from R-CNN to Faster R-CNN.

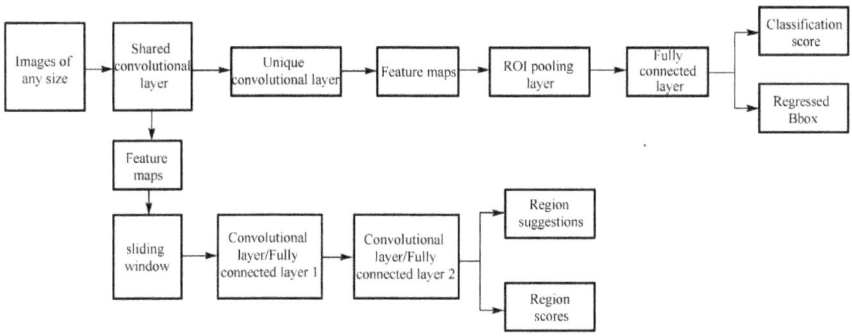

Figure 6.4. Network structure of Faster R-CNN.

Step 2: After forward propagation of the CNN network to the last shared convolutional layer, on the one hand, we get the feature maps for the input of the RPN network, and on the other hand, we continue to forward propagate to the unique convolutional layer to produce higher-dimensional feature maps.

Step 3: The feature map for RPN network input is passed through the RPN network to obtain the region suggestions and region scores, and the region scores are suppressed with non-extremely large values (with a threshold of 0.7), and the region suggestions with their *Top-N* scores are output to the ROI pooling layer.

Step 4: The high-dimensional feature map obtained in step 2 and the region suggestions output in step 3 are simultaneously input into the ROI pooling layer to extract the corresponding region suggestion features.

Step 5: After the proposed features of the region obtained in step 4 are passed through the fully connected layer, the classification score of the region and the regressed Bbox are output.

(2) Advantages and disadvantages analysis

Advantages: (1) It improves the detection accuracy and speed. (2) It realize an end-to-end target detection framework. (3) The generation of the suggestion box only takes about 10 ms.

Disadvantages: (1) It cannot achieve real-time detection of the target; (2) As it acquires the candidate box and then categorizes each candidate box, the computation is relatively large.

6.2.2.1.5 Mask R-CNN (1) Principle

Mask R-CNN is the masterpiece of Kaiming He, developed in 2017, which achieves excellent results by performing instance segmentation along with target detection. The design of its network is also relatively simple: Based on Faster R-CNN, a branch is added to the original two branches (classification + coordinate regression) for semantic segmentation, as shown in Figure 6.5.

The mask R-CNN framework parse is shown in Figure 6.6.

The steps of the mask R-CNN algorithm are as follows.

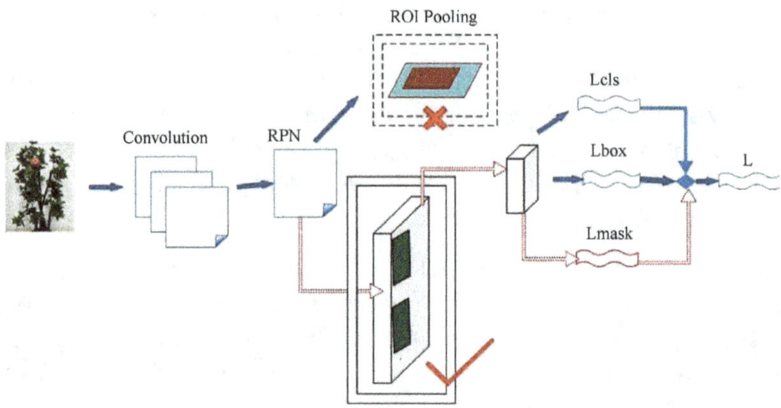

Figure 6.5. Semantic segmentation.

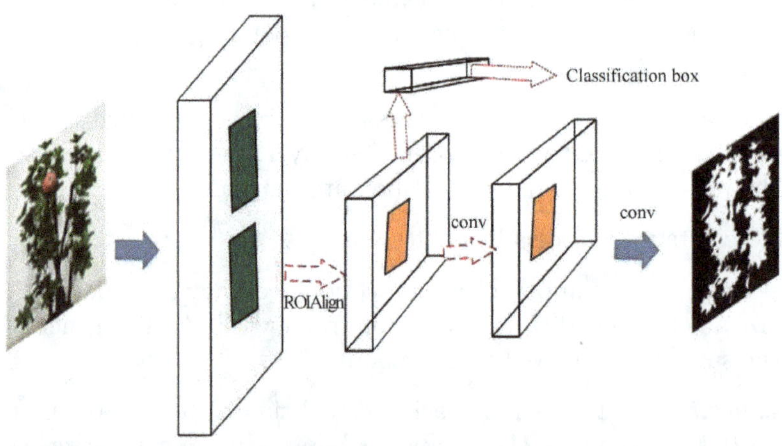

Figure 6.6. Explanation of the mask R-CNN framework.

Step 1: Input an image you want to process, and then perform the corresponding preprocessing operation.

Step 2: Feed it into a pre-trained neural network (ResNeXt, etc.) to obtain the corresponding feature map.

Step 3: Set an ROI for each point in this feature map to get multiple candidate ROIs.

Step 4: These candidate ROIs are fed into the RPN network for binary categorization (foreground or background) and BB regression to filter out some of the candidate ROIs.

Step 5: Perform a ROIAlign operation on these remaining ROIs (first correspond the pixels of the original map with those of the feature map, then correspond the feature maps with the fixed features).

Step 6: Perform classification, BB regression, and mask generation (FCN operation in each ROI) for these ROIs.

(2) Strengths and weaknesses analysis

Pros: (1) It analyzes the shortcomings of ROI Pooling, improves ROIAlign, and improves detection and instance segmentation at the same time. (2) It decomposes instance segmentation into two branches, classification and mask generation, and relies on the category labels predicted by the classification branch to select the corresponding masks for output. Meanwhile, the cross-loss entropy loss function is utilized instead of the multi-categorization loss function, which eliminates the competition between masks of different categories and generates accurate binary masks. (3) The classification and mask generation tasks are performed in parallel to speed up the model.
Disadvantages: Mask R-CNN is slightly slower than Faster R-CNN.

6.2.2.2 *Deep learning target detection algorithm based on regression methods*

Although Faster R-CNN is one of the mainstream object recognition algorithms, its speed does not meet real-time requirements. The subsequent emergence of algorithms such as YOLO and SSD has gradually highlighted its advantages and disadvantages. These algorithms make full use of the regression concept directly in the original image of multiple position to regress the target location border and target category.

6.2.2.2.1 YOLO algorithm YOLO stands for "You Only Look Once: Unified, Real-Time Object Detection", which basically summarizes the characteristics of the YOLO algorithm: "You Only Look Once" means that only one CNN operation is needed, "Unified" means a unified framework that provides end-to-end prediction, and "Real-Time" reflects the speed of the YOLO algorithm.

In 2016, Redmon *et al.* proposed the YOLO algorithm, a CNN that can predict the location and class of multiple prediction boxes at once. The network design strategy of the YOLO algorithm continues the core idea of GoogLeNet, which truly realizes end-to-end target detection and takes advantage of its speed, but its accuracy has decreased. However, the YOLO9000 algorithm proposed by Redmon *et al.* in 2016 improves its accuracy with the speed of the original YOLO algorithm. There are two main improvements: (1) a series of improvements on the original YOLO detection framework to make up for the lack of detection accuracy; (2) a method that combines target detection and target training into one is proposed. The training network of the YOLOv2 algorithm uses a downsampling method that can be dynamically adjusted under specific circumstances, a mechanism that allows the network to predict images with different sizes, enabling a balance between detection speed and accuracy. A comparison between YOLOv2 and other networks on VOC2007 is shown in Table 6.1.

Table 6.1. Comparison between YOLOv2 and other networks on VOC2007.

Detection Frameworks	Train	mAP	FPS
Fast R-CNN	2007 + 2012	70	0.5
Fast R-CNN VGG-16	2007 + 2012	73.2	7
Fast R-CNN ResNet	2007 + 2012	76.4	5
YOLO	2007 + 2012	63.4	45
SSD300	2007 + 2012	74.3	76
SSD500	2007 + 2012	76.8	19
YOLOv2 288 × 288	2007 + 2012	69	91
YOLOv2 352 × 352	2007 + 2012	73.7	81
YOLOv2 416 × 416	2007 + 2012	76.8	67
YOLOv2 480 × 480	2007 + 2012	77.8	59
YOLOv2 544 × 544	2007 + 2012	78.6	40

From Table 6.1, it can be seen that the YOLOv2 algorithm exceeds the real-time detection speed requirement in high-resolution image detection and reaches the advanced level.

(2) Advantages and disadvantages analysis

Advantages: (1) The YOLO algorithm converts the object detection task into a regression problem, which greatly accelerates the detection speed, allowing the YOLO algorithm to process 45 frames of images per second. Moreover, since each network predicts the target window using full-image information, the false detection rate is dramatically reduced. (2) The YOLO algorithm uses full-image information for prediction. Unlike sliding windows and candidate boxes, the YOLO algorithm utilizes full-image information during training and prediction. The Fast R-CNN algorithm incorrectly detects the background block as a target because it cannot see the global image during detection. Compared to the Fast R-CNN algorithm, the YOLO algorithm can reduce the background prediction error rate by half. (3) The YOLO algorithm can learn the generalized information of the target. It has much higher accuracy than other target detection algorithms.
Disadvantages: (1) For the detection of small objects, the detection of objects close to each other will be less effective. (2) Each grid can only predict one object, which is easy to cause missed detection, and is relatively sensitive to the scale of the object, and the generalization ability is weak when facing objects with large scale changes.

6.2.2.2.2 SSD algorithm The R-CNN family of algorithms (R-CNN, SPP-Net, Fast R-CNN, Faster R-CNN, mask R-CNN, etc.) with Proposal + Classification-based target detection have achieved very good results, but they are still far from real-time performance in terms of speed. The YOLO algorithm not only achieves real-time performance but also has a much higher mAP than the previously mentioned R-CNN family of algorithms. However, the YOLO algorithm also exhibits certain limitations. To address these shortcomings, the SSD network has made substantial improvements in both aspects while maintaining an optimal balance between mean average precision (mAP) and real-time processing efficiency.

(1) Principle

The SSD algorithm is a combination of the Faster R-CNN algorithm and the YOLO algorithm, which combines the regression idea in the YOLO algorithm and the prior boxes mechanism of the Faster R-CNN algorithm. The SSD algorithm will output a series of discretized prediction boxes, which are generated on different levels of features, and calculates the objects in each prediction box likelihood of belonging to each category, i.e., the score. At the same time, the shape of these prediction boxes is to be fine-tuned to make them conform to the outer rectangle of the object. There is also the fact that, in order to handle the case of different sizes of the same object, the SSD algorithm incorporates features of different resolutions. The SSD algorithm completely eliminates the stages of candidate box generation, pixel resampling, and feature resampling. This makes it easier for the SSD algorithm to go for optimized training and to incorporate the detection model into the system.

(2) Advantages and disadvantages analysis

Pros: Running speed exceeds that of the YOLO algorithm, while accuracy exceeds that of the Faster R-CNN algorithm (under certain conditions, for large targets in sparse scenes).
Disadvantages: (1) It needs to manually set the values of the initial scale and aspect ratio of the prediction box. The base size and shape of the prediction boxes in the network cannot be obtained directly through learning, but need to be set manually. And the size and shape of the prediction boxes used for each layer of features in the network can be different, leading to a debugging process that is highly dependent on experience. (2) The recognition of small-sized targets is still relatively poor and not up to the level of the Faster R-CNN algorithm. Because the SSD algorithm uses Conv4_3 low-level features to detect small objects, and the number of convolutional layers of low-level features is small, it suffers from the problem of insufficient feature extraction.

6.3 Integrated Design for Robotic Object Recognition

In recent years, the development of artificial intelligence is increasingly being perfected, with the application of deep learning in various scenarios emerging as an important trend. Notably,

Convolutional Neural Networks (CNNs) have become exceptionally advantageous in object detection tasks, distinguished by their high recognition accuracy, efficiency, rapid processing speed, and robust generalization capability. These attributes significantly enhance machine vision's adaptability to diverse and complex environments. CNN algorithms generally fall into two categories. The first comprises region proposal-based recognition algorithms, exemplified by R-CNN, Fast R-CNN, Faster R-CNN, and Mask R-CNN. These algorithms operate through a two-stage process that tackles target detection by initially identifying potential target-containing regions, known as regions of interest (ROIs), using candidate boxes. Following this, a classifier is employed to determine whether these boxes indeed contain the targets by calculating the likelihood of each target category's presence if detected. Despite their high detection accuracy and strong generalization ability, these algorithms tend to be slower, often falling short of real-time target capture requirements. The second category encompasses regression-based algorithms, with the SSD and YOLO series being prominent examples. These algorithms directly predict both the category and location of targets across the entire image using CNNs, offering a quicker detection speed while maintaining commendable accuracy. Among these, the YOLOv3 algorithm, introduced by Joseph Redmon in 2018, is selected for detailed exploration in this book. YOLOv3, a supervised learning-based machine learning algorithm, redefines object detection as a regression problem. This section delves into the YOLOv3 algorithm, covering its network structure, loss function definition, prediction box selection, and comparative analysis with other algorithms. The network structure of YOLOv3 is illustrated in Figure 6.7. The overall network setup file for the YOLOv3 algorithm is detailed in Appendix A.

6.3.1 *Boundary prediction*

The YOLOv3 algorithm first scales the original input image to a size of 416×416; it is then divided into a grid of $S \times S$ size. If the center point of a ground truth bounding box for the target object resides within a particular grid cell, that grid cell becomes specifically responsible for detecting that object. Each grid cell is required to predict B bounding boxes along with detection probabilities for C object categories. The final output consists of target bounding boxes for each category, with the system simultaneously computing

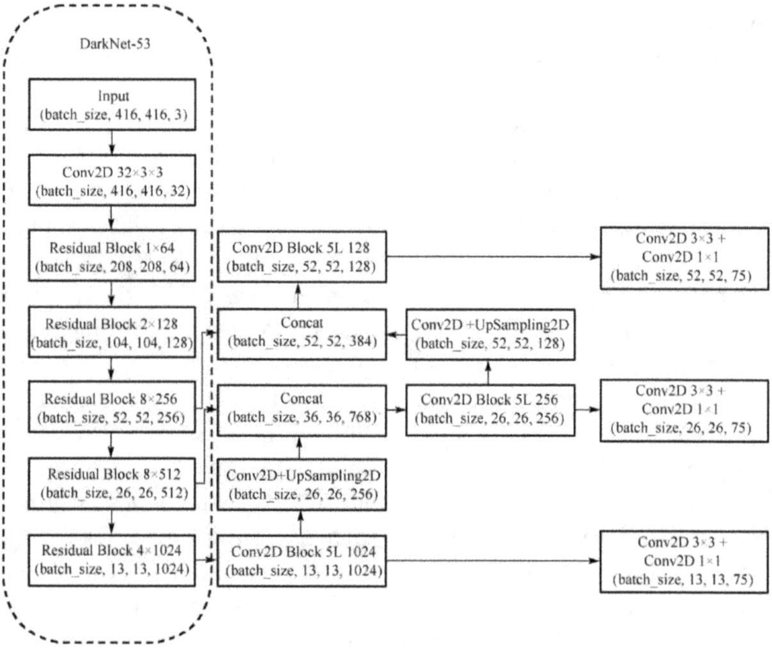

Figure 6.7. Network structure of YOLOv3.

confidence scores for each predicted bounding box. The confidence level is determined by two parts: (1) The probability that each grid contains a detected object; and (2) the accuracy of the bounding box. Among them, the accuracy of the bounding box is denoted by the symbol $t_{\text{confidence}}$, which is the intersection over union (IoU) of the prediction box and the truth box, and the formula of IoU is shown in Eq. (6.1):

$$\text{IOU} = \frac{\text{area of overlop}}{\text{area of union}} = \frac{\text{area}(B_p \cap B_{gt})}{\text{area}(B_p \cup B_{gt})} = \quad (6.1)$$

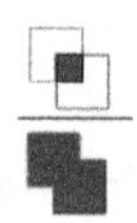

where B_p is the prediction box, B_{gt} is the truth box, and area denotes the area.

The formula for the accuracy of the bounding box $t_{\text{confidence}}$ can be obtained through Eq. (6.1):

$$t_{\text{confidence}} = \Pr(\text{object}) \times \text{IOU}_{\text{pred}}^{\text{truth}} \quad (6.2)$$

where Pr(object) indicates whether a certain type of detected object is present or not; if it is present, it is 1, otherwise it is 0.

The confidence of the predicted categories for each grid is

$$\Pr(\text{class}_i \,|\, \text{object}) \times \Pr(\text{object}) \times \text{IOU}_{\text{pred}}^{\text{truth}} = \Pr(\text{class}_i) \times \text{IOU}_{\text{pred}}^{\text{truth}} \tag{6.3}$$

where $\Pr(\text{class}_i \,|\, \text{object})$ denotes the probability that a certain type of detected object appears in the grid.

After understanding these parameters, the bounding box can be predicted. A flowchart depicting YOLOv3 bounding box prediction is shown in Figure 6.8.

When YOLOv3 predicts, each grid center may predict multiple bounding boxes, which requires further culling and screening. The

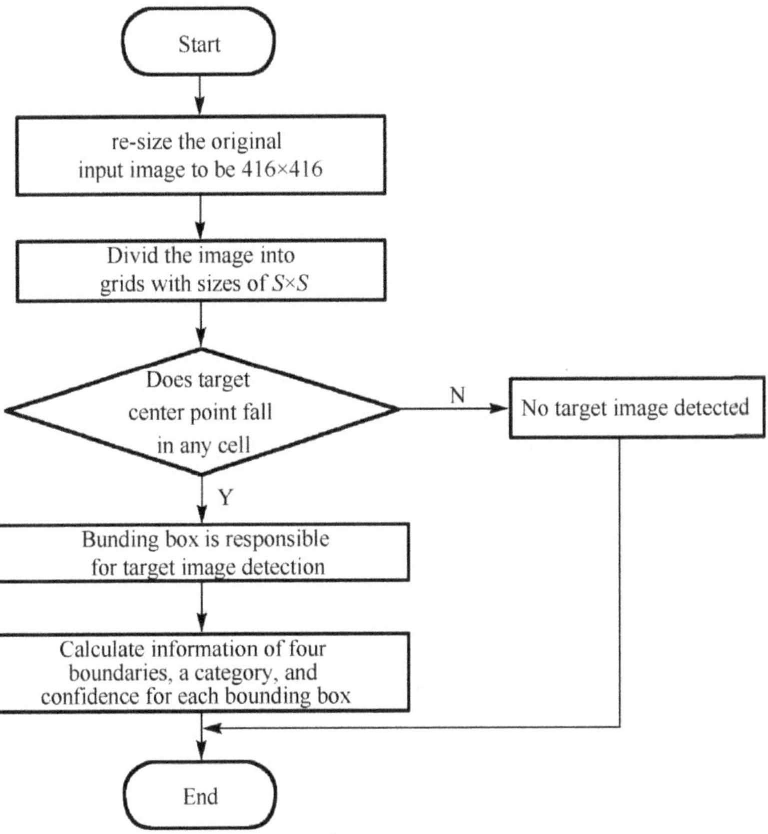

Figure 6.8. Flowchart of YOLOv3 bounding box prediction.

usual steps are as follows: Firstly, the bounding boxes with confidence lower than the threshold are eliminated by setting a suitable threshold, and the bounding boxes with confidence higher than the threshold are obtained; then, the obtained bounding boxes are subjected to non-maximum suppression operation, and the bounding box with the optimal confidence is selected as the final result. In YOLOv3, the prediction box has four parameters, x, y, w, and h. In order to speed up the operation of the network, these four parameters are normalized, as shown in Figure 6.9. Here, it is assumed that the input image is divided into a grid of size 7×7.

In Figure 6.9, the gray rectangular box represents the position of the predicted prediction box; (x_0, y_0) is the center point coordinates of the prediction box; the grid coordinates of the center point of the object to be detected are (r, c); w_{box} and h_{box} are the width and height of the prediction box, respectively; and w_{img} and h_{img} are the width and height of the input image, respectively. The normalization

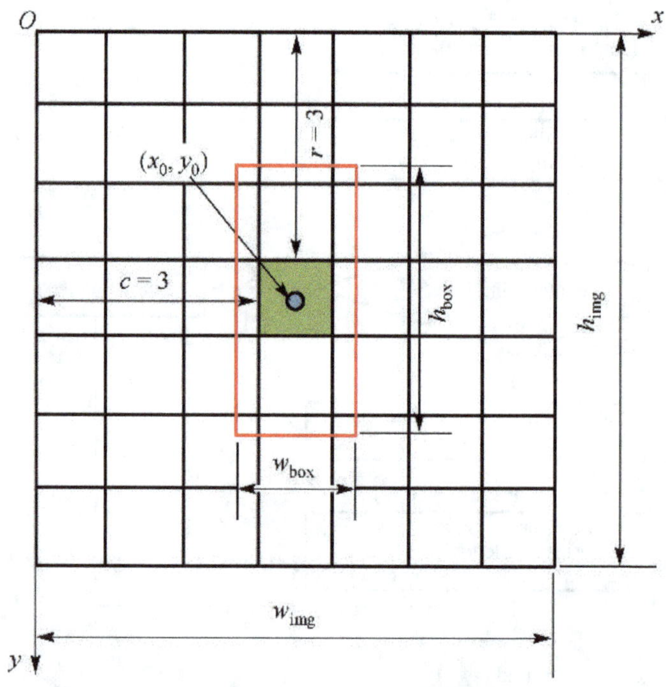

Figure 6.9. Schematic diagram of the prediction box.

process is shown in Eq. (6.4):

$$
\begin{aligned}
w &= \frac{w_{\text{box}}}{w_{\text{img}}} \\
h &= \frac{h_{\text{box}}}{h_{\text{img}}} \\
x &= x_0 \cdot \frac{S}{w_{\text{img}}} - c \\
y &= y_0 \cdot \frac{S}{h_{\text{img}}} - r
\end{aligned}
\tag{6.4}
$$

After normalization, all the information of YOLOv3 prediction output can be obtained, i.e., x, y, w, h, and $t_{\text{confidence}}$, as well as one of the C categories. Therefore, the overall output size of the network is $S \times S \times (5 \times B + C)$.

6.3.2 *Loss functions*

In terms of the loss function, YOLOv3 uses a binary cross-entropy loss function, which incorporates the coordinate error, the classification error, and the IoU error. Its calculation formula is

$$
\begin{aligned}
&\text{loss\_functions} \\
&= \lambda_{\text{coord}} \sum_{i=0}^{S^2} \sum_{j=0}^{B} I_{ij}^{\text{obj}} [(x_i - \hat{x}_i)^2 + (y_i - \hat{y}_i)^2] \\
&\quad + \lambda_{\text{coord}} \sum_{i=0}^{S^2} \sum_{j=0}^{B} I_{ij}^{\text{obj}} [(\sqrt{w_i} - \sqrt{\hat{w}_i})^2 + (\sqrt{h_i} - \sqrt{\hat{h}_i})^2] \\
&\quad - \sum_{i=0}^{S^2} \sum_{j=0}^{B} I_{ij}^{\text{obj}} [C_i \log(C_i) - (1 - \hat{C}_i) \log(1 - C_i)] \\
&\quad - \lambda_{\text{noobj}} \sum_{i=0}^{S^2} \sum_{j=0}^{B} I_{ij}^{\text{noobj}} [C_i \log(C_i) - (1 - \hat{C}_i) \log(1 - C_i)] \\
&\quad - \sum_{i=0}^{S^2} I_{ij}^{\text{obj}} \sum_{c \in \text{classes}} [\hat{p}_i(c) \log(p_i(c)) + (1 - \hat{p}_i) \log(1 - p_i(c))]
\end{aligned}
\tag{6.5}
$$

where S is the coefficient of dividing the grid; B is the number of prediction boxes to be predicted in a single grid; C is the total number of categories; p is the predicted category probability; w_i is the width of the prediction box in the ith grid; h_i is the height of the prediction box in the ith grid; x_i is the horizontal coordinate of the center of the prediction box in the ith grid; y_i is the vertical coordinate of the center of the prediction box in the ith grid; $\hat{*}$ is the data corresponding to the labeled dataset; I_{ij}^{obj} indicates that the jth ground truth box in the ith grid predicts a certain object; I_{ij}^{noobj} indicates that the jth ground truth box in the ith grid does not predict a certain object; the first term on the right-hand side of the equation is the error of the center coordinate; the second term on the right-hand side of the equation is the error of the width–height coordinate of the prediction box; and the third and fourth terms on the right-hand side of the equation are the errors of the prediction's confidence level.

6.3.3 *Feature extraction network*

DarkNet framework is a lightweight deep learning framework, which is developed based on the C language and CUDA; it can rely on both CPU and GPU for computation, with lower requirements for equipment, and it is easy to migrate, easy to install, more flexible, and, at the same time, easy to improve and extend from the underlying code, which has many advantages.

YOLOv3 has designed DarkNet-53, a backbone feature extraction network with better overall performance, based on the previous version. DarkNet-53, drawing inspiration from the concept of residual neural networks, merges DarkNet-19 with residual neural networks by adding a skip connection on top of the convolution and combination of convolutions. Its accuracy is close to that of the SOTA classification network, but it greatly improves the detection speed. The DarkNet-53 feature extraction network mainly consists of 3×3 and 1×1 convolutional layers. The residual unit and convolution module are shown in Figure 6.10.

In the residual unit of YOLOv3, the convolution layer first performs a convolution operation of 3×3 with a stride of 2 and then saves that convolution layer; then, it performs a convolution operation of 1×1 and 3×3 (used to change the number of channels); and at the same time, the result is summed up with the previously saved

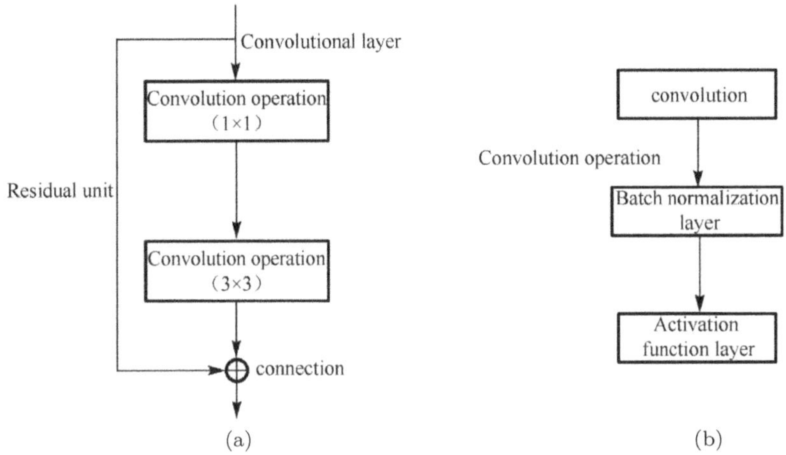

Figure 6.10. (a) Residual unit and (b) convolution module.

convolution layer to get the final result as the output of the residual unit. The purpose of the batch normalization layer in Figure 6.10 is to speed up convergence during training, and the purpose of the activation function layer is to avoid the problem of vanishing gradients due to increasing the depth of the network layers. Meanwhile, YOLOv3 utilizes the idea of Feature Pyramid Network (FPN) structure to extract different scale features on different feature layers and finally outputs three different feature maps for multi-scale prediction: small (13×13), medium (26×26), and large (52×52) scales. The model is more versatile and adapted to tasks of object detection at multiple scales; especially, the detection accuracy of small objects has been greatly improved.

6.3.4 *Comparison among different networks*

Viewing the ArXiv database for the year 2020 yields the data shown in Tables 6.2 and 6.3 (the experimental data were obtained by running on the Titan X platform, and the experimental image size was 320×320).

In Table 6.2, *Top-N* represents the object to be detected in the top N prediction results with the highest scores, which is equivalent to the accuracy of detection, and the larger its value, the higher the accuracy; Bn Ops and BFLOP/s can represent the computational efficiency of the neural network, and the larger their values, the higher

Table 6.2. Comparison of operational performance of various backbone networks.

Runtime performance	DarkNet-19	ResNet-101	ResNet-152	DarkNet-53
Top-1	74.1	77.1	77.6	77.2
Top-5	91.8	93.7	93.8	93.8
Bn Op (network operations require *n billions of* operations)	7.29	19.7	29.4	18.7
BFLOP/s (the network can run n billion calculations per second)	1246	1039	1090	1457
FPS	171	53	37	78

Table 6.3. Recognition accuracy of multiple backbone networks.

One-stage algorithm	Backbone network	AP	AP50	AP75
YOLOv2	DarkNet-19	21.6	44.0	19.2
SSD513	ResNet-101-SSD	31.2	50.4	33.3
DSSD513	ResNet-101-DSSD	33.2	53.3	35.2
RetinaNet	ResNet-101-FPN	39.1	59.1	42.3
RetinaNet	ResNetXt-101-FPN	40.8	61.1	44.1
YOLOv3 608 × 608	DarkNet-53	33.0	57.9	34.4

the computational efficiency; and the overall running speed of the backbone network can be directly described in frames per second (FPS), and the larger its value, the higher the running efficiency. As can be seen from Table 6.2, the operation efficiency of DarkNet-53 is lower than that of DarkNet-19, but its accuracy is significantly higher; the accuracy of ResNet-101 and ResNet-152 is almost equal to that of DarkNet-53, but the operation efficiency of DarkNet-53 is higher than that of ResNet-101 and ResNet-152 by a factor of 1.5 and 2, respectively.

In Table 6.3, AP denotes average precision, and AP50 and AP70 represent the average precision when the IoU is 0.5 and 0.7, respectively. Comparing the data in Table 6.3, it can be seen that RetinaNet has the most outstanding performance in terms of the metric of recognition accuracy; however, checking the database, it can be found that on Titan X, RetinaNet can achieve a mAP of 57.5 in 198 ms, while YOLOv3 can achieve a mAP of 57.9 in 51 ms. YOLOv3 achieves 57.9 mAP in 51 ms. YOLOv3 runs 3.8 times faster than RetinaNet with almost the same performance. In addition, YOLOv3 outperforms SSD and YOLOv2 in terms of recognition accuracy. To summarize, the YOLOv3 algorithm is undoubtedly superior under the criteria of operation performance, accuracy, and detection speed.

6.4 Integrated Practice for Robotic Object Recognition

Through experiments, it was discovered that the complexity of the environmental obstacles' background makes them harder to recognize, particularly when these obstacles are significantly obscured. This complexity hampers the camera's ability to capture comprehensive information about the obstacles. Often, obstacles appear in images as mere fragments or small sections, as illustrated in Figure 6.11.

As shown in Figure 6.11, more discernible obstacles are present in all the boxes, and these obstacles occupy only a minor portion of the entire image. The presence of such small objects complicates object recognition and challenges the precision of localizing prediction boxes for small objects. Failure to recognize these elements significantly

Figure 6.11. Areas where environmental obstacles are present.

affects the accuracy of subsequent branch reconstruction efforts. Additionally, the YOLOv3 algorithm has historically struggled with the recognition of small target objects. Consequently, this section introduces an enhancement to the YOLOv3 algorithm that focuses on two main aspects: enhancing the detectability of small objects and improving their localization accuracy. This improved neural network algorithm is designated as Improved-YOLOv3 (I-YOLOv3).

6.4.1 *Deepen network hierarchy*

Referring to the book *Deep Learning*, the effect of increasing the depth of the network on the neural network can be obtained, as shown in Figure 6.12.

It is also analyzed for the top-performing models in ImageNet, as shown in Figure 6.13.

Figures 6.12 and 6.13 are both derived from the relevant domain database. It can be seen that the deeper the network layers, the more expressive and better the performance of their network. Increasing the depth of the network can have the following effects on the performance of the neural network:

(1) It can better fit the features of the input image and enhance the expressive ability of the network structure. Deeper network structures can enhance the nonlinear expression ability and learning ability of the network and enrich the semantic information of the

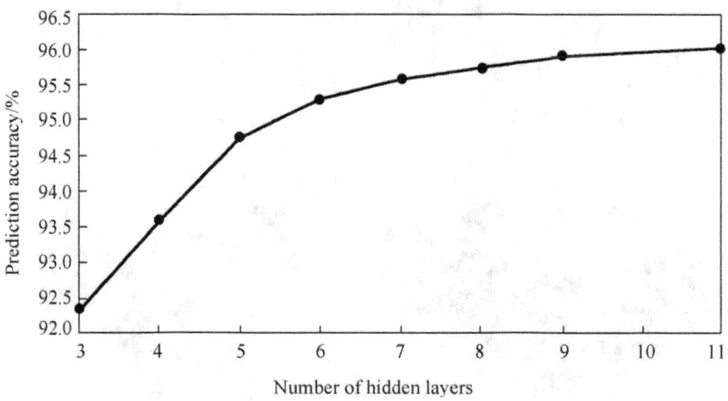

Figure 6.12. Prediction accuracy versus network layer depth curve.

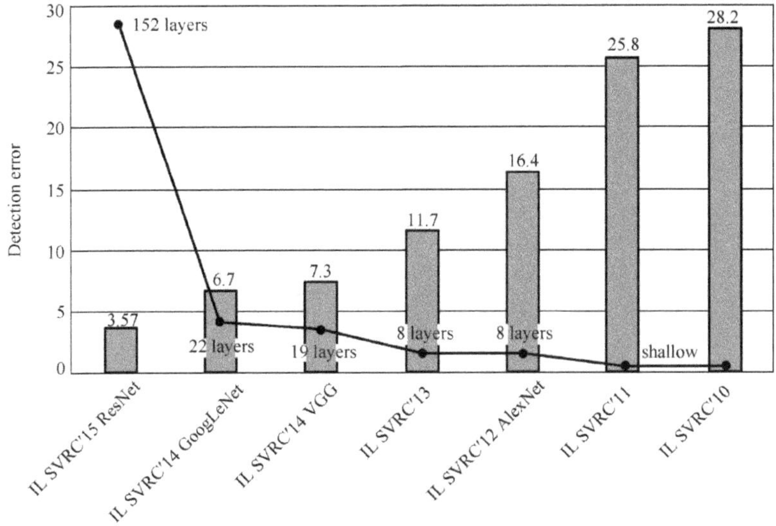

Figure 6.13. Performance comparison of top-performing models in ImageNet.

network. In other words, the network achieves good expressive ability with complex inputs. For this book, the environmental obstacle itself is a more complex object, and increasing the depth of the network is conducive to enhancing the model's ability to "recognize" the obstacle.

(2) Increasing the depth of the network means that, when expressing the same information, less information needs to be expressed in each layer. This effectively reduces the number of structural parameters of the network and facilitates layer-by-layer learning of the network.

(3) Deeper network layers make the effective receptive field of view of the deep convolutional kernel larger, which is favorable to the recognition of small objects. As already mentioned in this book, due to obstacle occlusion, obstacles appear in the form of small segmented objects, which has a greater impact on obstacle recognition, so increasing the depth of the network can obtain richer obstacle information, which is conducive to the reconstruction of obstacles.

However, over-deepening the network structure will also bring about the effects of gradient instability and network saturation, so the depth of the network cannot be increased excessively. In this book, we add

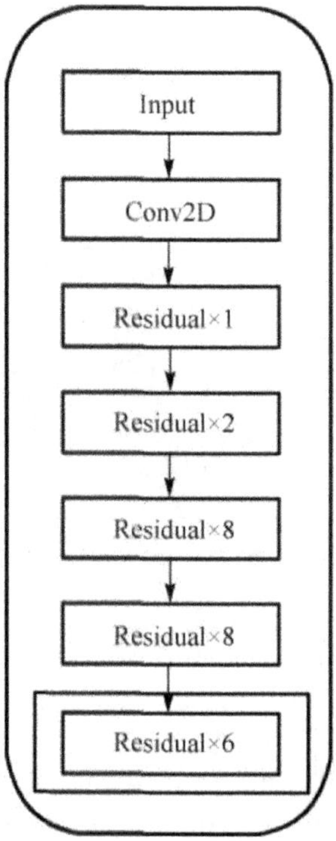

Figure 6.14. DarkNet-57 backbone feature extraction network structure (Residual indicates residual block).

two extra convolutional layers of sizes 1×1 and 3×3 on top of the YOLOv3 backbone network as the feature extraction network of-YOLOv3 and name it DarkNet-57, as show in Figure 6.14.

6.4.2 *Network reconstruction based on image pyramid*

In neural networks, fusion of feature layers at different scales is an effective means to improve segmentation performance. The output layers of seven different stages of the residual module in DarkNet-57 are displayed and scaled to the same size using feature visualization techniques, and the results are shown in Figures 6.15.

It can be seen that in the Layer 1–Layer 4 stages, the images have a higher resolution, and the contours of the obstacles and their locations can be seen more clearly. These layers are also commonly

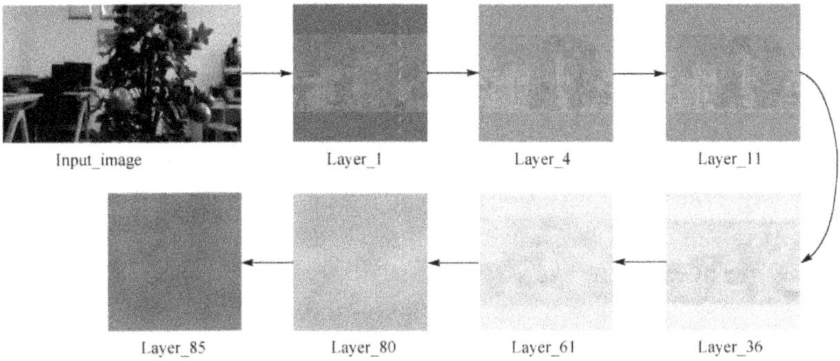

Figure 6.15. Output images of residual blocks at different stages.

referred to as low-level feature maps, and they tend to contain more specific detailed information, such as location. In the Layer 61–Layer 85 stage, on the other hand, these maps can only vaguely see black and white pixel dots, which express the category to which a specific object in the image belongs, i.e., semantic information. These layers are also commonly referred to as high-level feature maps. How to efficiently fuse the low-level feature maps with the high-level feature maps, effectively combine the position information with the semantic information, and utilize the advantages of both for prediction is a key step in improving the performance of the network.

Feature Pyramid Networks (FPNs) are capable of fusing features at different scales for prediction, and their structure is shown in Figure 6.16.

In the whole prediction process, the input original image is firs subjected to convolution operation to obtain the feature layers Layer_1, Layer_2, and Layer_3, and Layer_3 generates Layer_4; then,

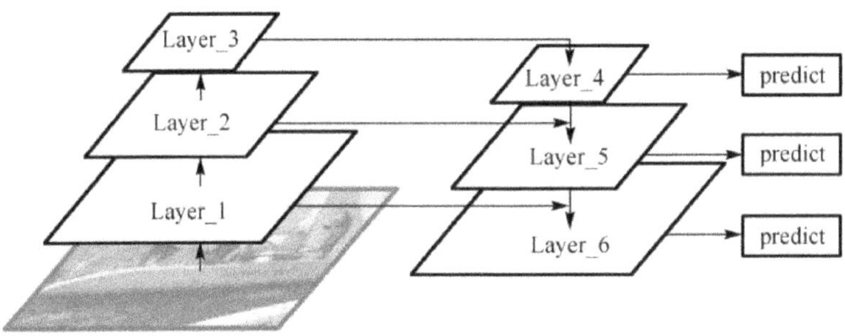

Figure 6.16. Feature Pyramid Network structure.

up-sampling operation is carried out on the feature layer Layer_4 to make it have the same dimensions as Layer_2. Next, Layer_2 and Layer_4 are subjected to the convolution operation, and the obtained result is output as Layer_5; then, Layer_4 is used as input for the prediction layer. And so on, thereby fusing the feature maps of different layers. By constructing such a deep network pyramid, the feature information of multiple layers is fused, and the prediction is performed on the feature maps of different scales.

This book also draws on such fusion ideas to optimize and improve the YOLOv3 algorithm. The high-level feature maps and low-level feature maps at different stages are fused, and finally three sets of feature maps with different scales are obtained for the network prediction output of I-YOLOv3, whose detailed information of network layer fusion is as follows.

This book sets the size of the input image to $608 \times 608 \times 3$. The I-YOLOv3 network structure is shown in Figure 6.17.

Step 1: The feature network is obtained using DarkNet-57, and Conv_80 is consecutively convolved with 1×1 and 3×3 to obtain the first set of initial YOLO layers, denoted as YOLO_init1.

Step 2: YOLO_init1 is subjected to one convolution operation of 1×1 and 3×3 to obtain the first set of final feature YOLO layers, denoted as YOLO_1.

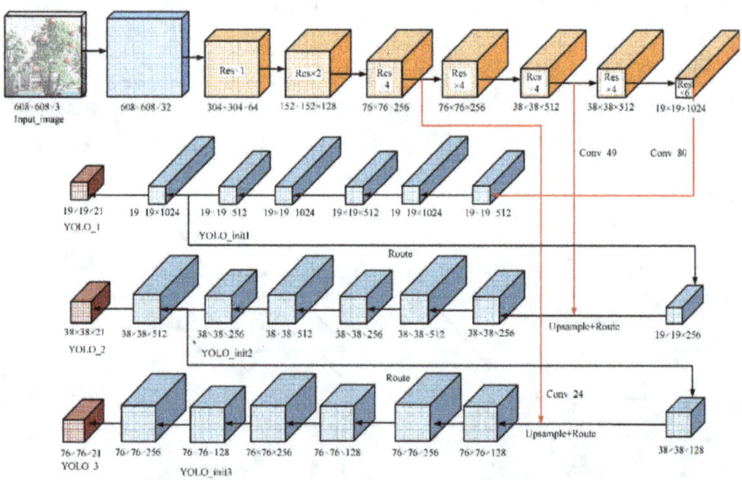

Figure 6.17. I-YOLOv3 network architecture.

Step 3: YOLO_init1 is subjected to a sampling operation, convolved with Conv_49 using DarkNet-53, and continued to use successive convolution operations of 1×1 and 3×3 to obtain a second set of initial YOLO layers, denoted as YOLO_init2.

Step 4: YOLO_init2 is subjected to one convolution operation of 1×1 and 3×3 to obtain the second set of final feature YOLO layers, denoted as YOLO_2.

Step 5: YOLO_init2 is subjected to an up-sampling operation and convolution operation with Conv_24 using DarkNet-53, and continued to use successive convolution operations of 1×1 and 3×3 to obtain the third set of initial YOLO layers, denoted as YOLO_init3.

Step 6: YOLO_init3 is subjected to one convolution operation of 1×1 and 3×3 to obtain the third set of final feature YOLO layers, denoted as YOLO_3.

By reconfiguring the network structure through the above steps, three sets of final feature YOLO layers with different scales will be obtained: YOLO_1, YOLO_2, and YOLO_3, and these three sets of feature YOLO layers will be utilized for the prediction of location and category.

6.4.3 *K-means prior frame clustering*

The YOLOv3 algorithm follows the anchor box mechanism of Faster R-CNN. Among them, the aspect ratio of the anchor box has a certain impact on the accuracy and speed of object detection. During the training process, the parameters of the candidate box are also continuously adjusted with the increase in the number of iterations to make them closer to the parameters of the ground truth. In the YOLOv3 algorithm, the parameters of the anchor box are trained on the COCO dataset, and the shapes of the objects in the COCO dataset are mostly square, and the target objects that need to be recognized in this book can be predicted with square boxes; however, if the environmental obstacles are elongated, the clustering operation needs to be carried out again to get a more accurate anchor box. Clustering is usually performed using the K-means clustering algorithm. In object detection, the IoU of the overlap between predicted boxes and anchor boxes, IoU (B, C), is commonly used as a standard for measuring distances during clustering, and the formula for IoU is shown in Eq. (6.1), and the formula for the measure is shown in Eq. (6.6):

$$d(B,C) = 1 - \mathrm{IOU}(B,C) \tag{6.6}$$

where B is the result of sample clustering and C is the center of all sample clusters.

The P–R curve is used to calculate the mAP value, and we can measure the model performance by plotting the P–R curve, which is detailed in Appendix B. Appendix C provides the code for the network training process, and the network in Appendix C needs to load in the code from Appendix A. The accuracy is verified with the code from Appendix B at the end of the training. The three codes in Appendix A, Appendix B, and Appendix C can be used to apply the concepts about neural networks for target recognition discussed in this book.

6.5 Summary

This chapter mainly addresses the problem of recognizing targets and branches, as well as the reconstruction of branches. Firstly, for the problem of recognizing small target regions, the network structure of the YOLOv3 algorithm is reconstructed, and the I-YOLOv3 algorithm is proposed for recognizing targets and branches; then, for the problem of too much background of the I-YOLOv3 prediction boxes, a more accurate prediction box is obtained by using the minimum outer rectangle; then the traversal search is conducted, and using the distance constraints and the angle constraints, it is judged that all the groups of prediction boxes belong to the same branch; finally, the polynomial fitting method is used to fit the curve of the branch and reconstruct the 3D spatial structure of the branch.

6.6 Exercises

6.1 Briefly describe the development of object recognition.
6.2 Design and practice traditional object recognition methods.
6.3 Design and practice deep learning-based object detection algorithms.
6.4 Design and practice the YOLO algorithm.
6.5 Design and practice common obstacle recognition methods.

Chapter 7

Integrated Design and Practice for Robot Path Planning

7.1 Introduction

Chapter 5 explored the design and implementation of the robot's visual perception system. Chapter 6 delved into the design and application of target and obstacle recognition, along with the reconstruction of obstacles in three-dimensional (3D) space. Building upon this foundation, the current chapter focuses on designing and implementing path planning for robots. This chapter addresses two main issues: Firstly, how the robot navigates around the target in two-dimensional (2D) space; and secondly, the route that the robot's end-effector takes to grasp the target in a multi-obstacle, 3D space where the target is situated.

7.2 Theory of Robot Path Planning

An octree map is obtained and subjected to a dimensionality reduction operation, resulting in a 2D occupancy grid map. This transformation facilitates path planning for the robot. Path planning can be broadly classified into two main categories based on whether the robot possesses a map: map-based and mapless path planning. Map-based path planning necessitates prior knowledge, whereas mapless path planning involves the mobile robot recognizing and tracking a specific target object to address various navigation challenges for

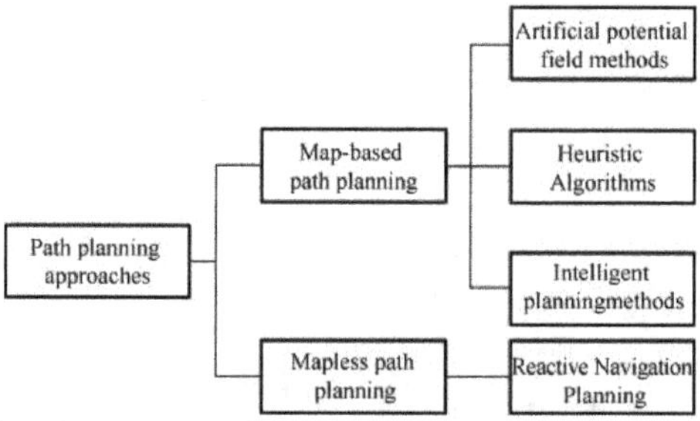

Figure 7.1. Route planning classification.

particular purposes. These two approaches to path planning are further divided into several subcategories, as depicted in Figure 7.1.

Besides, there are other ways of categorization; for example, according to the completeness of the path, the planning problem can be classified into two forms: resolution-complete and probabilistic-complete. The main methods of resolution-complete include the roadmap method, the cell decomposition method, and the potential field method; while the main methods of probabilistic-complete include Probabilistic Road Maps (PRM) and Rapidly-exploring Random Tree (RRT), which will not be described here. This book utilizes map-based path planning, so the focus will be on describing this approach. The three path planning approaches shown in Figure 7.1 are described as follows:

(1) *Artificial potential field method*[42]: The artificial potential field method is implemented on a 2D occupancy grid map using the Virtual Force Field (VFF) method. It is assumed that in the occupancy grid map, there is a virtual force on top of the grid containing obstacles that repels the mobile robot, which is used to repel the robot in order to go around the obstacles, and there is a virtual force on top of the grid containing the goal point that attracts the mobile robot, which is used to attract the robot toward the goal point location. The repulsive and attractive forces are represented as vectors in all the grids, and the combined force is calculated as the sum of the vectors and is used to propel the mobile robot. The combination of repulsive and attractive forces constitutes a virtual force potential field. However, the artificial potential field method suffers from the occurrence of "local minima" during the search, i.e., the robot does not move because the combined force is equal to zero in the

potential field. To solve this problem, more parameters need to be introduced.

(2) *Intelligent planning methods*: Intelligent planning methods include fuzzy logic control method, ant colony algorithm, particle swarm algorithm, genetic algorithm, neural network algorithm, and other emerging intelligent technologies. The fuzzy logic control method adopts an approximation of natural language, taking the currently known environmental obstacle information as the input of fuzzy reasoning and producing the movement mode of the mobile robot as the output by means of fuzzy reasoning. The ant colony algorithm[43] is a kind of bionic algorithm designed by simulating the behavior of ants searching for the shortest path to food. The particle swarm algorithm,[44] also known as the bird swarm algorithm, starts from a random solution and searches for the optimal solution by iterative means; this algorithm has a fast search speed and high accuracy. The genetic algorithm adopts a multi-point search method, which is capable of searching for the global optimal solution with a high probability.

(3) *Heuristic algorithms*: Heuristic algorithms are represented by the A* algorithm, which is an improvement of Dijkstra's algorithm. Later, some scholars also proposed the D* algorithm, which mainly utilizes the advantages of A* algorithm through heuristic evaluation function, which can solve the dynamic shortest path problem. Both algorithms can quickly correct and replan an optimal path according to the environmental information acquired by the robot in the process of movement, which can effectively reduce the local planning time and meet the requirements of real-time navigation planning. In addition, by modifying the heuristic evaluation function and the direction of map search of the A* algorithm, we can effectively improve its path planning speed, which has certain adaptability to the complex environment.

The various algorithms exhibit their respective advantages and limitations. Considering the ROS (Robot Operating System) software platform employed in this robotic system, the ROS navigation stack provides two grid-based path planning algorithms, among which the A* algorithm stands out for its direct integrability and relative implementation simplicity. Therefore, this book selects the A* algorithm for the integrated design of robotic path planning to fulfill comprehensive design requirements.

7.3 Integrated Design of Robot Path Planning

7.3.1 *Global path planning in 2D space*

Reducing a known 3D map into a 2D occupancy grid map, and performing path planning when the environment map is already fully known, is also referred to as the global path planning problem. The accuracy of global planning is determined by the precision of the environment map. This book sets the resolution of the 2D occupancy grid map to 1 cm, i.e., the width of each grid represents a distance of 1 cm in the actual scene, which is sufficient for path planning here. The principle of the A* algorithm and related improvements are described in detail as follows.

7.3.1.1 *A* algorithm*

The A* algorithm, co-proposed by Hart, Nilsson, and Raphael in 1968,[45] is a direct search algorithm for solving optimal path more efficiently. By combining the BSF algorithm and Dijkstra's algorithm,[46] it introduces a heuristic evaluation function, avoids blind search, improves search efficiency, and substantially improves the path planning accuracy. The A* algorithm consists of an estimate of the cost from the current node to the goal node, and the actual cost spent from the departure node to the current node. A mathematical model is used to express the estimation function of a certain optimal path, as shown in Eq. (7.1):

$$f(n) = g(n) + h(n) \tag{7.1}$$

where $f(n)$ is the minimum cost estimate from the departure node to the target node, which consists of two parts: $g(n)$ is the actual path cost from the departure node to the current node n; and $h(n)$ is the above-mentioned heuristic evaluation function, which represents the minimum path estimate cost from the current node n to the target node.

In the A* algorithm, the value of the cost is usually expressed by the distance value, and some of the commonly used distances are the Manhattan distance, Euclidean distance, and Chebyshev distance. At the same time, the A* algorithm needs to fulfill the following conditions:

(1) An optimal solution exists over the entire search space.
(2) The space for solving is limited.
(3) The search cost of each child node is greater than 0.
(4) $h(n) \leqslant h^*(n)$, where $h^*(n)$ is the actual cost.

Condition 4 shows that for the A* algorithm, the heuristic evaluation function $h(n)$ is very important. $g(n)$ is the current cost that has been consumed and its value is fixed, while the only factor that affects the overall estimated cost $f(n)$ is $h(n)$. For $h(n)$, the more the number of constraints set, the greater the possibility of excluding paths and, subsequently, the higher the corresponding search efficiency. When $h(n) = 0$, the algorithm in this case is Dijkstra's algorithm, where, undoubtedly the nodes of the search is increased, reducing the search efficiency, but the shortest path is always found. When $h(n) \leqslant h^*(n)$, the A* algorithm will definitely be able to search for an optimal path, while still maintaining a high search efficiency.

The search process of the A* algorithm is shown in Figure 7.2. Throughout the search process, the A* algorithm mainly utilizes two sets: close_set and open_set. open_set mainly stores the points that have not been detected, while close_set mainly stores the points that have been detected. Elements in open_set are arranged in ascending order according to the f-value, and each time the search is looped, the node with the smallest *f-value is* selected from open_set for detection, and the detected point is placed in close_set, while the f, g, and h of the nodes adjacent to the point are calculated at the same time. When the node in open_set is empty or the search reaches the target node, the search stops.

Take a grid of size 5×5 as an example, assuming that its starting point is S and its ending point is E, and the cost of moving in the horizontal and vertical directions is 10; additionally, it can only move along the horizontal or vertical direction. Then, the whole search process is shown in Figure 7.3.

In Figure 7.3, the black grid represents the obstacles, and the red arrows represent the optimal paths finally obtained. The meanings of the values above each small grid are as follows: The upper-right value represents the g value, the lower-right value represents the f value, and the lower-left value represents the h value.

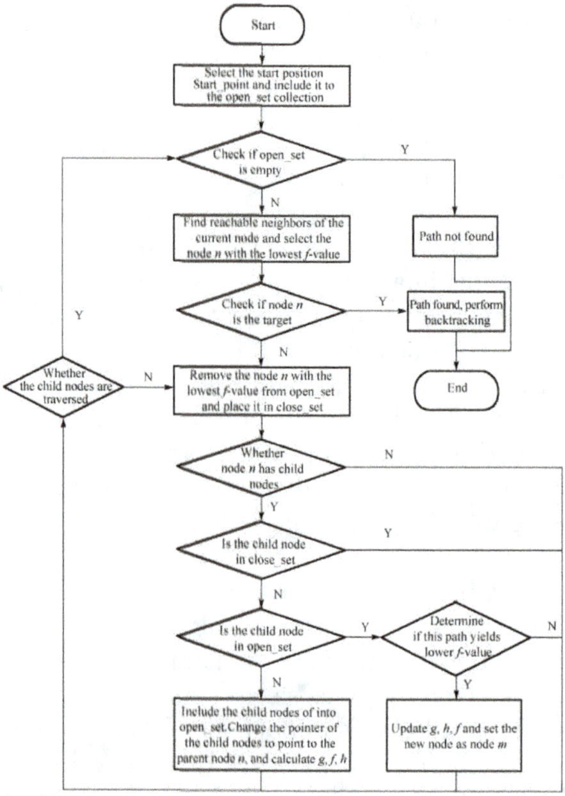

Figure 7.2. Search process of the A* algorithm.

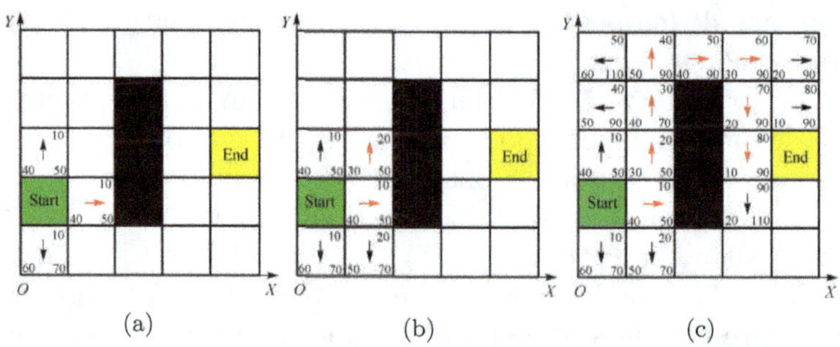

Figure 7.3. Schematic diagram of the A* algorithm search. (a) Starting Point adjacent node calculations, (b) (2,2) Node Cost Calculation and (c) Final Search Path.

7.3.1.2 *Improvements to the A* algorithm*

The traditional A* algorithm is a unidirectional search process, which can find an optimal path in a shorter time when the scene is relatively simple. However, since the maps constructed from actual scenes often cover a larger space and the conditions of the maps are more complex, this increases the time required for path search. To address this problem, this book improves the A* algorithm by using a bidirectional search to increase the search efficiency (only time optimization is used as a constraint).

The A* algorithm is a unidirectional search from the starting point to the goal point. The so-called bidirectional search, on the other hand, constructs two unidirectional searches by starting the search toward each other from the point of departure and the point of goal at the same time. The unidirectional search from the departure point to the target point is regarded as forward search, and the unidirectional search from the target point to the departure point is regarded as reverse search. In forward search, the current optimal node searched in reverse search is regarded as the target point for searching; in reverse search, the current optimal node searched in forward search is regarded as the target point for searching, and the two kinds of searches are carried out crosswise until the target point of the two kinds of searches becomes the same, and the search ends.

In some special cases, there will be no intersection of the two algorithms; then, it is sufficient to search using the unidirectional A* algorithm.

The implementation process of the bidirectional A* algorithm is as follows.

The A* algorithm utilizes open_set and close_set. Whereas, the bidirectional A* algorithm requires open_set1 and close_set1 for the forward search and open_set2 and close _set2 for the reverse search. Also, remember that the estimation function for the forward search is $f_1(n) = g_1(n) + h_1(n)$ and the estimation function for the reverse search is $f_2(n) = g_2(n) + h_2(n)$.

(1) Initialize open_set1 and open_set2.
(2) First, a forward search is performed to extend the current node S. The following operation is performed for node n after node S:

(a) Determine whether node n is an obstacle or a point that has already been searched. If so, skip it; if not, calculate $f_1(n) = g_1(n) + h_1(n)$.
(b) Store node n in open_set1.

(3) Determine whether open_set1 is an empty set. If yes, then the search fails, so terminate the search at the same time; if not, then arrange the nodes in open_set1 in ascending order according to the value of f_1, select the node with the smallest value of f_1 from open_set1 for detection, and put the detected point in close_set1. If the point is also in close _set2, then terminate the search and skip to step 5, otherwise proceed to the next step.
(4) Perform a reverse search according to steps 2 and 3, transforming the corresponding sets and cost functions accordingly.
(5) According to the estimation function, calculate the cost of the two search methods, and at the same time, determine whether there is an intersecting node N in the two close sets (the departure point and the target point are not counted). If there is one, then take node N as the intermediate connecting point, then backtrack to the forward search and the reverse search, calculate the two costs from the departure point to the node N and from the target point to the node N, and add up the two to get the total value of the surrogate. If there is not, then the search fails, and the single A* algorithm is used for the search.

Multiple sets of simulation experiments are conducted on the traditional A* algorithm and the bidirectional A* algorithm in occupancy grid maps of different sizes, and one set of simulation results is shown in Figure 7.4.

Figure 7.4 uses different colors to represent the various stages of the search. Among them, the blue grid represents the departure point; the cyan grid represents the goal point; the white grid represents the blank area, i.e., without any obstacles in the grid, and at the same time, it is the unsearched area; the black grid represents the existence of obstacles; the rosy red grid represents the node in the next search; the red grid represents the node that has been retrieved but is not the optimal node during the forward search; the yellow grid represents the node that has been retrieved but is not the optimal node during the reverse search; and the green grid represents the search path from departure point to the goal point.

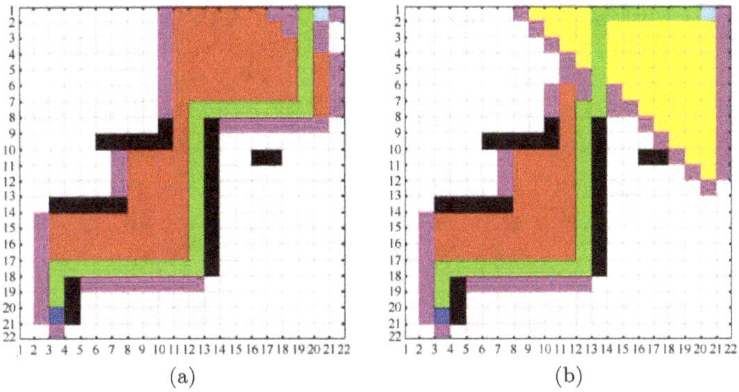

Figure 7.4. Simulation of the two search methods. (a) A* Search and (b) Bidirectional A* Search.

Table 7.1. Search time comparison.

Different groups of experiments	Experiment 1	Experiment 2	Experiment 3	Experiment 4	Experiment 5
Unidirectional search method consumption time/s	0.8732	1.5734	2.2537	3.4218	4.2138
Bidirectional search method consumes time/s	0.5732	0.9008	1.2523	1.8604	2.4356
Lowering time/s	0.3002	0.6726	1.0014	1.5614	1.7782
Percentage reduction/%	34.4	42.7	44.4	45.6	42.2

Data were recorded for five different sets of experiments, and the results obtained are shown in Table 7.1.

As can be clearly seen from Table 7.1, utilizing the bidirectional search method for path planning can reduce the consumption time by about 40%, and for environments with larger scenes, it will greatly reduce the overall path planning time of the robot.

7.3.2 *Global path planning in 3D space*

For the path planning problem of robots in a high-dimensional space, the commonly used solution is the RRT47 algorithm. This algorithm was proposed by Lavalle in 1998. It utilizes a tree structure instead of a graph structure with directions and random sampling of the state space for collision detection. It finally obtains a valid path from the start to the end by sampling and expanding the tree nodes. At the same time, the RRT algorithm is a probabilistic complete search algorithm, that is, in the free space, as long as there is a path that can be pointed in the direction of the starting point to the end point, at the same time, given enough sampling points, the RRT algorithm will certainly be able to find a passable path. The algorithm does not require accurate modeling of the spatial structure and is computationally small. Up to now, the RRT algorithm, which has been improved by many scholars, has been widely applied to the path planning problems of robots. In this section, 3D path planning is performed only for the end-effector of the operating arm of a humanoid mobile robot.

7.3.2.1 *RRT algorithm*

Taking the RRT algorithm in two-dimensional space as an example, as shown in Figure 7.5, $\boldsymbol{T}_k$ represents an extended random tree that already has k nodes; $\boldsymbol{X}_{\text{init}}$ is the root node of the tree, that is, the starting point of the search; and $\boldsymbol{X}_{\text{goal}}$ is the target point in the space of the extended random tree, that is, the end point of the search. When searching, first generate a random node $\boldsymbol{X}_{\text{rand}}$ in the reachable space of the random tree, then traverse each node on the whole extended random tree, search for the node $\boldsymbol{X}_{\text{nearest}}$ that is closest

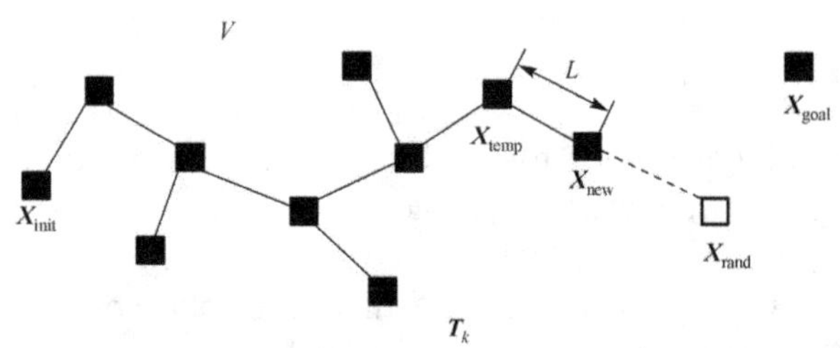

Figure 7.5. Schematic diagram of the RRT algorithm.

to $\boldsymbol{X}_{\text{rand}}$, and then search for a new node $\boldsymbol{X}_{\text{new}}$ in the direction of $\boldsymbol{X}_{\text{nearest}} \to \boldsymbol{X}_{\text{rand}}$ with a distance of L as the stride. Judge $\boldsymbol{X}_{\text{new}}$, and if it meets the conditions, then it will be added to the extended random tree $\boldsymbol{T}_k$. The metric distance is usually calculated using the Euclidean distance as shown in Eq. (7.2):

$$\text{dis}(\boldsymbol{x}_1, \boldsymbol{x}_2) = ||\boldsymbol{x}_2 - \boldsymbol{x}_1||_2 \tag{7.2}$$

where $|| - ||_2$ is the 2-norm of the vector.

Judge the distance between node $\boldsymbol{X}_{\text{new}}$ and node $\boldsymbol{X}_{\text{goal}}$; if $\boldsymbol{X}_{\text{new}}$ is in the neighborhood of $\boldsymbol{X}_{\text{goal}}$, i.e., it satisfies Eq. (7.3), then the search ends.

$$||\boldsymbol{X}_{\text{new}} - \boldsymbol{X}_{\text{goal}}||_2 \leqslant \text{dis}_{\text{goal}} \tag{7.3}$$

where dis_{goal} is the neighborhood distance of $\boldsymbol{X}_{\text{goal}}$ and the stop sign distance of the circular search.

The pseudo-code for the RRT algorithm is shown in Figure 7.6.

Several functions are introduced in the above pseudo-code, as described in the following:

(1) RandomSample function, which is a random sampling function, i.e., all sampling points obey a uniform distribution:

$$\begin{cases} \boldsymbol{X}_{\text{rand}} \cdot \text{x} = p * [\max(\text{X} \cdot \text{x}) - \min(\text{X} \cdot \text{x})] + \min(\text{X} \cdot \text{x}) \\ \text{X}_{\text{rand}} \cdot \text{y} = p * [\max(\text{X} \cdot \text{y}) - \min(\text{X} \cdot \text{y})] + \min(\text{X} \cdot \text{y}) \end{cases} \tag{7.4}$$

(2) SearchNearest function, which is a function to find the nearest point, as shown in Eq. (7.5):

$$\boldsymbol{X}_{\text{nearest}} = \min ||\text{X}_{\text{rand}} - \text{T}||_2 \tag{7.5}$$

(3) ExtendTree function, which is a function to extend the random tree. The extension formula is shown in Eq. (7.6):

$$\boldsymbol{X}_{\text{new}} = \text{X}_{\text{nearest}} + L\frac{\text{X}_{\text{rand}} - \text{X}_{\text{nearest}}}{||\text{X}_{\text{rand}} - \text{X}_{\text{nearest}}||_2} \tag{7.6}$$

(4) CollisionCheck function, which is a collision detection function and is a Boolean type function. All points between nodes $\boldsymbol{X}_{\text{nearest}}$

```
RRT algorithm
1. Function BuildRRT()                //Construct RRT function
2. Input (X_init,X_goa,L,X,N)         //Initialize the relevant parameters
3. Init T(X)_init  //Initialize the tree
4. For n=1 to N       //Enter the loop search
    X_rand = RandomSample(X)          //Generate a random sample point
    X_nearest = SearchNearest(T,X_rand)    //find the point with the closest distance
    X_new = ExtendTree(X_near,T,L)     //extend the nearest point to the random tree
      If CollisionCheck (X_nearest,X_new)==0     // Collision Check
          T.add(X)_new     //add the newly generated nodes to the random tree
      If( ||x_new - x_goal||_2 ⩽ dis_goal )     //Determine whether the neighborhood of the target point is reached or not
        Return(T)                      //Returns the extended randomized tree
      If(n>N)
          Return failed                //Search failed
5. Path = PATH(T)
6. Return (Path)                       //Returns the search path
```

Figure 7.6. Pseudo-code for the RRT algorithm.

and $\boldsymbol{X}_{\text{new}}$ are denoted as

$$\{\boldsymbol{X} : \boldsymbol{x} \in (\boldsymbol{X}_{\text{nearest}}, \mathrm{X}_{\text{new}})$$
$$\mathrm{x} = \mathrm{x}_{\text{nearest}} + t \times (\mathrm{x}_{\text{new}} - \mathrm{x}_{\text{nearest}}), \quad t \in [0, 1]\} \tag{7.7}$$

The RRT algorithm is simulated, and the results are shown in Figure 7.7.

In Figure 7.7, the red dots represent the starting points of the search, the green dots represent the target points, the black dots represent the newly added nodes during the search process, the blue lines represent the connections between the tree nodes, and the green broken line represents the path obtained from the final search. During the simulation process, some defects were found in the RRT algorithm:

(1) Different sampling strategies affect the overall convergence rate.
(2) The choice of the extent of the neighborhood space can have a significant impact on the accuracy of the algorithm.
(3) If there is too much information about obstacles in the space, it can lead to a slow solution.
(4) Randomized algorithms lead to more tortuous paths being planned.

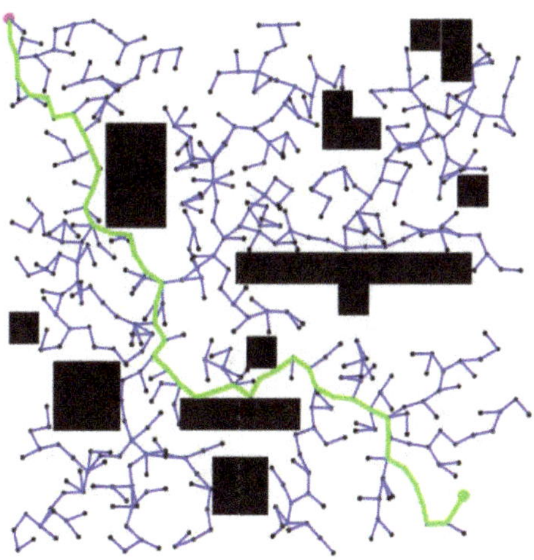

Figure 7.7. RRT 2D simulation diagram.

7.3.2.2 *RRT* algorithm*

Since the RRT algorithm is less efficient in path search in complex environments, Lavalle48 proposed a sampling model based on P probability in 2001, which improves the efficiency of the algorithm search by deciding the selection of random points in the sampling space through the probability value. Later, Lavalle and Kuff proposed a bidirectional search strategy to improve the efficiency of the algorithm, whose basic idea is similar to the bidirectional A* search method mentioned above, i.e., two trees are generated from the start point and the target point at the same time, and the bidirectional search is performed sequentially and alternately, which greatly improves the efficiency of the search. In this book, the three-dimensional spatial path planning is mainly carried out for the hand claw of an operating arm, and more requirements are needed for an optimal path (this book mainly considers the minimum consumption and the shortest path) under the guarantee of safe operation. The RRT* algorithm was proposed by Karaman49 in 2011, which introduces a cost function into the RRT algorithm to optimize the previous path through multiple iterations. While retaining the advantages of the RRT algorithm, the RRT* algorithm can also achieve rapid and effective path planning and obtain an optimal path that meets the criteria through the introduction of a cost function.

The RRT* algorithm mainly achieves path optimization by introducing a cost function. The general idea is similar to the RRT algorithm. The specific algorithm principle is as follows:

(1) Initialize the extended random tree $\boldsymbol{T}_k$ with k nodes. The root node of the tree is $\boldsymbol{X}_{\text{init}}$ and the target node is $\boldsymbol{X}_{\text{goal}}$.
(2) Pick a random point $\boldsymbol{X}_{\text{rand}}$ in the tree space and then traverse all the nodes on the whole extended random tree $\boldsymbol{T}_k$ to find the node $\boldsymbol{X}_{\text{nearest}}$ that is closest to $\boldsymbol{X}_{\text{rand}}$.
(3) Find and generate a new node $\boldsymbol{X}_{\text{new}}$ in the direction $\boldsymbol{X}_{\text{nearest}} \to \mathrm{X}_{\text{rand}}$ in steps of L.
(4) Collision detection is performed between $\boldsymbol{X}_{\text{nearest}} \to \mathrm{X}_{\text{rand}}$. If there is an obstacle, discard this node; otherwise, with node $\boldsymbol{X}_{\text{new}}$ as the center and R as the radius, find all the nodes in $\boldsymbol{T}_k$ in the neighborhood of node $\boldsymbol{X}_{\text{new}}$, all of which are stored in the point set $\{\boldsymbol{X}_{\text{near}}\}$.

(5) Iterate through all the nodes in $\{\boldsymbol{X}_{\text{near}}\}$ and find the $\boldsymbol{X}_{\min}$ node, which if satisfies the condition that the cost of $\boldsymbol{X}_{\min}$ to $\boldsymbol{X}_{\text{new}}$ is less than the cost of $\boldsymbol{X}_{\text{nearest}}$ to $\boldsymbol{X}_{\text{new}}$, then connect $\boldsymbol{X}_{\min}$ to $\boldsymbol{X}_{\text{new}}$ and disconnect $\boldsymbol{X}_{\text{nearest}}$ from $\boldsymbol{X}_{\text{new}}$.

(6) Traverse all nodes $\boldsymbol{X}_i$ $(i = 1, 2, \ldots, n)$ in $\{\boldsymbol{X}_{\text{near}}\}$ except node $\boldsymbol{X}_{\min}$ again to get the cost from the root node to $\boldsymbol{X}_{\text{new}}$ and then to $\boldsymbol{X}_i$. If the value of this generation is less than the cost from the root node to node $\boldsymbol{X}_i$ in the previous tree, then node $\boldsymbol{X}_i$ is disconnected from its parent node, and node $\boldsymbol{X}_i$ is connected with node $\boldsymbol{X}_{\text{new}}$; conversely, the relationship in the original tree is kept unchanged.

(7) Repeat the above steps until node $\boldsymbol{X}_{\text{new}}$ meets $||\boldsymbol{X}_{\text{new}} - \boldsymbol{X}_{\text{goal}}||_2 \leqslant \text{dis}_{\text{goal}}$ to stop searching.

Figure 7.8(a) depicts the search for the nearest node $\boldsymbol{X}_{\text{new}}$ based on the node $\boldsymbol{X}_{\text{rand}}$ obtained from random sampling; Figure 7.8(b) depicts the search for the node $\boldsymbol{X}_{\min}$ with the least costly distance to the node in the set of points $\{\boldsymbol{X}_{\text{near}}\}$, while disconnecting the node $\boldsymbol{X}_{\text{nearest}}$ from the node $\boldsymbol{X}_{\text{new}}$; and Figure 7.8(c) represents the traversal of the entire set of points again to find a new path with

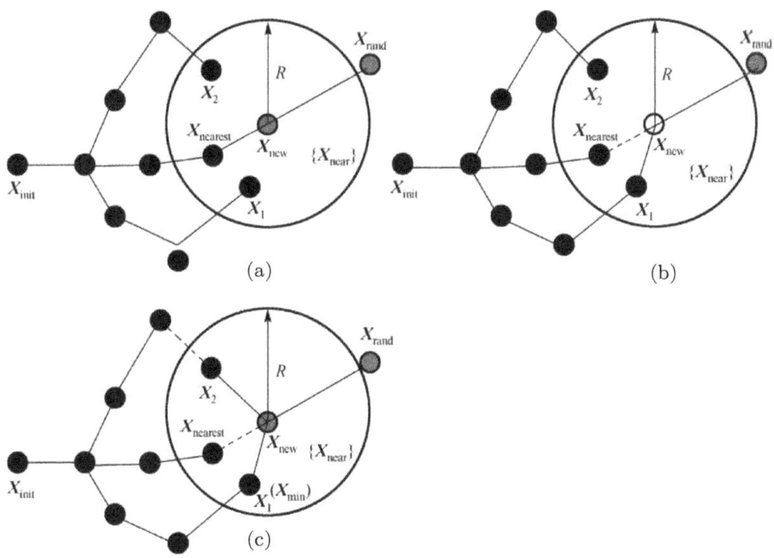

Figure 7.8. Schematic diagram of the RRT* algorithm.

the least total cost through the node $\boldsymbol{X}_{\text{new}}$ to the node $\boldsymbol{X}_2$ that reconstructs the entire extended random tree.

Compared with the RRT algorithm, the RRT* algorithm actually adds the judgment of the cost function and the process of reconstructing the tree. The pseudo-code of the RRT* algorithm is shown in Figure 7.9.

7.3.2.3 *RRT* algorithm improvement*

The RRT and RRT* algorithms can initially find a path, but that is not enough for our purposes in this book. Combined with the state of the operation arm movement in this book, the following are the needs for the path: (1) The path should be as short as possible; usually, the shorter the path, the lower the planning requirements for the operation arm during grasping. (2) The fewer the turning

```
RRT* algorithm

1. Function BuildRRT*()                          //Construct RRT

function

2. Input (Xinit,Xgoal,L,X,N,{Xnear},R)           //Initialize    the

relevant parameters

3. Init T(X)init   //Initialize the tree

4. For n=1 to N     //Enter the loop search

   Xrand = RandomSample(X)                              //

Generate a random sample point

   Xnearest = SearchNearest(T,Xrand)             //search  for  the
```

Figure 7.9. Pseudo-code for the RRT* algorithm.

```
point with the closest distance
    X_new = ExtendTree(X_near,T,L)                    //extend the
nearest point to the random tree
    If CollisionCheck(x_nearest,X_new )==0            //  Collision
Check
        {X_near} = AdjacentArea(T_k,X_new,R)          //find the set
of points in the immediate area of the new node
        X_min = CostMin({X_near},X_nearest,X)_new     //find the
least costly node in the neighborhood
        T.add(X_min,X)_new    //add the newly generated
nodes to the random tree
        TotalCostMin(T_k,{X_near|X_i≠X_min},X)_new    //find the
path that minimizes the total cost
        Rebuild(T_k,X_new,X_min,X)_i  //Reconstruct         the
extended random tree
    If( ||X_new − X_goal||_2 ≤ dis_goal )     //Determine whether the
neighborhood of the target point is reached or not
        Return(T)                        //Returns          the
extended randomized tree
    If(n>N)
        Return failed                //Search failed
5. Path = PATH(T)
6. Return(Path)          //Returns the search path
```

Figure 7.9. (*Continued*)

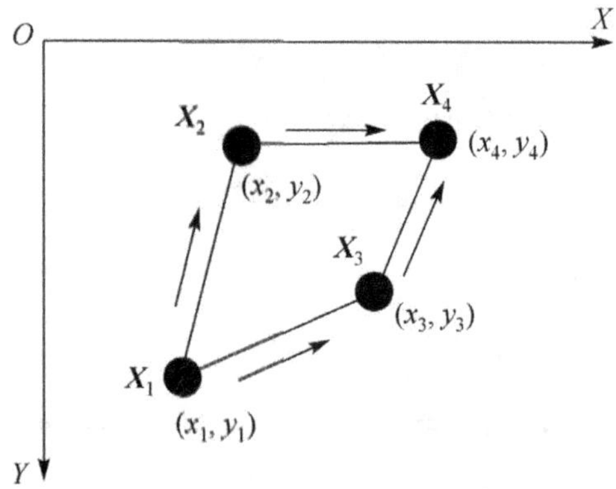

Figure 7.10. Schematic diagram of the RRT* cost function.

points in the path, the better, as it can simplify the movement of the arm. Therefore, an improved algorithm is proposed for the RRT* algorithm.

An example of the source of angular cost in two dimensions is shown in Figure 7.10.

(1) Reconstruct the cost function, Cost_function, which consists of two parts and is calculated as shown in Eq. (7.8):

$$\text{Cost\_function} = \alpha \min(\text{total\_distance}) + \beta \min(\text{total\_angle\_cos t}) \tag{7.8}$$

where min(total_distance) in the first term after the equal sign is the minimum cost of the path length (Euclidean distance) and min(total_angle_cost) in the second term after the equal sign is the minimum cost of the angular difference in the extended random tree.

There are two ways from node $\boldsymbol{X}_1$ to node $\boldsymbol{X}_4$ in Figure 7.10. Assuming that the distance generation value is the same, then it can be judged by the angle generation value, whose formula is shown in Eq. (7.9):

$$\begin{aligned} \text{total\_angle\_cost1} &= \beta(180^\circ - \arccos\langle \overrightarrow{\boldsymbol{X}_1\boldsymbol{X}_2}, \overrightarrow{\boldsymbol{X}_2\boldsymbol{X}_4}\rangle) \\ \text{total\_angle\_cost2} &= \beta(180^\circ - \arccos\langle \overrightarrow{\boldsymbol{X}_1\boldsymbol{X}_3}, \overrightarrow{\boldsymbol{X}_3\boldsymbol{X}_4}\rangle) \end{aligned} \tag{7.9}$$

At this point, by judging total_angle_cost1 and total_angle_cost2, you can get a more "straight" path.

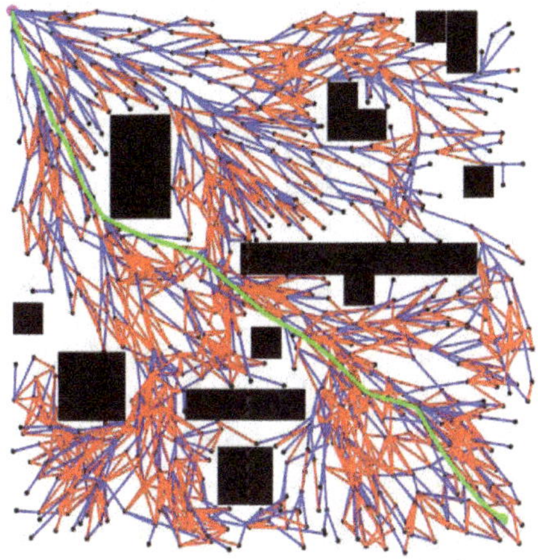

Figure 7.11. RRT* 2D simulation diagram.

(2) *Expansion of obstacles.* In order to reduce the risk of robot collision, the obstacle is inflated, so that although part of the working space of the working arm is sacrificed, the level of danger of robot operation is reduced.

A 2D simulation of the RRT* algorithm was performed, and the results are shown in Figure 7.11.

In Figure 7.11, the red dots represent the starting points of the search; the green dots represent the target points; the black dots represent the nodes in the tree; the blue lines represent the lines between $\boldsymbol{X}_{\text{new}}$ and $\boldsymbol{X}_{\text{nearest}}$; the red lines represent the lines between $\boldsymbol{X}_{\text{nearest}}$ and $\boldsymbol{X}_{\text{min}}$ after the judgment of the cost function; and the green broken line represents the final path obtained from the search. It can be seen that the path searched by the RRT* algorithm is obviously more "straight" than the path searched by the RRT algorithm. Furthermore, during the planning process, it is necessary to adopt the inverse kinematics solution used in the previous chapter to obtain a feasible solution without collision with obstacles.

7.4 Integrated Practice of Robot Path Planning

7.4.1 *Path planning practices in 2D space*

The following simulation scenario is built to perform 2D path planning and verified through experiments, as shown in Figure 7.12.

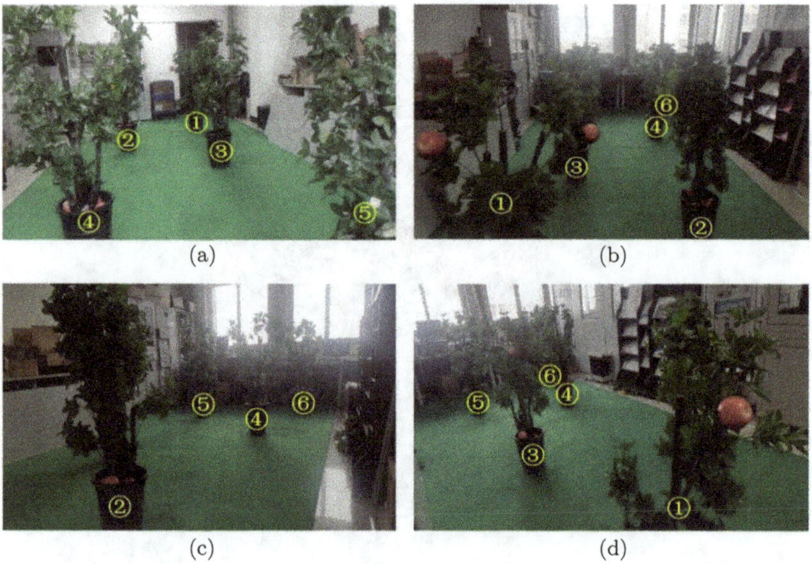

Figure 7.12. Experimental simulation diagram.

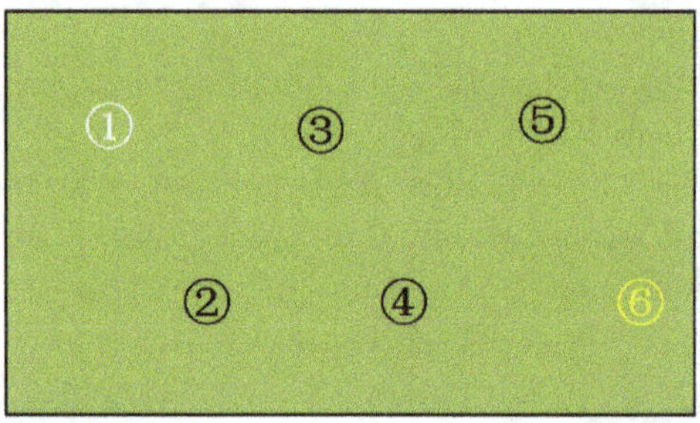

Figure 7.13. Simple diagram.
Note: The numbers represent the corresponding positions of the obstacles in the actual scene.

Figures 7.12(a)–7.12(d) show pictures of the actual scene taken from different viewpoints, and the circled serial numbers represent different numbers of the trees.

In order to represent the scene more intuitively, make simple diagrams, as shown in Figure 7.13.

In 2D estimation planning, the path planning is carried out with the starting point at obstacle 1 and obstacle 6 as the target point. In this scenario, the 3D map is downscaled to a 2D map with an

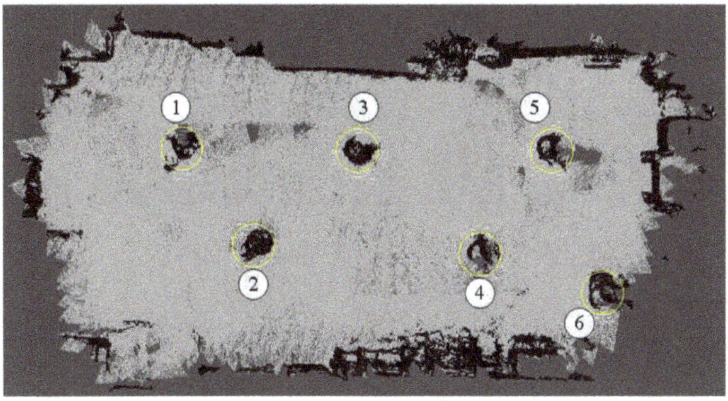

Figure 7.14. Two-dimensional occupancy grid map.
Note: The black areas inside the yellow circles correspond to the locations of the obstacles in the actual scene.

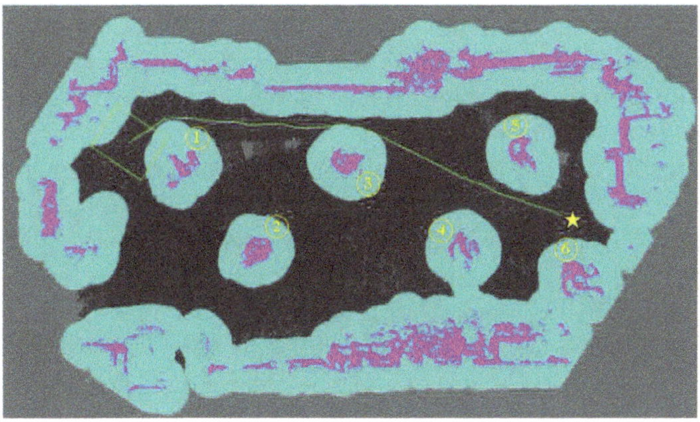

Figure 7.15. Planned path in the RVIZ simulation interface.

accuracy of 0.01 m, and the 2D occupancy grid map is shown in Figure 7.14.

The dimensions of the cart in this simulation are 0.77 m × 0.57 m, and the obstacles are inflated with an expansion radius of 0.285 m (half of the width of the cart) when performing the bidirectional A* algorithm. The path planned in the RVIZ simulation interface is shown in Figure 7.15.

In Figure 7.15, the purple areas represent the obstacle areas; the blue areas represent the expansion areas; the black areas represent

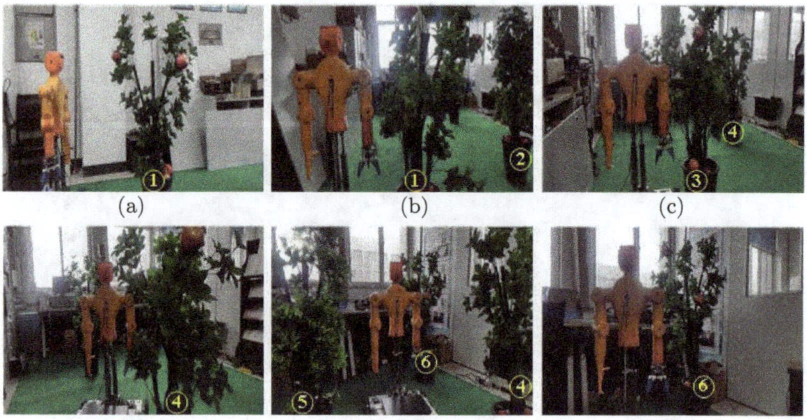

(a) (b) (c) (d) (e) (f)

Figure 7.16. Actual motion results of the cart.

the passable areas of the mobile cart; the yellow rectangle represents the starting position of the cart; the yellow pentagram represents the ending position of the cart; and the yellow curve represents the path obtained by searching through the bidirectional A* algorithm.

The result of the actual movement of the cart is shown in Figure 7.16.

Figure 7.16(a) shows the starting position of the cart; Figure 7.16(b) shows it passing to the left of Obstacle 1 after departure; Figure 7.16(c) shows it passing to the left of Obstacle 3; Figure 7.16(d) shows it going around the left of Obstacle 4; Figure 7.16(e) shows the path passing between Obstacles 4 and 5; and Figure 7.16(f) shows it arriving near Obstacle 6.

It can be seen that the cart can move more reasonably according to the path planned by the algorithm, indicating that the path planning algorithm proposed in this book is reasonable and effective.

7.4.2 *Path planning simulation experiments in 3D space*

7.4.2.1 *Simulation practice of the improved RRT* algorithm*

In order to visually compare the performance of the improved RRT* algorithm proposed in this book, the simulation is first analyzed and verified in a 2D plane. The 2D simulations of the RRT algorithm and the improved RRT* algorithm are carried out in three different

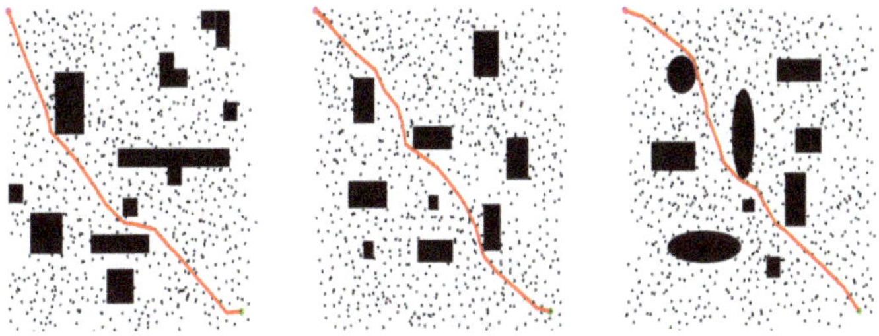

(a) (b) (c)

Figure 7.17. Simulation of the RRT algorithm. (a) simulation scenario 1; (b) simulation scenario 2; (3) simulation scenario 3.

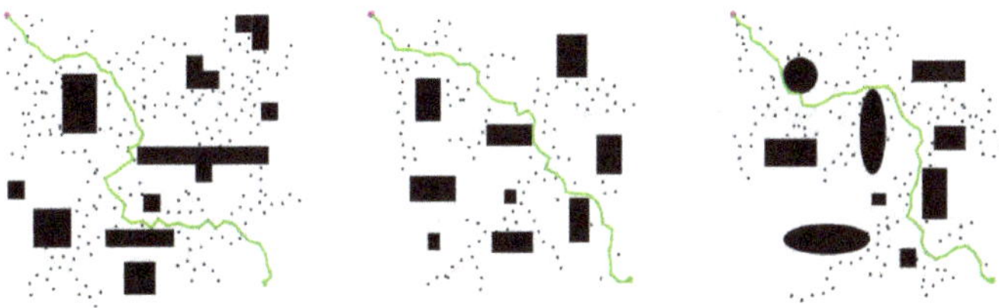

(a) (b) (c)

Figure 7.18. Simulation of the Improved RRT* algorithm. (a) simulation scenario 1; (b) simulation scenario 2; (3) simulation scenario 3.

scenarios with the same search stride, and the results are shown in Figures 7.17 and 7.18, respectively.

In Figures 7.17 and 7.18, the black dots represent the nodes to be searched by the two algorithms during the search process; the green and red lines represent the paths generated by the searches of the RRT algorithm and the improved RRT* algorithm, respectively.

The comparison shows that the RRT algorithm search is more simplified and straightforward and requires relatively fewer nodes to search. However, searching using the improved RRT* algorithm yields fewer path inflection points, shorter total path lengths, and a relatively smoother overall trend.

The path is smoothed using B-spline interpolation, as shown in Figure 7.19.

In Figure 7.19, the blue line represents the path searched by the improved RRT* algorithm; the cyan curve represents the path smoothed by three B-spline curves; and the control point of the B-spline curve is the inflection point of the blue line. Comparing these two curves, it can be seen that after three times of B-spline smoothing, the original "jagged" path has an obvious smoothing tendency, and such a path can make the working arm achieve smoother control.

7.4.2.2 *Three-dimensional spatial path planning*

In the Octave simulation environment, build the 3D attitude of the obstacles in the target environment, set the starting point and target point of the robot end-effector, plan the path by using the RRT* algorithm and fit it by using the cubic B-spline interpolation method, and do the simulation experiments in the following three scenarios, and the results are shown in Figures 7.20, 7.21, and 7.22.

In the above three scenarios, in general, it can be seen that on the basis of the RRT* algorithm, the use of angular difference and spatial

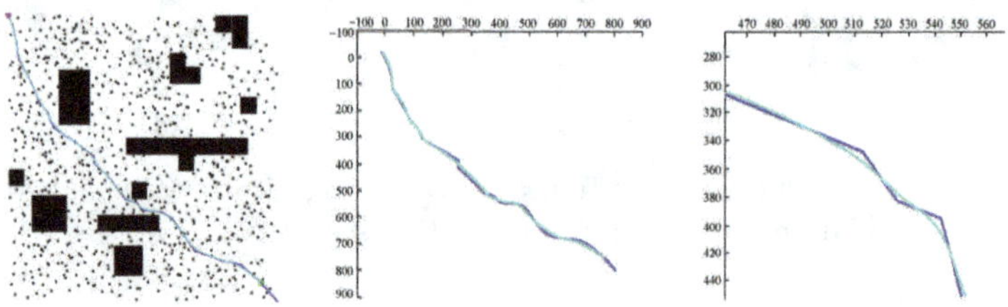

(a) (b) (c)

Figure 7.19. Schematic diagrams of the three B-spline interpolation smoothing RRT* paths. (a) 2D Plane Search Diagram, (b) Curve Comparison Chart and (c) Partial enlarged diagram.

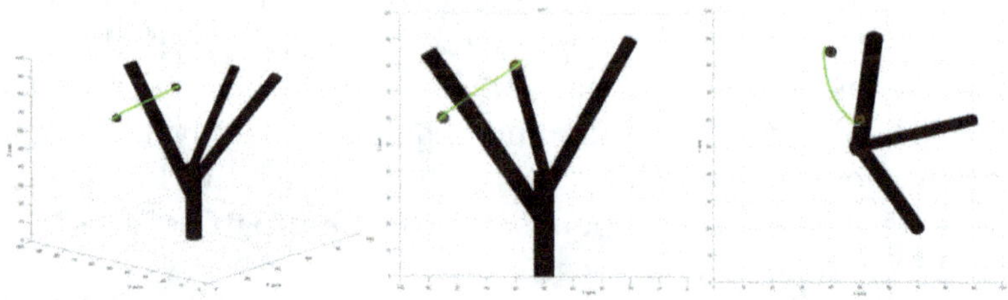

(a) (b) (c)

Figure 7.20. Target environment obstacle 3D simulation scene 1. (a) 3D environmental obstacle structure, (b) *YZ* plane projection and (c) *XY* plane projection.

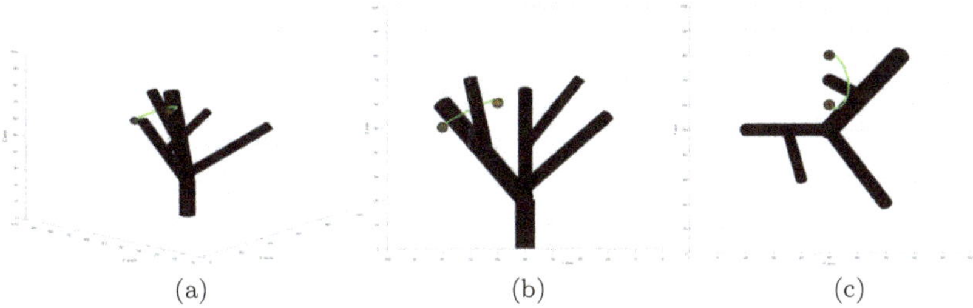

Figure 7.21. Target environment obstacle 3D simulation scene 2. (a) 3D environmental obstacle structure, (b) XY plane projection and (c) YZ plane projection.

Figure 7.22. Target environment obstacle 3D simulation scene 3. (a) 3D environmental obstacle structure, (b) YZ plane projection and (c) XY plane projection.

distance as the cost function, combined with the cubic B-spline interpolation, can provide a more reasonable path for the robot end-effector. The path is relatively smooth, giving the robot end-effector a preliminary foundation for path planning.

7.4.3 *Path smoothing*

Figures 7.17 and 7.18 illustrate that the paths found have issues such as folded corners, jagged edges, and overall lack of smoothness. A motion trajectory with numerous turning points can cause the robot to move with discontinuous velocity and acceleration, leading to rigid impacts and vibrations. This adversely affects the robot's movement stability and decreases its service life. Therefore, smoothing the searched trajectory curve is essential. Common techniques for this include the Bézier curve method, Hermite interpolation,

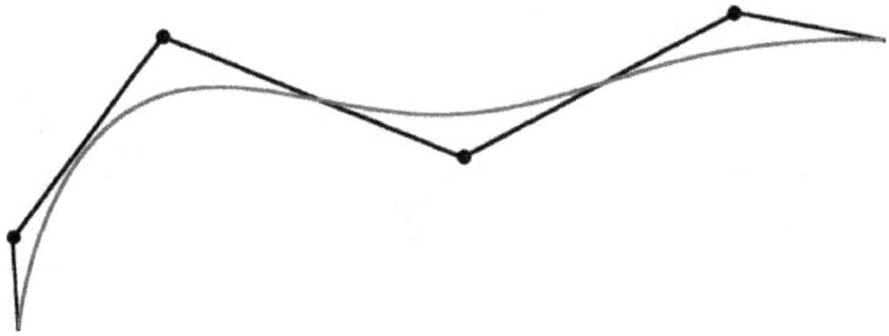

Figure 7.23. Schematic diagram of the effect of the cubic B-spline interpolation method.

Floyd's algorithm, and spline curve interpolation. In this book, the cubic B-spline interpolation method is utilized to smooth the planned path. For the path points identified by the RRT* algorithm presented here, there are various paths to choose from, enabling the selection of collision-free paths. As shown in Figure 7.23, the cubic B-spline interpolation method effectively smooths the "jagged" curve.

Assuming that there are n vertices in the plane or in space $\boldsymbol{P}_i$ $(i = 0, 1, \ldots, n)$, then the p-order B-spline function can be expressed as

$$B(t) = \sum_{i=0}^{n} \boldsymbol{P}_i \cdot M_{i,p}(t), \quad t \in [t_p, t_{p+1}] \tag{7.10}$$

where $\boldsymbol{P}_i$ is the vertex; t is a parameter; and $M_{i,p}(t)$ is the basis function of the p-order canonical B-spline interpolation, which is defined as shown in Eq. (7.11):

$$\begin{cases} M_{i,0}(t) = \begin{cases} 1, & t_i \leqslant t \leqslant t_{i+1} \\ 0, & \text{else} \end{cases} \\ M_{i,p} = \dfrac{t - t_i}{t_{i+1} - t_i} M_{i,p-1}(t) + \dfrac{t_{i+p+1} - t}{t_{i+p+1} - t_{i+1}} M_{i+1,p-1}(t) \\ \text{define } \dfrac{0}{0} = 0 \end{cases} \tag{7.11}$$

When $p = 3$, the curve is called a cubic B-spline interpolation curve, and the corresponding cubic basis function is

$$\begin{cases} M_{0,3}(t) = \frac{1}{6}(-t^3 + 3t^2 - 3t + 1) \\ M_{1,3}(t) = \frac{1}{6}(3t^3 - 6t^2 + 4) \\ M_{2,3}(t) = \frac{1}{6}(-3t^3 + 3t^2 + 3t + 1) \\ M_{3,3}(t) = \frac{1}{6}t^3 \end{cases} \quad t \in [0, 1] \tag{7.12}$$

At this point, the expression for the cubic B-spline interpolation curve is given by

$$B(t) = \frac{1}{6}[1, (t-i), (t-i)^2, (t-i)^3] \begin{bmatrix} 1 & 4 & 1 & 0 \\ -3 & 0 & 3 & 0 \\ 3 & -6 & 3 & 0 \\ -1 & 3 & -3 & 1 \end{bmatrix} \begin{bmatrix} p_i \\ p_{i+1} \\ p_{i+2} \\ p_{i+3} \end{bmatrix},$$

$$t \in [i, i+1], \quad i = 0, 1, \ldots, n-1 \tag{7.13}$$

7.5 Summary

This chapter addresses two primary issues: (1) the 2D global path planning for robots; and (2) the path planning for the robot's end-effector within the target environment space. Initially, the widely used A* algorithm is selected for path planning. To address the issue of prolonged planning time, the bidirectional A* algorithm is employed to shorten the duration of planning. Subsequently, the cost function of the RRT* algorithm is refined in 3D space to produce paths with fewer turning points. Finally, paths are smoothed using the cubic B-spline interpolation method. The path planning methodologies explored in this chapter lay a preliminary theoretical foundation for robot motion.

7.6 Exercises

7.1 For a four-wheeled vehicle, using obstacle avoidance technology, for each option, describe its advantages and disadvantages.

7.2 Consider a robot; design and implement its path planning algorithm and obstacle avoidance algorithm.

7.3 Discuss the similarities and differences between global and local path planning in 2D space.

7.4 Discuss the similarities and differences between global and local path planning in 3D space.

7.5 Briefly describe the similarities and differences between 2D path planning and 3D path planning.

7.6 Design and practice a 3D path planning algorithm for a tandem robot with six degrees of freedom.

Chapter 8

The Integrated Design and Practice of Robot Trajectory Planning

8.1 Introduction

Chapter 7 focused on the route formulation for the robot, detailing how the robot's end-effector was to navigate from a starting point to a designated target while avoiding any obstacles, without the consideration of time; this involves outlining the path that the robot's end-effector will follow.[41,42] It also explored the robot's trajectory formulation, which involves organizing the spatial and temporal relationship of the robot arm's position, speed, acceleration, and other kinetic parameters. This planning is essential for the seamless movement of the robot's end-effector and its various joints toward the intended target.[43]

The routes taken by the end-effector described in this chapter are selected with only its ability to avoid collisions in mind, neglecting the possibility of the robot arm's individual segments encountering obstacles as the end-effector moves. Building upon this, this book outlines a strategy for navigating around obstacles, establishing the robot's ultimate movement route based on a clear path for the end-effector. Additionally, to ensure the end-effector progresses smoothly from its starting point to the intended destination, this book details a method for planning the motion of the robot's dual arms. This approach aims for the robot to achieve its spatial navigation in a manner that is stable, precise, and swift.

8.2 Theory of Robot Trajectory Planning

Traditional articulated obstacle avoidance is based on the Work Space (W-space) assumption modification method, an early proposed method for robot obstacle avoidance. The basic idea is to assume a path from the initial point to the target point, and then carry out collision detection between the work arm and the obstacle. If a collision occurs, modify the path according to the information of the obstacle; continue to repeat the previous steps on the path that has been modified until a complete collision-free path is found before stopping.[44] The process continues to repeat the previous steps until a complete collision-free path is found and then stops.

In the 1980s, some scholars proposed an obstacle avoidance method based on C-space (Configuration Space),[45] which is also one of the more popular methods at present. In this method, the C-space is established by taking the center axis of each joint of the operating arm as the coordinate system and mapping the obstacles in the surrounding environment into the space to form the obstacle space. Therefore, the C-space is composed of the obstacle space and the free space, and by removing the obstacle space, the remaining space is the free space. The points in the free space represent that the operating arm does not collide with the obstacles, at which point the obstacle avoidance problem is transformed into avoiding the obstacle space to find a collision-free path in the C-space.[46]

The traditional W-space obstacle avoidance method is straightforward; however, its effectiveness diminishes in complex environments due to the constant need for collision detection, resulting in poor real-time performance. Furthermore, the drawback of C-space-based obstacle avoidance planning lies in its dependence on the number of robot joints for its dimensionality. Consequently, a robot with more joints entails a more complex calculation process.

Building on the above-mentioned discussion, obstacle avoidance planning for the manipulator arm presented in this book is executed for each joint within W-space, differing from conventional W-space approaches for articulated manipulator arms. The methodology outlined here focuses solely on generating collision-free paths for the manipulator arm's end-effector, omitting obstacle avoidance planning for each link of the arm. Thus, upon identifying a known collision-free path, this book initially establishes the final configuration of the manipulator arm's individual links within W-space as the end-effector reaches the target. Subsequently, it outlines the movement of the manipulator arm along this final configuration towards the target, ensuring obstacle-free motion throughout.

During the manipulator arm's operational process, leveraging the pre-planned collision-free path within the workspace, the arm aims to track this path accurately to fulfill the grasping task. That is, as the arm follows the collision-free trajectory, upon the end-effector's arrival at the target and execution of the grasp, the arm's configuration should ideally align closely with the trajectory of the planned path. However, given that the end-effector's obstacle avoidance path is derived from fitting a curve through numerous discrete spatial points, perfect alignment of the arm's configuration with the pre-planned path is unachievable post-grasp. Consequently, the final configuration of the arm as the end-effector reaches the target is established, defining the ultimate tracking path of the manipulator arm.

8.3 Integrated Design of Robot Trajectory Planning

8.3.1 *Trajectory planning in joint space*

Since trajectory planning in joint space does not limit the path of the robot end-effector, it is generally used in point-to-point tasks. Trajectory planning in joint space involves the following: Firstly, the initial position of the robot's end-effector in the workspace and the desired target position should be determined; secondly, after the inverse kinematics analysis, the rotation angle of each joint is solved when the robot end-effector reaches the final target position, and then the change of the rotation angle of each joint over time is used for planning. Trajectory planning in joint space generally uses polynomial functions such as cubic polynomial interpolation and higher-order polynomial interpolation to describe the trajectories of the operating arm joints as they move over time.

8.3.1.1 *Cubic polynomial interpolation*

One of the simpler methods commonly used in trajectory planning algorithms under joint space is the cubic polynomial interpolation method. However, the cubic polynomial interpolation method can only make the paths and velocities in the planning process smooth. Take one of the joints of the robot arm as an example. Let the start and stop moments of the robot's motion be t_0 and t_s, respectively, and the joint angle values at the start and stop moments be θ_0 and θ_s, respectively, and the joint velocities be $\dot{\theta}_0 = 0$ and $\dot{\theta}_s = 0$, respectively. Let the basic form of the cubic polynomial be shown in Eq. (8.1):

$$\theta(t) = a_0 + a_1 t + a_2 t^2 + a_3 t^3 \tag{8.1}$$

Equation (8.1) is a cubic polynomial designed for the variation of the joint angle over time. In Eq. (8.1), the initial value θ_0 and the target value θ_s of the joint angle are known. However, in order to determine the four unknown coefficients in Eq. (8.1) and in order to ensure the smooth movement of the joint to the endpoint, the constraints that need to be satisfied are shown in Eq. (8.2):

$$\begin{cases} \theta(t_0) = \theta_0 \\ \theta(t_s) = \theta_s \\ \dot{\theta}(t_0) = 0 \\ \dot{\theta}(t_s) = 0 \end{cases} \tag{8.2}$$

Based on the above-provided constraints, the time-dependent angular velocity $\dot{\theta}(t)$ and angular acceleration $\ddot{\theta}(t)$ of the robot joints during rotation can be obtained:

$$\begin{cases} \dot{\theta}(t) = a_1 + 2a_2 t + 3a_3 t^2 \\ \ddot{\theta}(t) = 2a_2 + 6a_3 t \end{cases} \tag{8.3}$$

Let the start moment $t_0 = 0$; the stop moment t_s is also known. Therefore, according to Eqs. (8.2) and (8.3), the following can be obtained:

$$\begin{cases} a_0 = \theta_0 \\ a_1 = 0 \\ a_2 = \frac{3}{t_s^2}(\theta_s - \theta_0) \\ a_3 = -\frac{2}{t_s^3}(\theta_s - \theta_0) \end{cases} \tag{8.4}$$

To summarize, the regular relationship between the angular displacement $\theta(t)$, angular velocity $\dot{\theta}(t)$, and angular acceleration $\ddot{\theta}(t)$ of the robot's joint rotation with time $t \in [t_0, t_s]$ can be obtained, respectively:

$$\begin{cases} \theta(t) = \theta_0 + \frac{3}{t_s^2}(\theta_s - \theta_0)t^2 - \frac{2}{t_s^3}(\theta_s - \theta_0)t^3 \\ \dot{\theta}(t) = \frac{6}{t_s^2}(\theta_s - \theta_0)t - \frac{6}{t_s^3}(\theta_s - \theta_0)t^2 \\ \ddot{\theta}(t) = \frac{6}{t_s^2}(\theta_s - \theta_0) - \frac{12}{t_s^3}(\theta_s - \theta_0)t \end{cases} \tag{8.5}$$

Similarly, when the start angle θ_0 and the end angle θ_s of each joint of the robot are known, trajectory planning for each joint can be performed based on the cubic polynomial interpolation method.

8.3.1.2 *Higher-order polynomial interpolation*

To ensure smoothness for robot motion process, it is not only necessary for us to determine those parameters, such as joint angle, angular velocity, and angular acceleration, etc., for both start and end-points of the robot joint trajectory, it is also necessary for us to obtain some other parameters, such as angular acceleration and angular acceleration, for joint angle change planning over time. Therefore, some higher-order polynomial interpolation equations, such as the fifth-degree and the sixth-degree polynomial interpolation equations, instead of cubic polynomial interpolation ones, have to be employed for such purposes. In this chapter, the fifth-degree polynomial interpolation is used as an example.

Let the function of a fifth-degree polynomial be as shown in Eq. (8.6):

$$\theta(t) = a_0 + a_1 t + a_2 t^2 + a_3 t^3 + a_4 t^4 + a_5 t^5 \tag{8.6}$$

The planning time of the joint angle from the start point is the start time t_0, and the end time of the joint angle to the endpoint is t_s. To facilitate the calculation, the start time is set to 0. When the start angle of the joint angle is set to θ_0, the end angle is set to θ_s, while the start joint angular velocity is set to be $\dot{\theta}_0$, the end joint angular velocity is set to $\dot{\theta}_s$, and the start joint angular acceleration is set to be $\ddot{\theta}_0$, and the end joint angular acceleration is set to $\ddot{\theta}_s$, the following polynomial interpolation equations could be obtained for trajectory planning,

$$\begin{cases} \theta(t_0) = \theta_0 = a_0 \\ \theta(t_s) = \theta_s = a_0 + a_1 t_s + a_2 t_s^2 + a_3 t_s^3 + a_4 t_s^4 + a_5 t_s^5 \\ \dot{\theta}(t_0) = \dot{\theta}_0 = a_1 \\ \dot{\theta}(t_s) = \dot{\theta}_s = a_1 + 2a_2 t_s + 3a_3 t_s^2 + 4a_4 t_s^3 + 5a_5 t_s^4 \\ \ddot{\theta}(t_0) = \ddot{\theta}_0 = 2a_2 \\ \ddot{\theta}(t_s) = \ddot{\theta}_s = 2a_2 + 6a_3 t_s + 12a_4 t_s^2 + 20a_5 t_s^3 \end{cases} \tag{8.7}$$

When solving the above polynomial interpolation equations with the initial values given, the coefficients of those equations could be obtained as follows, which could finally determine the trajectory planning equations,

$$\begin{cases} a_0 = \theta_0 \\ a_1 = \dot{\theta}_0 \\ a_2 = \frac{\ddot{\theta}_0}{2} \\ a_3 = \frac{20\theta_s - 20\theta_0 - (8\dot{\theta}_s + 12\dot{\theta}_0)t_s - (3\ddot{\theta}_0 - \ddot{\theta}_s)t_s^2}{2t_s^3} \\ a_4 = \frac{-30\theta_s + 30\theta_0 + (14\dot{\theta}_s + 16\dot{\theta}_0)t_s - (3\ddot{\theta}_0 - 2\ddot{\theta}_s)t_s^2}{2t_s^4} \\ a_5 = \frac{12\theta_s - 12\theta_0 - (6\dot{\theta}_s + 6\dot{\theta}_0)t_s - (\ddot{\theta}_0 - \ddot{\theta}_s)t_s^2}{2t_s^5} \end{cases} \tag{8.8}$$

8.3.2 *Trajectory planning in workspace*

For trajectory planning under joint space, the path trajectory of the end-effector of the work arm is not described. When the end-effector is required to move along a given path, it is necessary to plan the path trajectory of the end-effector under the workspace. For trajectory planning under the workspace, the end-effector of the work arm is required to move steadily, accurately, and quickly along the planned path. According to the start position and target position of the end-effector under the workspace, the position path can be fitted using straight line segments and circular arcs. Therefore, trajectory planning in the workspace involves trajectory planning for straight line segments and circular arcs in that space. The main idea of straight line segment and circular arc planning in the space is to discretize the fitted path to get the positional state of the end-effector by combining the orientation information completed by each positional state planning, i.e., combine the attitude information of each completed position status into a series of position and orientation information of the end-effector, and then try to solve the angle values for each operating joint arm corresponding to the position and orientation state via inverse kinematics analysis. In this section, the pose of the end-effector is divided into position and orientation for planning, i.e., the positional path of the end-effector is planned when interpolating straight line segments and circular arcs in the workspace; the orientation of the end-effector is converted into quaternion and then planned.

8.3.2.1 *Positional linear interpolation*

When the path of the end-effector of the operating arm is a straight line segment, and its start point and target point are known, it is necessary to interpolate the path of the straight line segment to ensure that the end-effector moves along the straight line segment to the target point. As shown in Figure 8.1, the start point of the straight line segment path is $P_o(x_o, y_o, z_o)$ and the target point is $P_f(x_f, y_f, z_f)$.

From the two points in space, you can determine the equation of the function of this straight line segment as shown in Eq. (8.9):

$$\frac{x - x_o}{x_f - x_o} = \frac{y - y_o}{y_f - y_o} = \frac{z - z_o}{z_f - z_o} \tag{8.9}$$

The linear distance between two points can be obtained from the coordinates of the two points, which will be denoted as d_p, and the

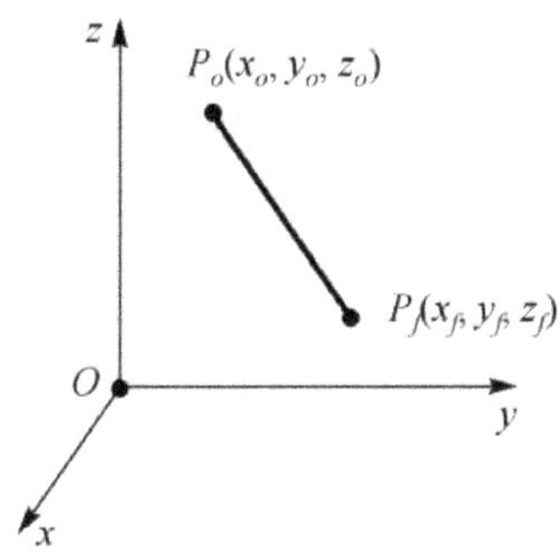

Figure 8.1. Start and target points of end-effector in space.

value of d_p is shown in Eq. (8.10):

$$d_p = \sqrt{(x_f - x_o)^2 + (y_f - y_o)^2 + (z_f - z_o)^2} \tag{8.10}$$

Next, the interpolation operation is performed on the straight line segment P_oP_f, i.e., the discretization of the straight line segment P_oP_f. If N points are inserted into the straight line segment between the position points P_o and P_f at equal intervals, and the distance between every two points is d, then we can get Eq. (8.11):

$$d = \frac{d_p}{N+1} \tag{8.11}$$

After inserting N points at equal intervals on the straight line segment P_oP_f, including its start point and target point, there are $N+2$ discrete points on the straight line segment, and these discrete points are recorded as P_i, $i = 1, 2, 3, \ldots, N+1, N+2$. The start point is recorded as the first discrete point and the target point is recorded as the first $\overrightarrow{AB} = \lambda\overrightarrow{P_oP_f}$ $(i = 1, 2, 3, \ldots, N+1, N+2)$ discrete point; then, we can get the relationship between the direction vector $\overrightarrow{P_oP_i}$ from the start point P_o to the interpolation point P_i and the direction vector $\overrightarrow{P_oP_f}$ from the start point P_o to the target point P_f, i.e.,

$$\overrightarrow{P_oP_i} = \lambda\overrightarrow{P_oP_f},\ i = 1, 2, 3, \ldots, N+1, N+2 \tag{8.12}$$

where $\lambda \in (0, 1)$ and $\lambda = \frac{i-1}{N+1}$. When $\lambda = 0$, it refers to the position of the end-effector at the starting point; when $\lambda = 1$, it means the end-effector has reached the target point. Therefore, the coordinate values of the ith discrete point on the straight line segment in the directions of x, y, z axes $P_i(x_i, y_i, z_i)$ can be obtained as shown in Eq. (8.13):

$$\begin{aligned} x_i &= x_o + \lambda(x_f - x_o) \\ y_i &= y_o + \lambda(y_f - y_o) \\ z_i &= z_o + \lambda(z_f - z_o) \end{aligned} \tag{8.13}$$

As a result, the position coordinates of a series of interpolation points are solved, and then after inverse kinematics analysis, the joint angle value of the operating arm corresponding to each interpolation point is obtained. The joint angle values can then be planned in joint space to ensure that while the end-effector of the operating arm moves along the specified linear path, the other joints can also move smoothly to the target angle.

8.3.2.2 *Spatial position arc interpolation*

Similar to the straight line interpolation in the workspace, for the circular arc in the workspace, the end-effector of the operating arm is required to move to the target position along the specified spatial circular arc path. In the spatial arc interpolation algorithm, the calculation of the arc length between any two initial points and the endpoint on the arc curve can be converted to the calculation of the increment of the central angle of the circle corresponding to the two points. Let there be three points in the spatial coordinate system $A(x_a, y_a, z_a)$, $B(x_b, y_b, z_b)$, and $C(x_c, y_c, z_c)$, and the steps for planning the spatial curve determined by these three points are as follows:

(1) Determine the center O and radius of the arc r

From the three points in space $A(x_a, y_a, z_a)$, $B(x_b, y_b, z_b)$, and $C(x_c, y_c, z_c)$, a plane M can be determined. Since all three points are in this plane, the normal vector $\vec{n}$ of the plane M can be found as shown in Eq. (8.14):

$$\begin{aligned}\vec{n} = \overrightarrow{AB} \times \overrightarrow{AC} &= \begin{vmatrix} \vec{i} & \vec{j} & \vec{k} \\ x_b - x_a & y_b - y_a & z_b - z_a \\ x_c - x_a & y_c - y_a & z_c - z_a \end{vmatrix} \\ &= [(y_b - y_a)(z_c - z_a) - (y_c - y_a)(z_b - z_a)]\vec{i} \\ &= [(x_c - x_a)(z_b - z_a) - (x_b - x_a)(z_c - z_a)]\vec{j} \\ &= [(x_b - x_a)(y_c - y_a) - (x_c - x_a)(y_b - y_a)]\vec{k}\end{aligned} \tag{8.14}$$

Therefore, the equation of the plane M can be obtained as shown in Eq. (8.15):

$$\begin{aligned}&[(y_b - y_a)(z_c - z_a) - (y_c - y_a)(z_b - z_a)](x - x_a) \\ &\quad + [(x_c - x_a)(z_b - z_a) - (x_b - x_a)(z_c - z_a)](y - y_a) \\ &\quad + [(x_b - x_a)(y_c - y_a) - (x_c - x_a)(y_b - y_a)](z - z_a) = 0\end{aligned} \tag{8.15}$$

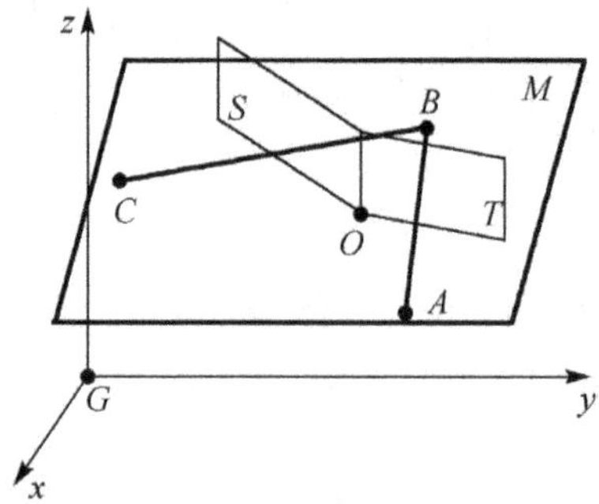

Figure 8.2. Space plane relationship diagram.

The plane M is the plane determined by the three points A, B, and C. To determine the centers of the arcs determined by these three points, place the perpendicular bisectors T and S of the line segments AB and BC, respectively, over the midpoints of the segments as shown in Figure 8.2.

From this, the normal vectors of the perpendicularly bisecting planes T and S are $\overrightarrow{AB}$ and $\overrightarrow{BC}$, so the equations of the two planes T and S can be determined as shown in Eqs. (8.16) and (8.17), respectively:

$$(x_b - x_a)[x - (x_a + x_b)/2] + (y_b - y_a)[y - (y_a + y_b)/2] + (z_b - z_a)[z - (z_a + z_b)/2] = 0 \tag{8.16}$$

$$(x_c - x_b)[x - (x_b + x_c)/2] + (y_c - y_b)[y - (y_b + y_c)/2] + (z_c - z_b)[z - (z_b + z_c)/2] = 0 \tag{8.17}$$

Therefore, the point where the three planes M, T, and S *intersect* is the center O of the arc $\widehat{ABC}$ determined by the points A, B, and C. By combining the three plane equations, the coordinate value of the center of the circle $O(x_o, y_o, z_o)$ can be determined. Also, solve for its radius r, as shown in Eq. (8.18):

$$r = \sqrt{(x_a - x_o)^2 + (y_a - y_o)^2 + (z_a - z_o)^2} \tag{8.18}$$

(2) Coordinate transformation

Based on the three planes determined previously, a new spatial coordinate system O-UVW *is* established, as shown in Figure 8.3.

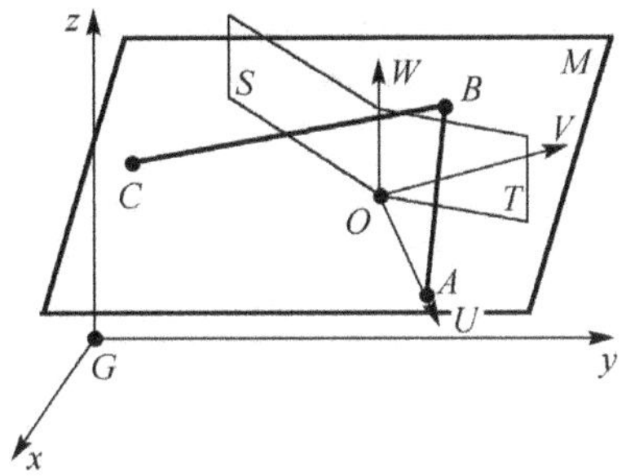

Figure 8.3. *O-UVW coordinate* system.

In Figure 8.3, the center O of the circular arc $\widehat{ABC}$ is taken as the origin of the coordinate system O-UVW, the U-axis is along the direction of the vector $\overrightarrow{OA}$, the W-axis is along the direction of the intersection of the two perpendicular bisectors T and S, i.e., the direction of the normal vector $\vec{n}$ of the plane M, and the direction of the V-axis is determined according to the right-hand rule. The unit vectors of direction of each axis of the coordinate system O-UVW are

$$\begin{aligned} \vec{u} &= \frac{\overrightarrow{OA}}{\left|\overrightarrow{OA}\right|} \\ \vec{w} &= \frac{\vec{n}}{|\vec{n}|} \\ \vec{v} &= \vec{u} \times \vec{w} \end{aligned} \tag{8.19}$$

It can be shown that the homogeneous transformation matrix ${}_O^G\boldsymbol{T}$ of the initial base coordinate system G-XYZ transformed to the coordinate system O-UVW is

$${}_O^G\boldsymbol{T} = \begin{bmatrix} u_x & v_x & w_x & x_o \\ u_y & v_y & w_y & y_o \\ u_z & v_z & w_z & z_o \\ 0 & 0 & 0 & 1 \end{bmatrix} \tag{8.20}$$

The coordinates of the four points O, A, B, and C in the initial base coordinate system G-XYZ transformed to the coordinate system O-UVW *are* denoted as $O(u_o, v_o, w_o)$, $A(u_a, v_a, w_a)$, $B(u_b, v_b, w_b)$, and $C(u_c, v_c, w_c)$, so that the transformation relationship between the two coordinate systems can be obtained, as shown in Eq. (8.21):

$$\begin{bmatrix} u_o \\ v_o \\ w_o \\ 1 \end{bmatrix} = {}_O^G\boldsymbol{T}^{-1} \begin{bmatrix} x_o \\ y_o \\ z_o \\ 1 \end{bmatrix} \quad \begin{bmatrix} u_a \\ v_a \\ w_a \\ 1 \end{bmatrix} = {}_O^G\boldsymbol{T}^{-1} \begin{bmatrix} x_a \\ y_a \\ z_a \\ 1 \end{bmatrix}$$
$$\begin{bmatrix} u_b \\ v_b \\ w_b \\ 1 \end{bmatrix} = {}_O^G\boldsymbol{T}^{-1} \begin{bmatrix} x_b \\ y_b \\ z_b \\ 1 \end{bmatrix} \quad \begin{bmatrix} u_c \\ v_c \\ w_c \\ 1 \end{bmatrix} = {}_O^G\boldsymbol{T}^{-1} \begin{bmatrix} x_c \\ y_c \\ z_c \\ 1 \end{bmatrix} \tag{8.21}$$

where ${}_O^G\boldsymbol{T}^{-1}$ denotes the inverse matrix of the homogeneous transformation matrix ${}_O^G\boldsymbol{T}$. Meanwhile, since the center of the circle O is the origin in the coordinate system O-UVW, and A, B, and C *are* in the UOV plane, $u_o = v_o = w_o = 0$, $w_a = w_b = w_c = 0$, and $u_a = r$.

(3) Calculation of the central angle of the circle corresponding to the arc of the circle

Since, in the newly established O-UVW coordinate system, the arc $\overset{\frown}{ABC}$ only lies on the two-dimensional UOV plane, it is analyzed within this 2D plane, as shown in Figure 8.4. In Figure 8.4, the central angle corresponding to the arc is denoted by θ, with the starting point A of the arc located on the U axis, and the central angle is determined by judging the position of the endpoint C on the arc.

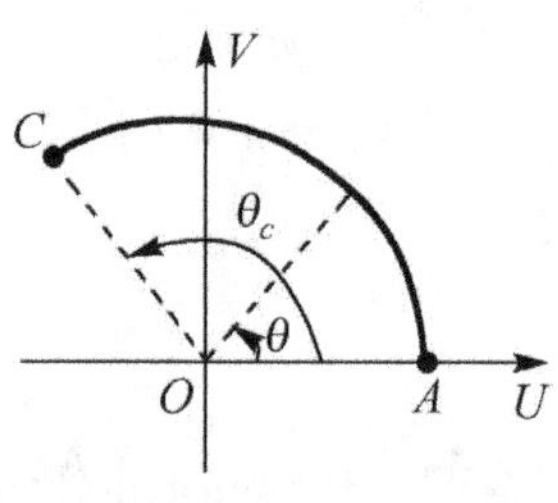

Figure 8.4. Circular trajectory in the UOV plane.

When $v_c > 0$, it means that the endpoint C is located in the first or second quadrant of this 2D coordinate system, i.e., $\theta_c = a\tan 2(v_c, u_c) < \pi$.

When $v_c < 0$, it means that the endpoint C is located in the third or fourth quadrant of this 2D coordinate system, i.e., $\pi < \theta_c = 2\pi + a\tan 2(v_c, u_c) < 2\pi$.

(4) Calculation of the coordinates of points on the arc of a circle

If the radius r of the arc and the corresponding central angle θ of the circle are known, then the length l of the arc is obtained:

$$l = r\theta \tag{8.22}$$

Setting the linear velocity to v and the time interval between two interpolated points on the arc trajectory to T_s, the arc length of two neighboring interpolated points on the arc can be obtained Δl:

$$\Delta l = vT_s \tag{8.23}$$

The total number of start, end, and interpolation points can be calculated over the entire arc $\widehat{ABCN}$:

$$N = \left[\frac{l}{\Delta l}\right] + 1 \tag{8.24}$$

In the formula, $[\frac{l}{\Delta l}]$ means that by rounding $\frac{l}{\Delta l}$, we can get the corresponding central angle of the arc between two neighboring interpolation points on the arc $\Delta\theta$:

$$\Delta\theta = \frac{\theta}{N-1} \tag{8.25}$$

To summarize, the ith interpolation point on the arc is denoted as $P_i(u_i, v_i)$ and the corresponding rotated central angle θ_i of the circle can be obtained when moving from the initial point to this interpolation point:

$$\theta_i = (i-1)\Delta\theta, \quad i = 1, 2, 3, \ldots, N-1, N \tag{8.26}$$

where, when $i = 1$, the central angle of the circle is $0°$, which means the starting point on the arc; when $i = N$, the central angle of the circle is θ, which means the endpoint on the arc.

As a result, the coordinate value of an interpolated point $P_i(u_i, v_i)$ on the arc can be obtained in the 2D planar coordinate system:

$$\begin{aligned} u_i &= r\cos\theta_i \\ v_i &= r\sin\theta_i \end{aligned} \tag{8.27}$$

(5) Transform the obtained interpolation points into the coordinate system G-XYZ.

Combining the above-mentioned steps yields the coordinate representation of the interpolated point P_i on the spatial arc in the coordinate system:

$$P_i(u_i, v_i, w_i) = P_i(r\cos\theta_i, r\sin\theta_i, 0) \tag{8.28}$$

Therefore, a series of interpolated points in the coordinate system O-UVW *are* transformed to the coordinate system G-XYZ:

$$\begin{bmatrix} x_i \\ y_i \\ z_i \\ 1 \end{bmatrix} = {}^{G}_{O}\boldsymbol{T} \begin{bmatrix} u_i \\ v_i \\ w_i \\ 1 \end{bmatrix} \tag{8.29}$$

In summary, the positions of all interpolation points on the spatial arc in the base coordinate system can be obtained, that is, the planning of the spatial arc is completed. According to the position of the interpolation points on the arc, combined with the inverse kinematics solution, we can get the angle value of each joint of the robot when the end-effector is at each interpolation point position.

8.3.3 *Robot arm coordinate motion planning*

The motion of the robot's dual arms is under the same workspace; therefore, the common workspace of the arms needs to be analyzed to avoid a collision of the arms. The workspace of the robot is the maximum reachable range of the end-effector in three-dimensional space.

In this book, the common workspace of the main and auxiliary operating arms of the robot is analyzed after the lower body movement of the robot, so the common base coordinate system of the robot arms is the chest-centered coordinate system $\{O\}$. The position of the end-effector in the base coordinate system $\{O\}$ is obtained by performing forward kinematics calculations at each joint of the operating arm, where the position of the end-effector is noted to be $P(p_x, p_y, p_z)$, and the relationship between it and the angle of each joint of the operating arm is obtained:

$$P(p_x, p_y, p_z) = \text{fkine}(\theta_1, \theta_2, \ldots, \theta_n) \tag{8.30}$$

where (p_x, p_y, p_z) denotes the positional coordinates of the end-effector in the base coordinate system O; fkine() denotes the forward kinematics function, the independent variable of which is the angle of the joints of the operating arm, and the value of the angle of each joint is within the limit of its joints, i.e., $\theta_i \in [\theta_{i\,\min}, \theta_{i\,\max}](i = 1, 2, \ldots, n)$; and n denotes the number of joints of the robot.

Thus, the robot's workspace can be represented as

$$\begin{aligned} W[P] = \{&\text{fkine}(\theta_1, \theta_2, \ldots, \theta_n); \theta_i \in [\theta_{i\,\min}, \theta_{i\,\max}]\} \subset R^3, \\ &i = 1, 2, \ldots, n \end{aligned} \tag{8.31}$$

The idea of the Monte Carlo method is to first apply the mathematical method of random sampling and randomly traverse the values within the limits of each joint angle, where the more the points that are taken, the higher the accuracy of the workspace. Then, the selected joint angle values are subjected to forward kinematics computation, and finally the set of all random points is the workspace of the robot. The main steps of applying the Monte Carlo method in this book are as follows:

(1) According to Eqs. (3.21) and (4.20), obtain the position expressions $P_L(p_{lx}, p_{ly}, p_{lz})$ and $P_R(p_{rx}, p_{ry}, p_{rz})$ of the end-effector of the robot's auxiliary operation arm and the end-effector of the main operation arm in the coordinate system $\{O\}$, respectively:

$$\begin{aligned} p_{lx} = {}& d_5(s_4(c_3s_1 + c_1s_2s_3) - c_1c_2c_4) - d(s_6(s_5(s_1s_3 - c_1c_3s_2) \\ &- c_5(c_4(c_3s_1 + c_1s_2s_3) + c_1c_2s_4)) - c_6(s_4(c_3s_1 + c_1s_2s_3) \\ &- c_1c_2c_4)) - c_1c_2d_3 \\ p_{ly} = {}& d_5(s_4(c_1c_3 - s_1s_2s_3) + c_2c_4s_1) + d(c_6(s_4(c_1c_3 - s_1s_2s_3) \\ &+ c_2c_4s_1) - s_6(s_5(c_1s_3 + c_3s_1s_2) - c_5(c_4(c_1c_3 - s_1s_2s_3) \\ &- c_2s_1s_4))) + c_2d_3s_1 \end{aligned}$$

$$p_{lz} = d(s_6(c_5(s_2s_4 - c_2c_4s_3) - c_2c_3s_5) - c_6(c_4s_2 + c_2s_3s_4)) \\ - d_5(c_4s_2 + c_2s_3s_4) - d_3s_2d_1 \quad (8.32)$$

$$p_{rx} = d_5(c_4(c_3s_1 - c_1s_2s_3) + c_1c_2s_4) - c_1c_2d_3 \\ p_{ry} = -d_5(c_4(c_1c_3 + s_1s_2s_3) - c_2s_1s_4) - c_2d_3s_1 \\ p_{rz} = d_1 + d_5(s_2s_4 + c_2c_4s_3) - d_3s_2 \quad (8.33)$$

(2) Restrict each joint variable of the robot's dual arms to its minimum to maximum joint range, and utilize the rand$(N, 1)$ function to randomly extract N values on the interval $(0, 1)$ to obtain the angle values of each joint:

$$\theta_i = \theta_{i\,\min} + (\theta_{i\,\max} - \theta_{i\,\min}) \times \text{rand}(N, 1) \quad (8.34)$$

(3) Combined with the forward kinematics of the robot, each set of joint angle values of the main and auxiliary operating arms obtained is substituted into Eqs. (3.21) and (4.20), respectively, to calculate the position coordinates of the end-effector and draw the simulation diagram of the robot workspace, as shown in Figure 8.5.

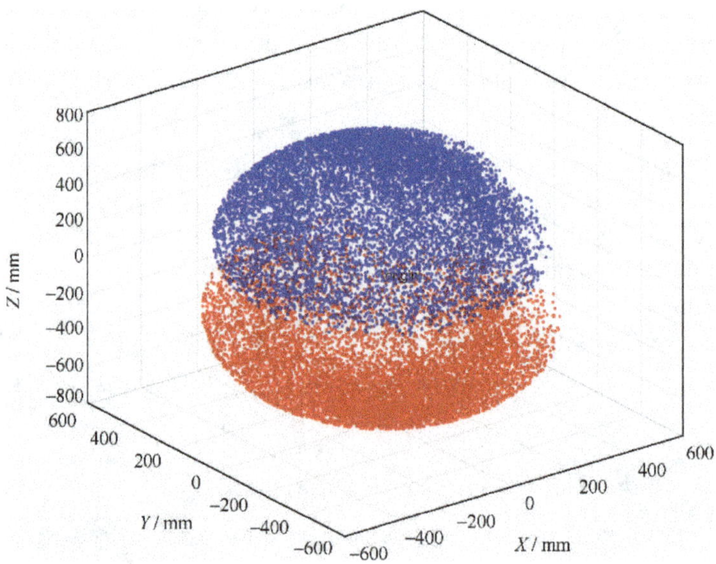

Figure 8.5. Robotic workspace simulation.

In Figure 8.5, the value of N is 10000, the red color indicates the reachable space of the end-effector of the robot auxiliary operation arm and the blue color indicates the reachable space of the end-effector of the robot main operation arm. The simulation map in 3D space is projected onto the XOY plane, XOZ plane, and YOZ plane, as shown in Figure 8.6.

Analyzing Figure 8.6, it can be noted that in the XOY plane (corresponding to the machine side-view plane), the reachable space range of the end-effector of the robot auxiliary working arm is larger than that of the end-effector of the main working arm, and the range is added to the reachable space of the end-effector of the main working arm by an annular space with a width of approximate 30 mm. The common reachable workspace for the main and auxiliary robotic arms on the XOZ and YOZ planes is approximated as an ellipsoidal shape. On the XOZ plane, which corresponds to the front side of the robot, the shared workspace of the main and auxiliary arms extends from −400 mm to 290 mm in the X-axis direction and from −120 mm to 150 mm in the Z-axis direction. On the YOZ plane, representing the top view of the robot, the shared workspace ranges from −420 mm to 415 mm in the Y-axis direction and from −120 mm to 150 mm in the Z-axis direction. In summary, during the task operations of robots with dual arms positioned directly in front of the robot's body, the maximum vertical, horizontal, and depth distances within this shared workspace are approximately 690 mm, 270 mm, and 420 mm, respectively. Furthermore, beyond this shared workspace, there is no interference between the end-effectors of the robot's arms.

8.4 Integrated Practice of Robot Trajectory Planning

8.4.1 *Joint space trajectory planning*

Setting the auxiliary operation arm and main operation arm to move from the initial state $(0°, 0°, 0°, 0°, 0°, 0°)$, $(0°, 0°, 0°, 0°, 0°)$ to the target position state $(-80°, -10°, 20°, -30°, 12°, 2°)$, $(-80°, -10°, 20°, -30°, 12°)$, respectively, the trajectory planning of the operation arm under the joint space is carried out by using cubic polynomials and quintic polynomials, as shown in Figures 8.7 and 8.8, respectively.

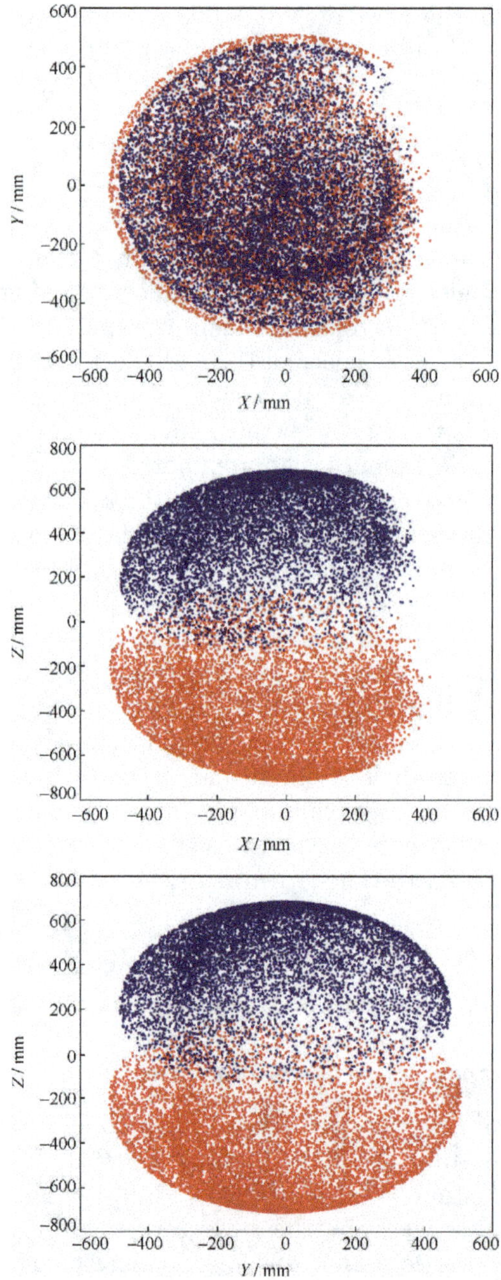

Figure 8.6. Projection of the robot simulation workspace onto a two-dimensional plane.

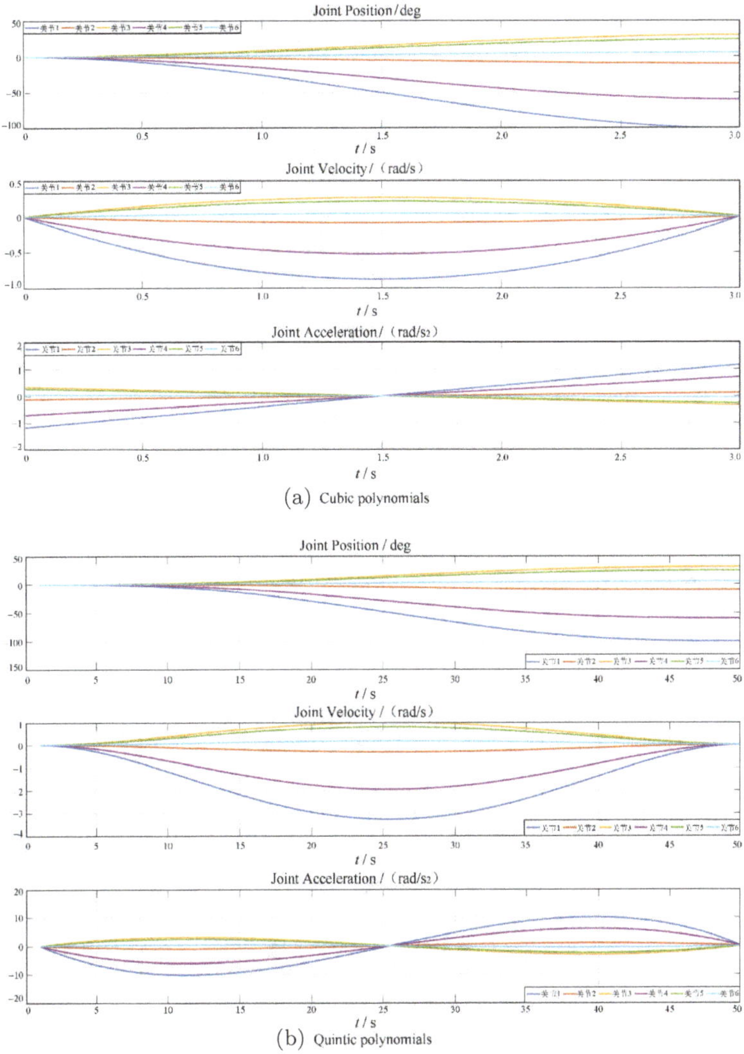

Figure 8.7. Auxiliary work arm trajectory planning in joint space.

8.4.2 *Workspace trajectory planning*

Under the workspace, when using the spatial linear interpolation method for planning, the initial position of the auxiliary work arm end-effector is $A_L(-513, 0, -200)$ and the target position is $B_L(298.8, -272.4, -0.0416)$; the initial position of the main work

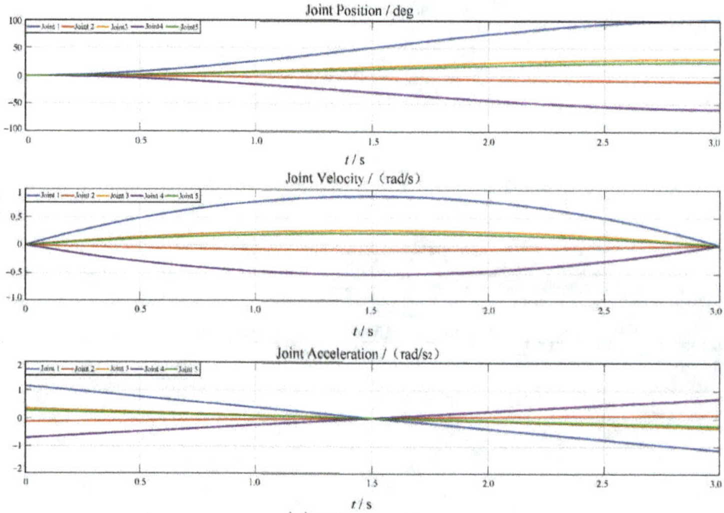

(a) Cubic polynomials

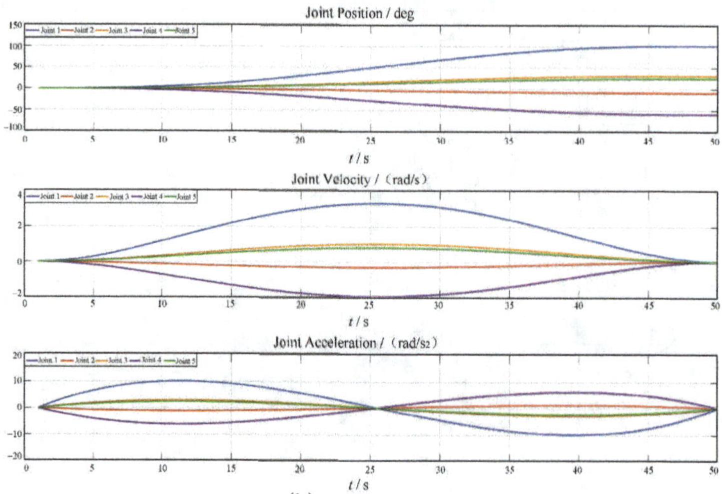

(b) Quintic polynomials

Figure 8.8. Master work arm trajectory planning in joint space.

arm end-effector is $A_R(-183, -330, 200)$ and the target position is $B_R(218.4, -415.7, 362.7)$. Dual-arm spatial linear trajectory planning is shown in Figure 8.9.

Under the workspace, when the spatial arc interpolation method is used for planning, the initial position of the end-effector of the

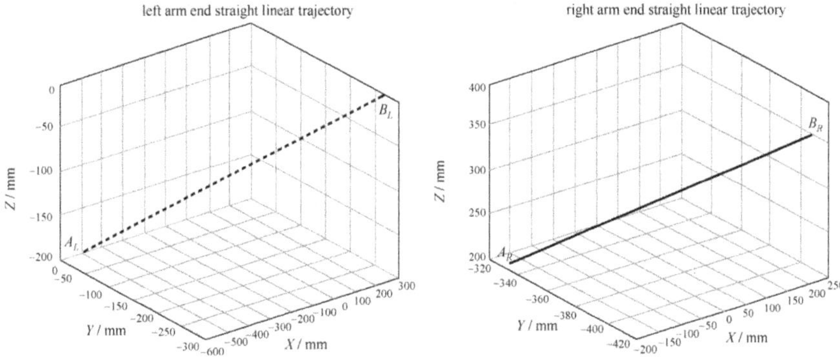

(a) (b)

Figure 8.9. Linear trajectory planning in dual-arm space. (a) Auxiliary operation arm end-effector linear trajectory and (b) Main arm end-effector linear trajectory.

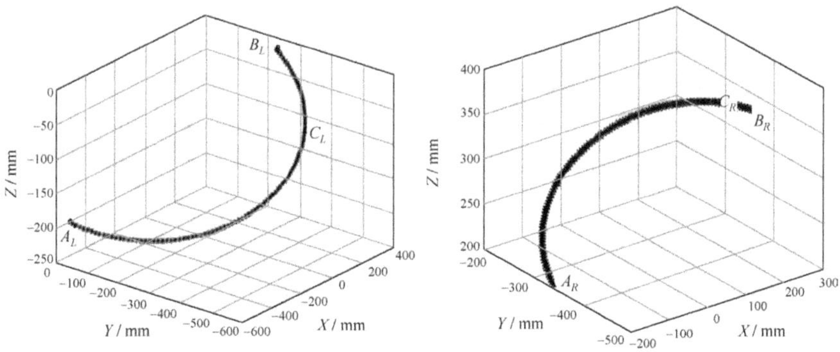

(a) (b)

Figure 8.10. Dual-arm space arc trajectory planning. (a) Auxiliary operation arm end actuator arc trajectory and (b) Main arm end actuator arc trajectory.

auxiliary operation arm is $A_L(-513, 0, -200)$, the angle value of each joint is $(-80°, -10°, 20°, -30°, 12°, 2°)$, the corresponding path point position is $C_L(73.37, -472.4, -63.44)$, and the target position is $B_L(298.8, -272.4, -0.0416)$. The initial position of the end-effector of the main operation arm is $A_R(-183, -330, 200)$, the angle value of each joint is $(-80°, -10°, 20°, -30°, 12°)$, the corresponding path point position is $C_R(207.9, -367.4, 356.7)$, and the target position is $B_R(218.4, -415.7, 362.7)$. The spatial arc trajectory planning of the two arms is shown in Figure 8.10.

8.4.3 *Dual-arm coordinated grab task*

Currently, in many complex operational tasks, such as grasping irregular objects, assembling complex workpieces, and handling large objects, compared with single-arm robots, dual-arm robots have stronger applicability. As for robot dual-arm coordinated control, the main task is to realize motion trajectory planning for the two arms to drive the operation arms to reach the targeted states without any collisions between the two arms themselves. During the planning process, the arms can reach the target point according to the planned trajectory only when the arms meet different constraints for different objective tasks. In the planning process, only by satisfying different constraints for different target tasks between the two arms can the arms be controlled to reach the target point according to the planned trajectory movement.

At present, the dual-arm coordination planning of robots is mainly categorized into loose coordination and tight coordination. Loose coordination means that in the same workspace, the two arms independently complete their respective tasks; tight coordination means that in the same workspace, the two arms work in accordance with constraints to maintain a certain regularity along the preset trajectory movement. The dual-arm robot is strongly coupled in the tight coordination task, such as when the dual-arms carry the same object, the motion law of the dual-arms completely depends on the desired motion law of the target, and does not have relative motion with the target.

Based on the functions to be realized by the robot in the laboratory, i.e., when the auxiliary operating arm assists the main operating arm in completing the task of grasping the fruits in the canopy, the main operating system does not need assistance, or the two arms can independently complete the task of grasping the target object, this book investigates and analyzes the loose coordination task for the robot's dual arms. According to the above-described tasks of the robot's dual arms, the robot's auxiliary operation arm end-effector and the main operation arm end-effector do not have the same target object; the main operation arm end-effector's target objective is to grasp the object, while the auxiliary operation arm's target objective is an obstacle. Therefore, there is no force constraint or relative motion between the main and auxiliary operation arm end-effectors.

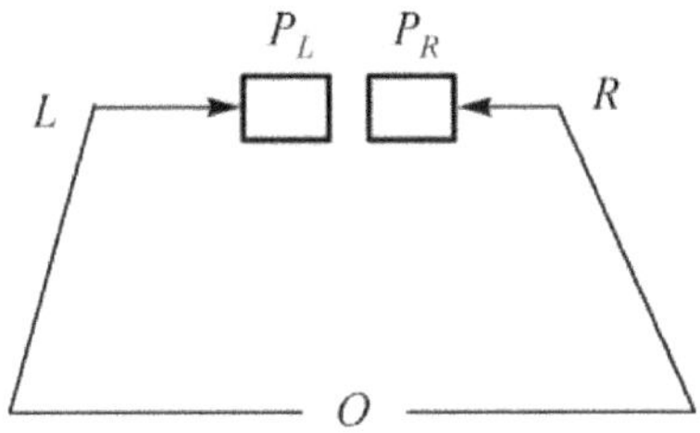

Figure 8.11. Schematic diagram of dual-arm operation.

This analysis of the workspace of the robot's dual arms shows that there are positional constraints between the end-effectors of the main and auxiliary operation arms.

According to the DH model of the main and auxiliary arms of the robot, the base coordinate system of the two arms of the robot is selected to be the center coordinate system of the robot's chest, $\{O\}$, and its origin is noted to be O. Here, the origins of the attached coordinate system of the end-effector of the auxiliary arm and the main arm are noted to be L and R, respectively, and the origins of the attached coordinate system of the end-effector of the two arms to grasp the target is noted to be P_L and P_R, respectively, as shown in Figure 8.11.

As shown in Figure 8.11, when there is no constraint between the main and auxiliary operation arms of the robot, the position and orientation matrices of the targets of the main and auxiliary operation arms in the coordinate system $\{O\}{}_{P_L}^{O}\boldsymbol{T}$, ${}_{P_R}^{O}\boldsymbol{T}$ can be obtained as shown in Eq. (8.35):

$$\begin{aligned} {}_{P_L}^{O}\boldsymbol{T} &= {}_{L}^{O}\boldsymbol{T}{}_{P_L}^{L}\boldsymbol{T} \\ {}_{P_R}^{O}\boldsymbol{T} &= {}_{R}^{O}\boldsymbol{T}{}_{P_R}^{R}\boldsymbol{T} \end{aligned} \tag{8.35}$$

In the formula, ${}_{P_L}^{L}\boldsymbol{T}$ and ${}_{P_R}^{R}\boldsymbol{T}$ represent the position and orientation matrices of the target of the main and auxiliary operating arms in the attached coordinate system of the end-effector, respectively; ${}_{L}^{O}\boldsymbol{T}$ and ${}_{R}^{O}\boldsymbol{T}$ represent the position and orientation matrices of the end-effector of the main and auxiliary operating arms in the base coordinate system, respectively.

Splitting the homogeneous transformation matrix into a 3×3-order orientation matrix $\boldsymbol{R}$ and a 3×1-order position vector $\boldsymbol{P}$, the following can be obtained by equating ${}_{P_L}^{L}\boldsymbol{T}$ with $({}_{L}^{P_L}\boldsymbol{T})^{-1}$ and ${}_{P_R}^{R}\boldsymbol{T}$ with ${}_{P_R}^{R}\boldsymbol{T}({}_{R}^{P_R}\boldsymbol{T})^{-1}$:

$$
\begin{aligned}
{}_{P_L}^{L}\boldsymbol{T} &= \left({}_{L}^{P_L}\boldsymbol{T}\right)^{-1} = \begin{pmatrix} {}_{L}^{P_L}\boldsymbol{R}_{3\times3} & {}_{L}^{P_L}\boldsymbol{P}_{3\times1} \\ \boldsymbol{0}_{1\times3} & 1 \end{pmatrix}^{-1} = \begin{pmatrix} {}_{P_L}^{L}\boldsymbol{R} & -{}_{P_L}^{L}\boldsymbol{R}{}_{L}^{P_L}\boldsymbol{P} \\ \boldsymbol{0}_{1\times3} & 1 \end{pmatrix} \\
{}_{P_L}^{R}\boldsymbol{T} &= \left({}_{R}^{P_R}\boldsymbol{T}\right)^{-1} = \begin{pmatrix} {}_{R}^{P_R}\boldsymbol{R}_{3\times3} & {}_{R}^{P_R}\boldsymbol{P}_{3\times1} \\ \boldsymbol{0}_{1\times3} & 1 \end{pmatrix}^{-1} = \begin{pmatrix} {}_{P_R}^{R}\boldsymbol{R} & -{}_{P_R}^{R}\boldsymbol{R}{}_{R}^{P_R}\boldsymbol{P} \\ \boldsymbol{0}_{1\times3} & 1 \end{pmatrix}
\end{aligned} \tag{8.36}
$$

The position expressions ${}_{P_L}^{O}\boldsymbol{P}$ and ${}_{P_R}^{O}\boldsymbol{P}$ of the main and auxiliary operating arm end-effectors P_L, P_R in the base coordinate system $\{O\}$ can be obtained by using Eqs. (8.35) and (8.36), respectively, as shown in Eq. (8.37):

$$
\begin{aligned}
{}_{P_L}^{O}\boldsymbol{P} &= {}_{L}^{O}\boldsymbol{P} - {}_{P_L}^{O}\boldsymbol{R}{}_{L}^{P_L}\boldsymbol{P} \\
{}_{P_R}^{O}\boldsymbol{P} &= {}_{R}^{O}\boldsymbol{P} - {}_{P_R}^{O}\boldsymbol{R}{}_{R}^{P_R}\boldsymbol{P}
\end{aligned} \tag{8.37}
$$

In the formula, ${}_{L}^{O}\boldsymbol{P}$ and ${}_{R}^{O}\boldsymbol{P}$, respectively, indicate the positions of the origin of the coordinate system of the end-effector of the main and auxiliary operation arms in the base coordinate system $\{O\}$; ${}_{P_L}^{O}\boldsymbol{R}$ and ${}_{P_R}^{O}\boldsymbol{R}$, respectively, indicate the transformation matrices of the origin of the coordinate system of the target to be grasped by the end-effector of the main and auxiliary operation arms in the base coordinate system $\{O\}$; and ${}_{L}^{P_L}\boldsymbol{P}$ and ${}_{R}^{P_R}\boldsymbol{P}$, respectively, indicate the positions of the origin of the coordinate system of the end-effector of the main and auxiliary operation arms in the coordinate system of the corresponding target.

Based on the above-mentioned relationship, it can be noted that when there is no constraint between the robot arms, and the orientation and position information of the target in the base coordinate system $\{O\}$ and the position information of the end-effector in the coordinate system of the target are acquired, the position expression of the end-effector in the base coordinate system $\{O\}$ can be obtained.

According to Eqs. (8.35) and (8.36), the orientation transformation matrices ${}_{P_L}^{O}\boldsymbol{R}$ and ${}_{P_R}^{O}\boldsymbol{R}$ of the end-effector of the main auxiliary operation arm P_L and P_R, respectively, in the base coordinate system $\{O\}$ can be obtained, as shown in Eq. (8.38):

$$\begin{aligned} {}^{O}_{P_L}\boldsymbol{R} &= {}^{O}_{L}\boldsymbol{R}{}^{L}_{P_L}\boldsymbol{R} \\ {}^{O}_{P_R}\boldsymbol{R} &= {}^{O}_{R}\boldsymbol{R}{}^{R}_{P_R}\boldsymbol{R} \end{aligned} \tag{8.38}$$

In the formula, ${}^{O}_{L}\boldsymbol{R}$ and ${}^{O}_{R}\boldsymbol{R}$ denote the respective transformation matrices of the end-effector coordinate system origins for the master and auxiliary operation arms within the base coordinate system $\{O\}$, while ${}^{L}_{P_L}\boldsymbol{R}$ and ${}^{R}_{P_R}\boldsymbol{R}$ represent the respective transformation matrices of the origins of the target coordinate system to be grasped by the end-effector of the master and auxiliary operating arm in the end-effector coordinate system.

According to Eq. (8.38), when the orientation transformation matrices of the end-effector of the main and auxiliary operating arms in the base coordinate system $\{O\}$ and the end-effector coordinate system, respectively, are obtained, the orientation information of the end-effector in the base coordinate system $\{O\}$ can be obtained.

In summary, when there are no constraints between the robot's two arms, the operating arm's inverse solution can be derived from the end-effector position and orientation information obtained earlier.

This book uses a dual-arm robot; based on the analysis of its workspace, the relative position and orientation between the two arms when the dual-arm end-effector finally reaches the target to be grasped are constrained. The constraints are shown in Eq. (8.39):

$$\begin{aligned} {}^{O}_{P_R}\boldsymbol{P} - {}^{O}_{P_L}\boldsymbol{P} &= \boldsymbol{P}_1 \\ {}^{O}_{P_R}\boldsymbol{R} &= {}^{O}_{P_L}\boldsymbol{R}\boldsymbol{R}_1 \end{aligned} \tag{8.39}$$

where $\boldsymbol{P}_1$ is a 3×1-order matrix representing the position constraints at the final state of the dual-arm end-effector and $\boldsymbol{R}_1$ is a 3×3-order matrix representing the orientation constraints at the final state of the dual-arm end-effector.

According to the constraints in Eqs. (8.38) and (8.39), the following can be obtained:

$$\begin{aligned} {}^{O}_{L}\boldsymbol{P} &= {}^{O}_{R}\boldsymbol{P} + {}^{O}_{P_L}\boldsymbol{R}{}^{P_L}_{L}\boldsymbol{P} - \boldsymbol{P}_1 - {}^{O}_{P_R}\boldsymbol{R}{}^{P_R}_{R}\boldsymbol{P} \\ {}^{O}_{L}\boldsymbol{R} &= {}^{O}_{R}\boldsymbol{R}{}^{R}_{P_R}\boldsymbol{R}\boldsymbol{R}_1{}^{P_L}_{L}\boldsymbol{R} \end{aligned} \tag{8.40}$$

This book does not consider the constraint relationship between the orientations of the end-effector of the auxiliary operation arm and the end-effector of the main operation arm, and only considers the positional constraints. When the auxiliary operation arm coordinates with the main operation arm, or both auxiliary operation arm and the main operation arm complete their grasping tasks independently, in order to avoid collisions between the two arms, it has to be guaranteed that, in the front space of the robot body, the practical distances between the two-arm end-effectors along the direction of the x-axis or the y-axis or the z-axis of the base coordinate system $\{O\}$ should not be less than 50 mm, i.e., only one of the absolute values of x-, y- or z- coordinate values of $\boldsymbol{P}_1$ in Eq. (8.39) should be no less than 50 mm.

Therefore, according to the collision-free path planned earlier, when the end-effector of the main operation arm moves along the path, the position information of the end-effector in the base coordinate system $\{O\}$ can be determined in real time; when the position constraints $\boldsymbol{P}_1$ with the end-effector of the auxiliary operation arm are also known, the position information of the auxiliary operation arm's end-effector in the base coordinate system $\{O\}$ can be obtained according to Eq. (8.40); and the inverse kinematic method of the auxiliary operation system can be applied to obtain the angle value when each joint moves to the target point.

8.5 Summary

In this chapter, based on the known collision-free path of the end-effector of the working arm, the final tracking path of the end-effector of the working arm when the end-effector reaches the target point is determined, obstacle avoidance planning is also carried out for the other links of the working arm, and according to the determined collision-free path of the working arm for the various joint corners, trajectory planning is carried out by using cubic polynomials and quintuple polynomials in the joint space. In the working space, trajectory planning is carried out using positional linear interpolation and spatial positional circular interpolation methods. The Monte Carlo method was used to analyze the workspace of the dual arms, and the coordinated operation of the dual arms was studied based on the loose coordination task.

8.6 Exercises

8.1 What is trajectory planning? Briefly describe the methods of trajectory planning and characterize them.

8.2 Suppose a robot has six rotating joints whose joint motions are all planned as cubic polynomials and are required to stop at a goal position after passing through two intermediate path points. How many independent cubic polynomials are required in total to describe the motion of the joints of this robot? How many coefficients are needed to determine these polynomials?

8.3 A rotating joint of a single-link robot, which starts moving from a standstill at $\theta = -5°$, is to be brought to a stop at the position $\theta = 180°$ in 4 s in a smooth motion. Try to determine the trajectory of the motion according to the following requirements:

(1) Design and practice joint motions to be planned in a cubic polynomial interpolation manner.

(2) Design and practice the planning of joint motions as parabolic transitions with linear interpolation.

8.4 Briefly describe both similarities and differences between joint space and workspace trajectory plannings for wheeled robots.

Chapter 9

Integrated Design and Practice for Robot Control

9.1 Introduction

The previous chapters have explored robot perception systems, target recognition, motion planning, and trajectory planning, laying the groundwork for robot control discussed in this chapter. Robot control, as suggested by its name, encompasses an array of controls utilized by the robot's control system to enable the completion of various tasks and actions. The robot control system, serving as the robot's brain, is responsible for guiding the robot to achieve a predetermined state in the desired manner through the interpretation of detection signals from sensors. The effectiveness of the robot control system significantly influences the robot's performance and, to some extent, impacts the advancement of robotics. Thus, possessing a perfectly functional, sensitive, and reliable control system is crucial for the coordinated action between the robot and its equipment, facilitating their collaborative effort to accomplish tasks.

This chapter delves into the comprehensive design and practice of robot control, covering the design and implementation of control in the robot's driver space, joint space, and workspace. Specifically, Section 9.2 discusses the theoretical underpinnings of robot control; Section 9.3 focuses on the integrated design of robot control, including the control of the robot's effector space, joint space, and workspace; Section 9.4 presents the integrated practice of robot control; and Section 9.5 summarizes the key points covered in this chapter.

9.2 Theory of Robot Control

Robotic systems are redundant, multivariable nonlinear control systems as well as complex coupled dynamical systems, and each control task has its own dynamical task in itself. In research, robot control systems are usually described by simplifying them into a number of low-order subsystems, including several aspects such as controllers, control variables, control levels, and control techniques.

9.2.1 *Robot controller*

Controlling a robot is analogous to managing the control and transmission within its subsystems. From the perspective of joints (or links), robot controllers can be classified into two types: single-joint (link) controllers and multi-joint (link) controllers. The former must incorporate compensation for steady-state errors, while the latter should address the compensation for coupled inertia. The effectiveness of robot control hinges on the sophistication of its "brain", namely, the processor. Given the diverse real-world operating conditions, various control methods can be applied, ranging from straightforward programmed automation and small-computer control to microprocessor control. The architecture of the robot control system can significantly vary as well, extending from single-processor control systems to complex multi-processor hierarchical distributed control systems. In the case of the latter, each processor is assigned a specific task or is directly connected to a particular part of the robot, such as a degree of freedom or an axis.

9.2.2 *Robot control variables*

As an example, the control variables for each joint of the operating arm in the robot platform are shown in Figure 9.1. If the robot is to be allowed to grasp a target, A, then it is necessary to know the

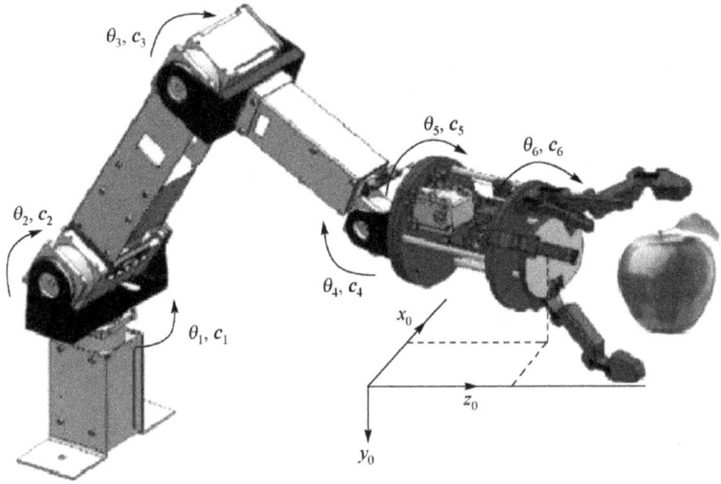

Figure 9.1. Control variables for each joint of the robot operating arm.

state of the end-effector relative to the target Aat any given moment, such as the position, orientation, and open/close state. The position of target A is given by a set of coordinate axes in the absolute system of the environment where it is located; let this set of axes be called the task axes (R). The state of the end-effector is represented by the values or parameters of this set of axes, which are components of the vector $\boldsymbol{X}$. To control the robot for operations, our task is to control the vector $\boldsymbol{X}$ over time, i.e., $\boldsymbol{X}(t)$, which represents the real-time position of the end-effector in space. $\boldsymbol{X}$ changes only when the joints θ_1–θ_6 move. We use the vector $\boldsymbol{\Theta}(t)$ to represent the joint variables ranging from θ_1 to θ_6.

The joints move under the action of torques $\boldsymbol{c}_1$–$\boldsymbol{c}_6$, which form the vector $\boldsymbol{C}(t)$. The vector $\boldsymbol{C}(t)$ is delivered to the joints by the torque vector $\boldsymbol{T}(t)$ of the drive motors through the speed changer, and these motors are controlled by one or more microprocessors to generate the torque vector $\boldsymbol{T}(t)$ under the power provided by the current or voltage vector $\boldsymbol{V}(v)$. Thus, the control of a robotic system is essentially the control of the transformation of the following bi-directional equation:

$$\boldsymbol{V}(t) \leftrightarrow \boldsymbol{T}(t) \leftrightarrow \boldsymbol{C}(t) \leftrightarrow \boldsymbol{\Theta}(t) \leftrightarrow \boldsymbol{X}(t) \tag{9.1}$$

9.2.3 *Robot control hierarchy*

According to the realization of the robot's function, its main control level can be divided into three control levels, i.e., the artificial intelligence (AI) level, the control mode level, and the servo system level, as shown in Figure 9.2.

9.2.3.1 *Level 1: Artificial intelligence level*

If a robot is commanded to "grasp target A", how is this task performed? First, it must be determined that the successful execution of the command is due at least to the fact that the robot generates the vector $\boldsymbol{X}(t)$ for the command, i.e., the motion of the end-effector with

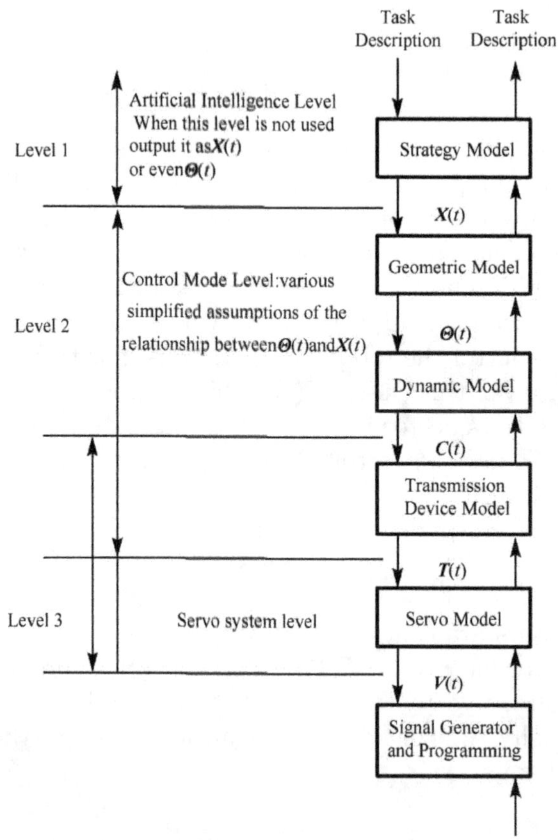

Figure 9.2. Main control levels of the robot.

respect to the object A. This level of control encompasses all possible issues related to AI, such as lexical and natural language understanding, plan generation, and task description. However, AI-level control is still in the research stage, and there are not enough applications to robots, and many practical problems remain to be solved.

9.2.3.2 *Level 2: Control mode level*

The control mode stage can establish a bidirectional relationship between the control vector $\boldsymbol{X}(t)$ and the motor torque vector $\boldsymbol{T}(t)$, and there are a variety of control modes that can be used in this control stage. The relationship between the two vectors can be expressed by the relationship expressed in Eq. (9.2):

$$\boldsymbol{T}(t) \leftrightarrow \boldsymbol{C}(t) \leftrightarrow \boldsymbol{\Theta}(t) \leftrightarrow \boldsymbol{X}(t) \tag{9.2}$$

The above-mentioned four vectors correspond to different structural system models of the robot: $\boldsymbol{T}(t)$ corresponds to the robot transmission device model; $\boldsymbol{C}(t)$ corresponds to the robot model of an articulated mechanical system; $\boldsymbol{\Theta}(t)$ corresponds to the model of the relationship between the articulated variables and the controlled values in the task space; and $\boldsymbol{X}(t)$ corresponds to the robot model in the workspace. However, they correspond to different problems in engineering practice.

The first issue at the control model level pertains to system dynamics. This level faces several challenges. Initially, accurately modeling mechanical errors in various interconnected components, such as dry friction and joint deflection, is challenging. Furthermore, even if it were feasible to account for these errors, the resulting model would encompass thousands of parameters, rendering the processor incapable of executing all necessary operations in real time at the appropriate speed. Additionally, predicting the control system's response to model transformations is difficult. Moreover, complexity exacerbates the challenge of transforming the model, particularly when dealing with nonlinear models.

9.2.3.3 *Level 3: Servo system level*

The focus within the servo system level is on the practical aspects of robotics. It's crucial to understand that the first and second levels of control are not always clearly separated. Moreover, it's essential to determine if the transmission and deceleration mechanisms are incorporated into the second level. Currently, there's a trend toward exploring motors that integrate deceleration mechanisms, allowing direct mounting on robot joints. However, this necessitates addressing issues related to the moment of inertia and the reduction ratio. Another point of interest is that servo systems have traditionally been analog. However, analog systems are increasingly being supplanted by digitally controlled servo systems. Therefore, the assumptions for deriving a satisfactory method might vary. These assumptions are influenced by the operator's depth of knowledge on the subject matter and the specific application of the robot.

9.2.4 *Robot control technology*

For the desired geometric path of the robot's end-effector, inverse kinematics can be used to calculate the joint motions, which can then be substituted into the equations of motion to know the effector's commands, which are then applied to make the robot ideally move the end-effector along the desired path. However, the robot may not move along the desired path because of external disturbances or because the system is not modeled. Therefore, here we would like the robot control technique to minimize or eliminate the path movement deviation.

In general, robot control techniques can be categorized into two main types: open-loop control and closed-loop control. First, if the dynamics model is accurate and complete, and there is no "noise" or other disturbances, the robot can be controlled by the control system after the desired trajectories $\boldsymbol{\Theta}_{\mathrm{d}}$, $\dot{\boldsymbol{\Theta}}_{\mathrm{d}}$, and $\ddot{\boldsymbol{\Theta}}_{\mathrm{d}}$ which are given by the trajectory generator. Let the desired path of $\boldsymbol{\Theta}_{\mathrm{d}} = \boldsymbol{\Theta}(t)$ be a function of time, then the torque required to make the robot obey the desired motion can be calculated using Eq. (9.3):

$$\boldsymbol{\tau} = \boldsymbol{M}(\boldsymbol{\Theta}_{\mathrm{d}})\ddot{\boldsymbol{\Theta}}_{\mathrm{d}} + \boldsymbol{V}(\boldsymbol{\Theta}_{\mathrm{d}}, \dot{\boldsymbol{\Theta}}_{\mathrm{d}}) + \boldsymbol{G}(\boldsymbol{\Theta}_{\mathrm{d}}) \tag{9.3}$$

The robot can work stably based on Eq. (9.3). Then, along the desired trajectory continuous torque solution equation, the effector control torque $\boldsymbol{\tau}$ can produce the desired path $\boldsymbol{\Theta}_{\mathrm{d}}$, which is the open-loop control algorithm, that is, based on a known desired path and the

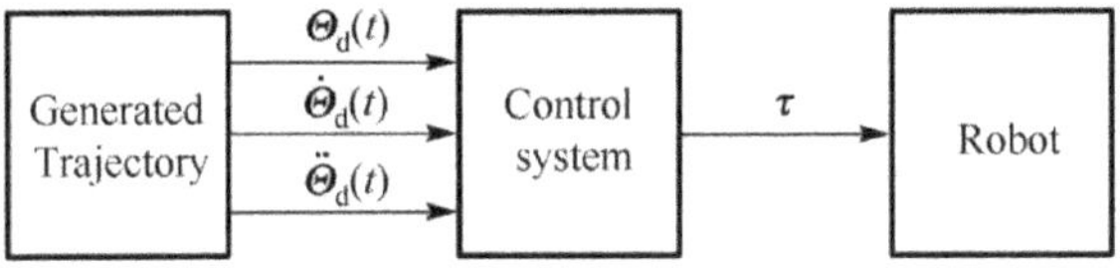

Figure 9.3. Block diagram of high-order open-loop control system for robots.

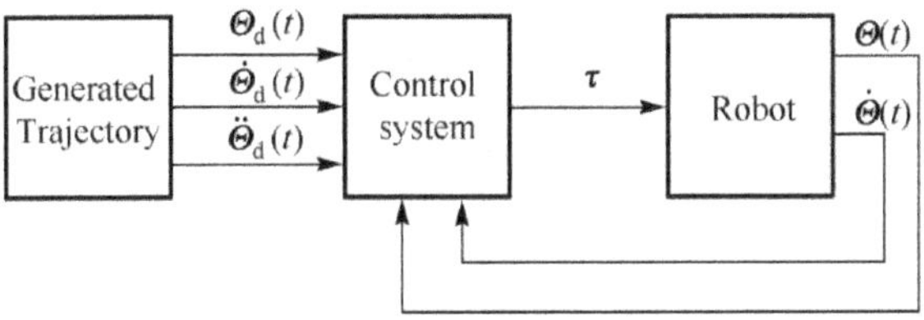

Figure 9.4. Block diagram of the robot high-order closed-loop control system.

equation of motion to compute the control commands. Finally, the control commands acting on the system will produce the desired path. The block diagram of the robot's high-order open-loop control system is shown in Figure 9.3.

However, this control approach does not utilize the feedback from the joint sensors, i.e., Eq. (9.3) is a function of the desired trajectories $\mathbf{\Theta}_\text{d}$ and $\dot{\mathbf{\Theta}}_\text{d}$ rather than the actual trajectory $\mathbf{\Theta}$. As a result, the dynamics model is incomplete, and there are unavoidable external disturbances, which makes the open-loop control approach impractical. To solve this problem, the control system needs to calculate the torque required by the effector based on the servo error function, i.e., closed-loop control is required.

Closed-loop control refers to controlling by utilizing feedback from the system. The core of its design is to ensure that the design of the closed-loop system meets specific performance requirements, and the most basic criterion is the system stability requirements, that is, when the robot faces a number of "moderate" interferences, in accordance with a variety of desired trajectories, the system always maintains "small" errors. In reality, robots can use position sensors, velocity sensors, acceleration sensors, or force sensors to directly or indirectly measure joint variables, which are converted into the joint space to provide information about the joint motion of the links with respect to the coordinate system. The relationship between the trajectory generator and the robot is given in Figure 9.4, where the robot receives a joint torque vector from the control system $\boldsymbol{\tau}$, and the sensors allow the controller to read the joint position vector and the joint velocity vector for closed-loop control.

The only method of closed-loop control for high-performance robots is to utilize feedback from the robot's sensors for control. This feedback typically calculates the servo error by comparing the difference between desired and actual positions and the difference between desired and actual velocities:

$$\boldsymbol{E} = \boldsymbol{\Theta}_{\mathrm{d}} - \boldsymbol{\Theta}, \dot{\boldsymbol{E}} = \dot{\boldsymbol{\Theta}}_{\mathrm{d}} - \dot{\boldsymbol{\Theta}} \tag{9.4}$$

This allows the control system to calculate the torque required by the driver based on the servo error function and reduce the servo error by calculating the torque of the driver. Some of the more common robot control methods are as follows.

9.2.4.1 *Feedback linearization or computerized torque control techniques*

This type of technique is a model-based control method that focuses on obtaining linear differential equations for deviation commands by defining control laws, which are then designed using linear control. Although feedback linearization techniques have been successfully applied to robots, the control laws designed are based on a nominal model of the robot. However, the method does not guarantee robustness of robot control due to parameter uncertainties or disturbances.

9.2.4.2 *Linear control techniques*

The simplest robot control technique is the design of linear controllers based on the linearization of the equations of motion at the operating point. Proportional, integral, differential, or any combination of these are commonly used linear control techniques, but this class of techniques can only locally determine the stability of the robot.

9.2.4.3 *Adaptive control techniques*

Adaptive control technique is a technique used to control uncertain or time-varying robots and is usually more effective for low-degree-of-freedom robots.

9.2.4.4 *Robust adaptive control techniques*

This class of control methods is designed based on a nominal model plus some uncertainty, where the uncertainty can appear in any parameter, such as the end-effector load.

9.2.4.5 *Gain scheduling control techniques*

This class of control methods applies linear control techniques to robot nonlinear dynamics. In gain scheduling, a large number of control points are usually selected to cover the robot operation range, and then for each control point, a linear controller is designed by making a linear time-varying approximation of the robot dynamics, and finally the controller parameters are inserted or scheduled among the control points.

9.3 Integrated Design of Robot Control

To enhance readers' comprehension of the robot control process, this section leverages the robot platform introduced in Chapter 2 as a case study. It elucidates the integrated design process of the robot through three dimensions: the design of the drive space control system, the joint space control system, and the workspace control system. The robot platform comprises a humanoid dual-arm system, a 3-RPS parallel system, an omnidirectional mobile chassis, and a sensing system. Meanwhile, its control system includes a host computer, an embedded control system, a motor driver module, DC motors, digital servos, sensors, and additional components. To facilitate omnidirectional movement, the chassis is equipped with four McNamee wheels. Our in-house developed humanoid robot is depicted in Figure 9.5.

In order to achieve the purpose of controlling the robot, after equipping the robot platform with various hardware devices, it is also necessary to design the corresponding control system and software system to cooperate with this hardware system to realize robot control. Taking this robot platform as an example, the control system of the robot is designed based on the embedded system, and its overall design framework is shown in Figure 9.6.

As shown in Figure 9.6, the control system design framework incorporates two distinct power supply specifications. The 24 V

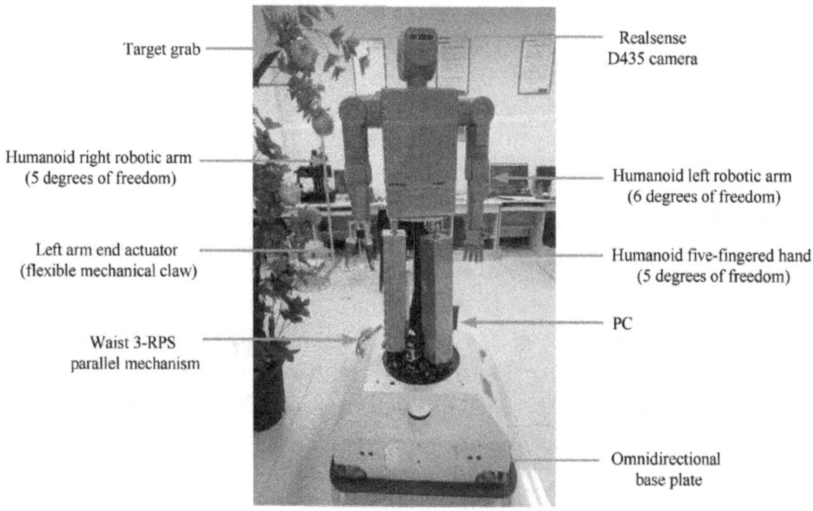

Figure 9.5. In-house developed humanoid robot at the laboratory.

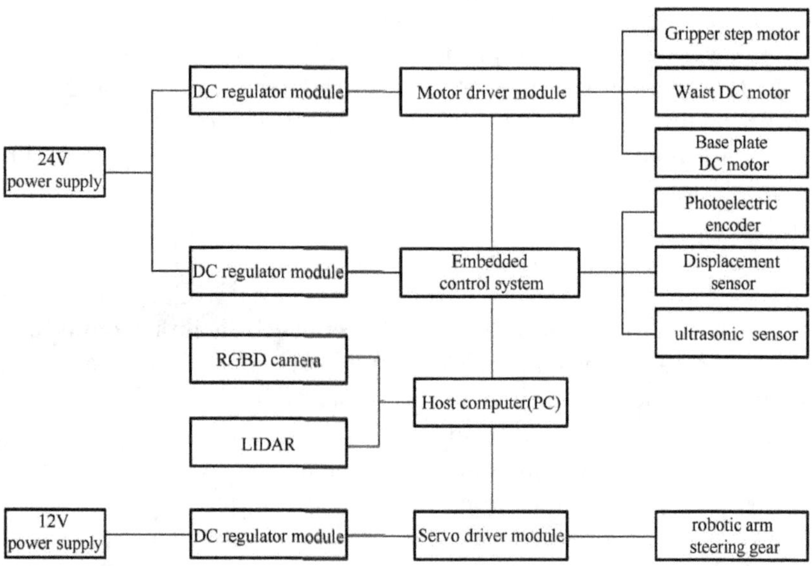

Figure 9.6. Control system design framework for the robot platform.

power supply primarily powers the motor driver module and the embedded control system, as well as the corresponding motors and sensors via the DC regulator module. Conversely, the 12 V power supply is dedicated to powering the digital servo of the operating arm and the drive control module, also through the DC regulator module. The DC regulator module's role is to adjust the power supply voltage to levels suitable for other modules. The motor driver module is responsible for receiving commands from the embedded control system and managing motor operations. Moreover, the system's host computer, a PC, sends commands to the embedded control system and the servo driver module through serial communication. Additionally, since the PC is equipped with its own power supply, it is directly connected to the RGB_D camera and the LIDAR via a USB 3.0 interface. This connection not only powers these devices directly but also simplifies the reception and processing of image and laser point cloud data.

The following sections introduce the integrated design of robot control, elaborating on the realization of the robot's anticipated goals and functionalities.

9.3.1 *Integrated design of robot drive spatial control*

The purpose of driver spatial control is to accomplish some kind of driver action by controlling the driver and through stable communication between the driver and the PC. Among them, the drivers of the robot platform can be categorized into three main types, which are the servo drivers used for the main and auxiliary operating arms and heads, the stepper motor drivers used for the robot's waist, foot, and end-effector, etc., and the DC motor drivers equipped for the omnidirectional mobile chassis. This section describes the control design for each of these three driver spaces.

9.3.1.1 *Spatial control of the robot's main and auxiliary operating arms and head servo driver*

Servos are commonly used in the rotary joints of robots, particularly in smaller models. Although servos from different brands and models may appear similar in shape, their output axes tend to be offset due to the internal gear sets' mounting configuration. A servo

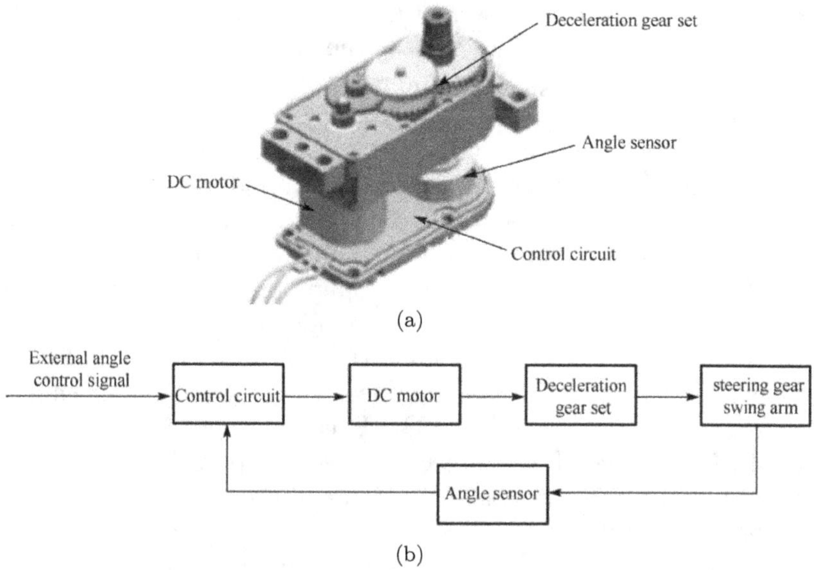

Figure 9.7. Robot servo driver. (a) The core device of common steering gears and (b) The working principle flow of the steering machine.

functions as an automatic control system that incorporates angular feedback, as illustrated in Figure 9.7(a). Beyond its casing, a servo's essential components include a DC motor, a gear reduction set, an angle sensor, and a control circuit. The operational process of a servo is depicted in Figure 9.7(b). Initially, the control circuit receives an external angle control signal and determines the rotation direction for the DC motor. Following this, the motor rotates, transmitting power to the servo's pendulum arm via the gear reduction set. Subsequently, the angle sensor measures the current angle and relays this information back to the control circuit. The circuit then assesses whether the target angle has been achieved based on the external angle control signal. If the desired angle is reached, the circuit halts the servo's rotation. Otherwise, the rotation continues until the specified angle is attained.

The robot platform mainly senses the external environment information through various types of sensors. In order to better perceive the environment and its own situation, the robot platform is equipped with a variety of sensors as its perception system, including visual sensors, displacement sensors, photoelectric encoders, and LIDAR.

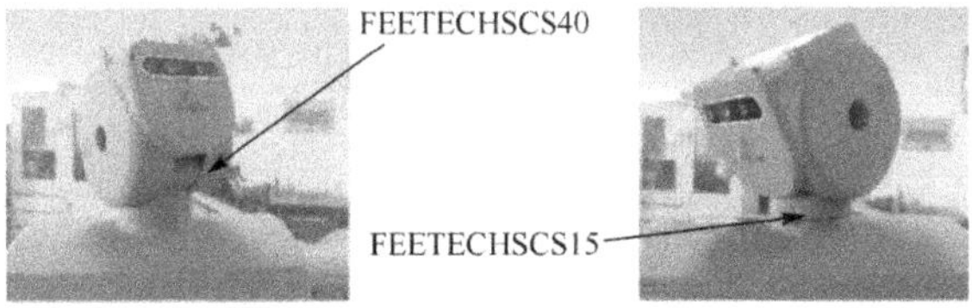

Figure 9.8. Robot servo driver.

In order to accurately acquire the position of the target and complete the grasping, the head of the robot platform is installed with a camera as a visual sensor to collect the color image information and depth image information from the outside world; at the same time, two servos are installed in its head as shown in Figure 9.8, so as to make its head have two degrees of freedom of left-right and up-down rotation. The angular range of up and down rotation is $\pm 30^\circ$, while the angular range of left and right head rotation is $\pm 90^\circ$.

In addition, in order to increase the working space of the robot, the main and auxiliary working arms and head joints of the robot platform are equipped with a total of 13 degrees of freedom, and each joint needs motion control. However, because the overall structure of the robot is very compact and the internal space is small, we also use digital servos as the power output of each joint; due to the different output moments required by different joints of the operating arm, here we choose a variety of servos as the joint power unit to meet the power required by the robot joints and, at the same time, reduce the waste of the joint power problem and to improve the endurance of the robot. The robot platform consists of the serial bus intelligent servo shown in Figure 2.7 and the ARM 32-bit microcontroller main control board shown in Figure 2.8, and the position sensing adopts the 360° 12-bit precision magnet sensing angle program, as well as the RS485/CAN mode communication level with strong anti-interference ability. In the actual control, all the servos can be connected in series, the ID number for each servo can be set, and a data line can be used to communicate with each servo in both directions through serial communication, and while controlling the servos, we can also obtain the servo's position, temperature, speed, and voltage feedback information, so as to realize the stable control of the robot system.

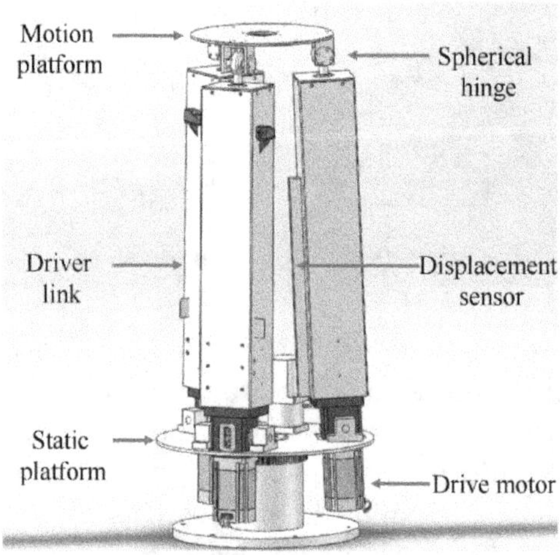

Figure 9.9. 3-RPS parallel structure used in the waist of a robot.

9.3.1.2 *Spatial control of stepper motor drivers for robot waist, foot and end-effector*

A stepping motor is an effector that converts electrical impulses into angular displacements characterized by simple and reliable control and low cost, and is mainly used in open-loop position control systems. It is mainly used in the robot platform for joint parts such as the waist, foot, and claw of the robot. Take the waist of the robot platform as an example, as shown in Figure 9.9, which adopts the 3-RPS parallel structure, the bottom of the driver link is fixed to the static platform, and the top of the link is fixed to the moving platform through the ball–hinge joint. The three driver links are 120° symmetrically distributed, with an internal ball screw nut structure, and the length of the links can be controlled by the drive motor. This design allows the 3-RPS parallel mechanism to be viewed as a tandem branch with three degrees of freedom (two rotation and one movement), so that the three tandem branches of the 3-RPS parallel mechanism can be controlled by stepper motors, thus realizing the robot's forward and backward bending, left and right bending, and lifting and lowering functions.

There are three types of stepper motors: permanent magnet (PM) type, variable reluctance (VR) type, and hybrid (HB) type. Among them, the PM-type stepper motor rotor is a permanent magnet and the stator is a winding, which rotates under the repulsive and attractive forces between the stator electromagnet and the rotor permanent magnet, with a step angle of generally 7.5°–90°. The VR-type stepper motor uses a gear-like core as the rotor, and the stator is an electromagnet. In the stator magnetic field, the rotor always turns to the position with the smallest magnetic resistance, and the step angle is generally 0.9° $\sim$ 15°, and the VR-type stepping motor is basically obsolete. The HB-type stepping motor is a composite form of PM-type and VR-type stepping motors, which has a number of axial grooves machined on the surfaces of the rotor permanent magnets and the stator electromagnet. Its torque-generating principle is the same as that of the PM-type stepping motor, and the shapes of rotor and stator are similar to those of the VR-type stepping motor, and the step angle is 0.5–15. The shape of the rotor and stator is similar to that of the VR stepper motor, and the step angle is generally 0.9°–15°. The HB stepper motor combines the advantages of the PM and VR stepper motors, and it is most widely used at present.

Because stepper motors can directly receive digital signals, they are widely used in digital control systems. When the stepper driver receives a pulse signal, it drives the stepper motor to rotate a fixed angle (stepping angle) in the set direction, so the number of pulses can be controlled by controlling the motor to control the amount of angular displacement, so as to achieve accurate positioning; at the same time, it can be controlled by controlling the pulse frequency to control the motor rotation speed and acceleration, so as to achieve the purpose of speed regulation. Simple control circuits for stepper motors can be implemented through various logic circuits, such as pulse distributors composed of gate circuits and flip-flops. This control method uses hardware, and once determined, it is difficult to change the control scheme. To change the control function of the system generally, we have to redesign the hardware circuit, making it less flexible. And in recent years, with the rapid development of software technology, microcomputers as the core of the computer control system for the control of stepper motors have opened up a new way, so the use of computer software or a combination of software and hardware methods has become the second choice, which also greatly enhances the function of the stepper motor system and improves the flexibility and reliability of the system.

9.3.1.3 *Spatial control of DC motors for robot omnidirectional moving chassis*

DC motor consists of two parts: motor body and driver. When operational, it mainly converts DC into AC in the armature winding through commutator, so as to make the armature produce electromagnetic torque in a constant direction. DC motors can be continuously variable, are convenient, and provide a wide range of speed regulation, good low-speed performance, smooth operation, and easy control of torque and speed. In robot platforms, DC motors are mainly used in the motion control of the robot's omnidirectional moving chassis, thus ensuring that the robot can work autonomously in indoor environments flexibly, stably, and with high precision.

The robot's omnidirectional moving chassis employs McNamee wheels, as depicted in Figure 9.10(a), serving as the driving wheels. To achieve precise control over the robot's movement, the chassis was modeled and kinematically analyzed, as illustrated in Figures 9.10(b) and 9.10(c). The blue line in Figure 9.10(b) signifies the drum roller, with its orientation indicating the mounting direction of the drum roller. The solid red line in Figure 9.10(c) represents the friction generated between the drum roller and the ground during the chassis's movement, while the dashed line illustrates the force decomposition. Viewing the cart as a rigid body enables the decomposition of its motion into three linear components: y-axis translation (forward and backward), x-axis translation (sideways),

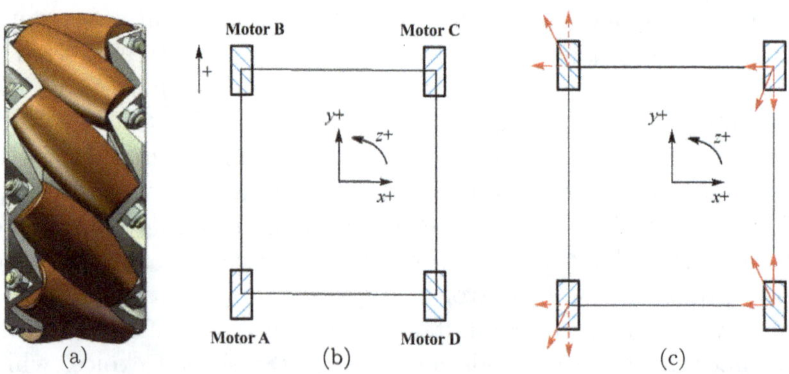

Figure 9.10. Robot chassis driver. (a) McNamee wheels, (b) Trolley physical model and (c) Chassis shift left.

and z-axis rotation (spin). Consequently, each wheel's motor module operates independently, allowing the four wheels' independent drivers to facilitate the robot's omnidirectional movement and to feedback the encoder values. Communication is achieved through the CAN bus, to which a USB or WIFI module can be connected, enabling the host computer to send and receive data. Four servo drive sub-modules are mounted independently on the CAN bus, incorporating the embedded STM32F405 chip as the primary control chip, motors, corresponding encoders, and servo drivers to create a closed-loop control system. The omnidirectional driver module of the robot chassis is presented in Figure 9.11.

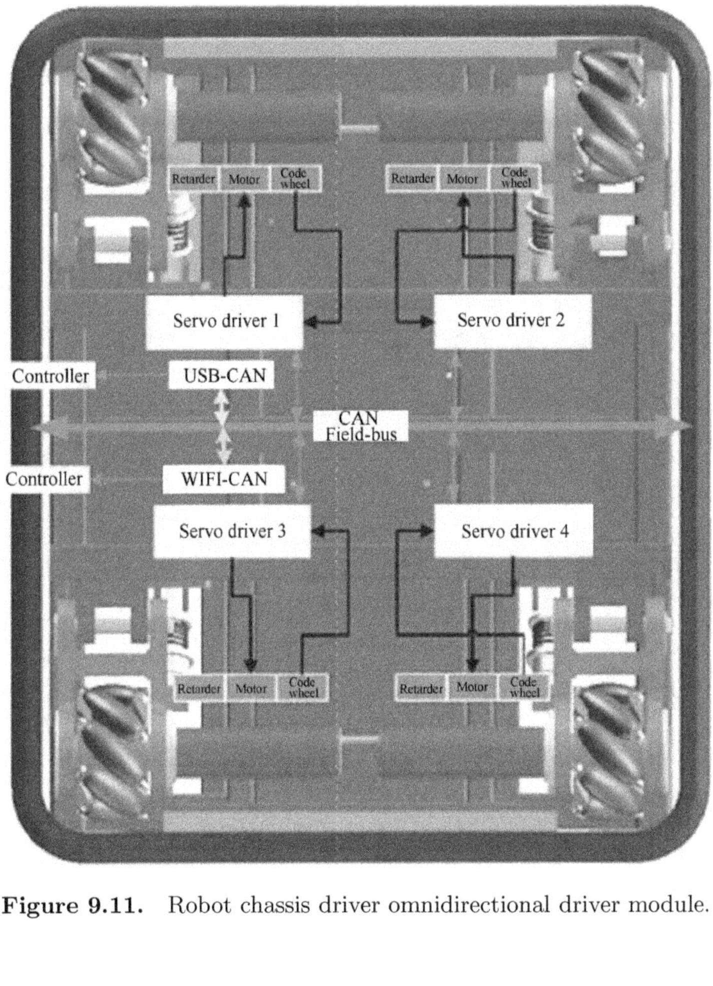

Figure 9.11. Robot chassis driver omnidirectional driver module.

Assuming that the length of the robot chassis is a and its width is b, the rotational speeds of the four McNamee wheels are v_{A}, v_{B}, v_{C}, and v_{D}, and the rotational speeds of the four McNamee wheels can be solved for based on the information about the motion of the cart on the plane:

$$\begin{cases} v_{\mathrm{A}} = -v_x + v_y - \omega\left(\dfrac{a+b}{2}\right) \\ v_{\mathrm{B}} = +v_x + v_y - \omega\left(\dfrac{a+b}{2}\right) \\ v_{\mathrm{C}} = -v_x + v_y + \omega\left(\dfrac{a+b}{2}\right) \\ v_{\mathrm{D}} = +v_x + v_y + \omega\left(\dfrac{a+b}{2}\right) \end{cases} \tag{9.5}$$

The target speed of the motor is sent to the underlying controller for control, as shown in Figure 9.12.

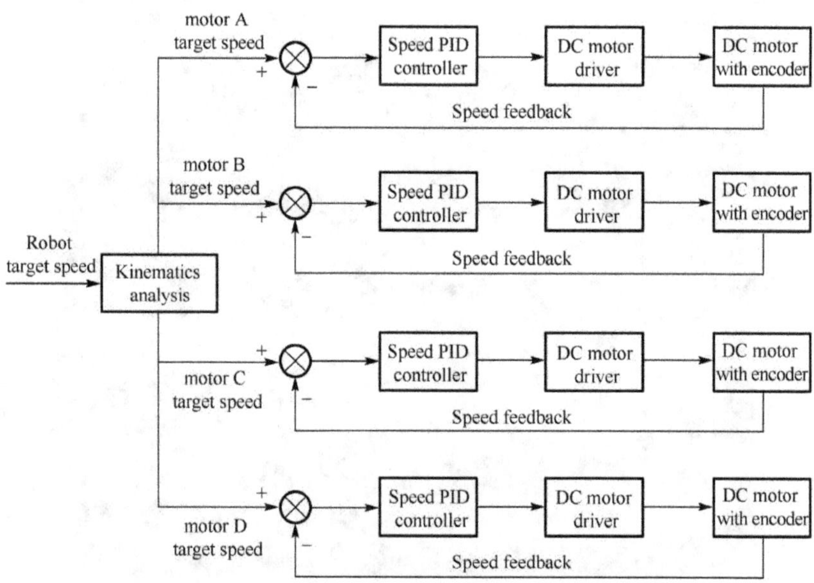

Figure 9.12. Robot chassis driver speed control.

In the figure, motors A–D represent the four MD60 planetary DC geared motors controlling four McNamee wheels in the robot chassis, and each DC motor is equipped with a 500-line photoelectric encoder. In the whole speed control process of the robot system, the host computer sends the robot target speed instruction to the embedded system; the control system analyzes the kinematics and decomposes the instruction into the target speed instruction of each motor; the instruction is sent to the DC motor driver through the speed proportional-integral-derivative (PID) controller. Using the photoelectric encoder to detect the motor angle, angular speed, and other related information, and according to the forward kinematics, the position and speed of the cart are obtained and then sent to the host computer through the serial port or wireless module.

9.3.1.4 *PC control and communication of robot driver*

In order to realize stable and efficient communication between the PC and each motor driver, and to send or read each motor and servo information in time, combining the communication characteristics of each control part, two schemes are used to design the robot motor and servo communication scheme. The PC communication model of the robot driver is shown in Figure 9.13. First of all, for the robot dual-arm and head servo communication needs, a direct connection scheme between the signal expansion board and the servo

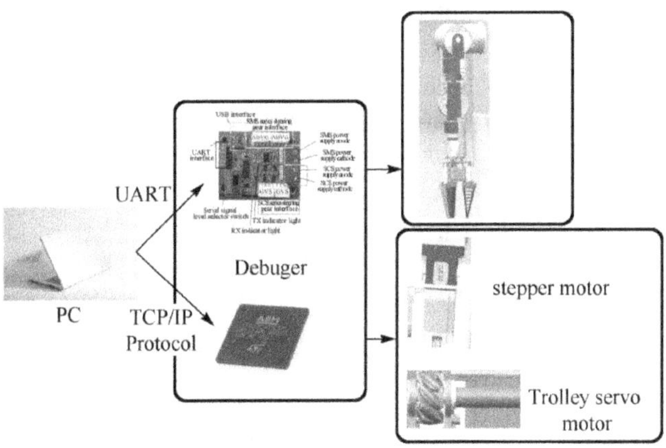

Figure 9.13. PC communication model for robot driver.

control board is adopted to reduce redundant signal lines and minimize wiring costs. Using a serial bus program for the servo and host computer communication protocol, the PC or controller sends a command packet, and the servo returns an answer packet. Multiple servos are allowed in a communication network, and each servo is assigned a unique ID number. The control command contains ID information, and only the servos matching the ID number can receive the command in full and return the answer message. A frame of control instruction data is divided into 1 start bit, 8 data bits, 1 stop bit, and no parity bit, making a total of 10 bits. After testing, the communication frequency can be up to 38400 bit/s – 1 Mbit/s, which meets the communication requirements.

Secondly, the embedded main control board is designed for the robot waist, foot, end-effector stepping motors, and cart servo motors. All stepper motors and servo motors control and communication interactions with the host computer occur through the embedded main control board. In order to reduce the line requirements, the communication between the main control board and the host computer (or PC) adopts the TCP/IP protocol. With the stable and reliable communication architecture of the TCP/IP protocol, the host computer can communicate stably with the embedded main control board connected to each stepping motor and cart servo motor through the underlying socket.

To realize the control system of the robot platform, the STM32 series controller shown in Figure 9.14 is used as the control unit chip for the robot embedded control system. The chip has a high operating frequency, built-in high-speed memory, and multiple USART, USB, and CAN interfaces. Meanwhile, the STM32 series controllers are widely used, easy to program, and can conveniently communicate with the host computer through the UART serial port.

As for the motor driver module of the robot platform, the dual-channel isolated DC motor driver shown in Figure 9.15 is used. Each interface of this motor driver module supports three-wire control of enable, forward and reverse rotation. and braking. The enable signal can be externally connected to PWM, and the forward and reverse rotation control signals can be connected in series with limit switches, which is characterized by a small size, a wide range of motor voltages, and a dual-circuit motor interface, which is well suited to the operating characteristics of the robot platform.

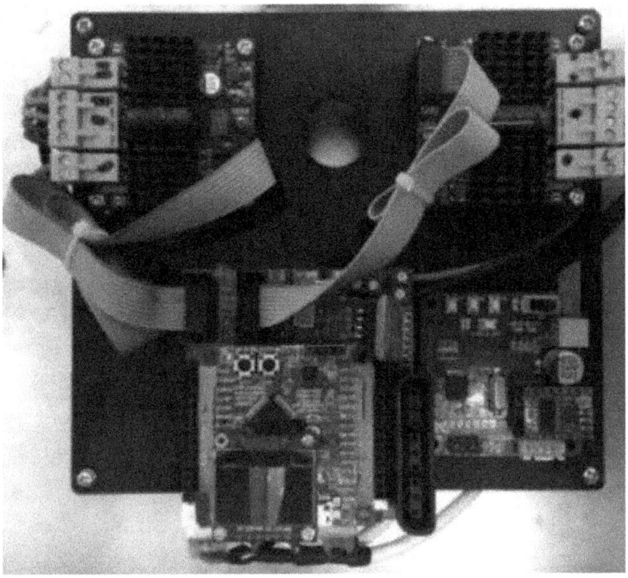

Figure 9.14. Controller mounted on the robot platform chassis.

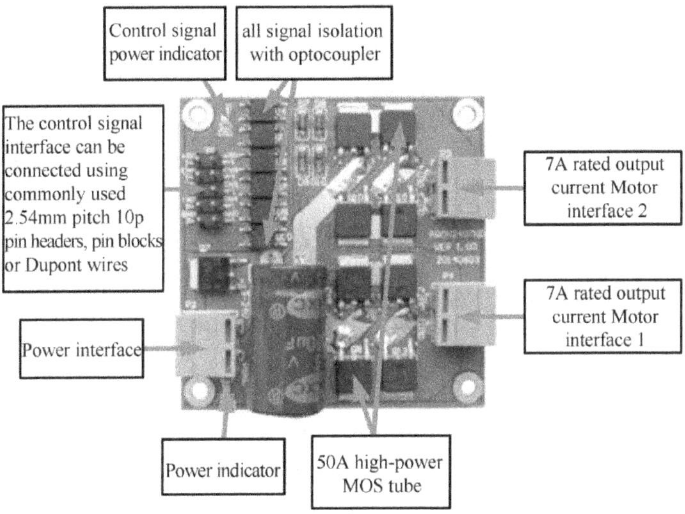

Figure 9.15. Motor driver module used in robot operation.

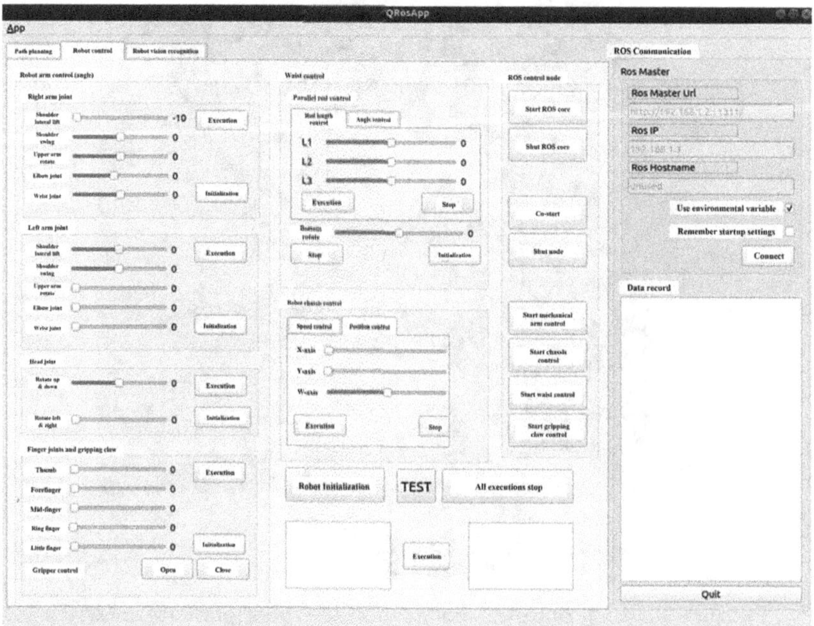

Figure 9.16. Local software interaction interface of the robot.

To enhance the intuitiveness and convenience of controlling the designed humanoid robot, we developed the human–computer interaction software using Qt, based on the host computer's control program, and implemented local control of the robot. This book covers the robot platform design, completed on the Ubuntu 18.04 system using the Qt 5.12 development framework, specifically focusing on the design of the human–computer interface. The "ROS Communication" section, illustrated in Figure 9.16, allows users to configure relevant information independently and manage the connection and disconnection between the software and the ROS node manager. Additionally, the interface's "Data Logging" section records user operation commands and provides feedback on control results. The "ROS Control Node" section manages the start and stop of each ROS node on the host computer. The "Waist Control" section is dedicated to adjusting the link length/angle of the 3-RPS parallel system. The "Robot Chassis Control" section governs the robot chassis's speed/position, and the "Robot Arm Control (Angle)" section manages the angles of each joint in the robot arm.

9.3.2 *Integrated design joint space control of robot*

The execution of robot joint motions is performed by motors/servos that are capable of realizing the desired motion of the system, and the most commonly used motors/servos in robotics applications are servo motors. Among servo motors, brushless DC servo motors are widely used due to their high control flexibility and low losses. Unlike permanent magnet DC servo motors, brushless DC servo motors consist of a rotor, stator, and static rectifier. The rotor of a brushless DC servo motor is the rotating coil that generates the magnetic flux and is a permanent magnet made of magnetic ceramic or rare earth. The stator of the brushless DC servomotor is a fixed armature made of polyphase coils. The static rectifier generates a feeding sequence for the armature coil phase based on the signal provided by the position sensor on the motor shaft. In a brushless DC servomotor, the rotor position sensor allows the coil orthogonal to the magnetic field to be found, and then the magnetic field acts on this coil to produce rotation of the rotor. After the rotation of the rotor, the electronic control module causes the magnetic field to interact with the windings of the stator in turn, in such a way that the armature magnetic field is always orthogonal to the stator magnetic field. The components of the joint actuation system are shown in Figure 9.17.

In terms of global input/output relationships, P_{c} characterizes the (usually electrical) power associated with the control law signals, while P_{u} characterizes the mechanical power required by the joints to drive the motion. The intermediate links characterize the power supplied to the engine (electric, hydraulic, or pneumatic) P_{a}, the power P_{p} supplied by the primary energy source with the same physical properties as P_{a}, and the mechanical power generated by

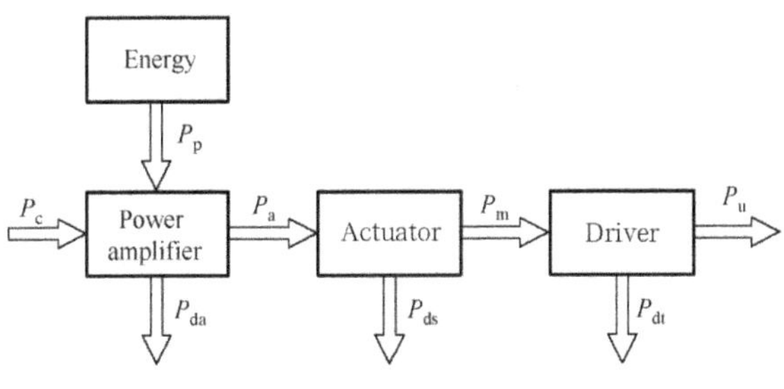

Figure 9.17. Components of the joint execution system.

the engine P_{m}. In addition, P_{da}, P_{ds}, and P_{dt} denote the power losses consumed by the amplifiers, the engine, and the transmission, respectively, for the execution of the transformation. In order to select the composition of the effector system, it is valuable to start from the forces and velocities required to characterize the joint movements by the composition of the mechanical power P_{u}.

The power amplifier has the task of regulating, in the presence of a control signal, the power flow provided by the primary energy source and which must be transmitted to the effector to perform the desired movement. In other words, the amplifier obtains from the energy source a portion of the available power proportional to the control signal and then transmits this power to the engine according to the appropriate force and flow. The input to the amplifier is the power obtained from the primary energy source P_{p}, which is associated with the control signal P_{c}. Part of the total power is transmitted to the effector (P_{a}) and part is dissipated (P_{da}). For brushless DC servomotors, the voltage (current) is AC, and the voltage and frequency values are determined by the control signals from the amplifier to make the motor perform the desired motion. For the power ranges (kilowatts) typically required for articulated motions, transistorized amplifiers that are properly switched by pulse width modulation (PWM) techniques are used. They enable a power conversion ratio $P_{\mathrm{a}}/(P_{\mathrm{p}}+P_{\mathrm{c}})$ of more than 0.9, as well as a power gain $P_{\mathrm{a}}/P_{\mathrm{c}}$ up to the 10^6 level. In this case, a DC–AC converter (inverter) is used to control the brushless DC servo motor. The task of the energy source is to provide the amplifier with one level of power, which is necessary for the operation of the actuating system. For servo motors, the power supply usually consists of a transformer and a typical uncontrolled bridge rectifier. They allow the conversion of the alternating current supplied by the distributor into a properly sized direct current that needs to be fed into the power amplifier.

From a modeling point of view, a brushless DC servo motor configured with a position sensor and a permanent magnet DC servo motor configured with a commutator can be described by the same differential equations, with the armature circuit shown in Figure 9.18. In the s-complex domain, the voltage equation is

$$V_{\mathrm{a}} = (R_{\mathrm{a}} + sL_{\mathrm{a}})I_{\mathrm{a}} + V_{\mathrm{g}} \tag{9.6}$$

$$V_{\mathrm{g}} = k_{\mathrm{v}}\Omega_{\mathrm{m}} \tag{9.7}$$

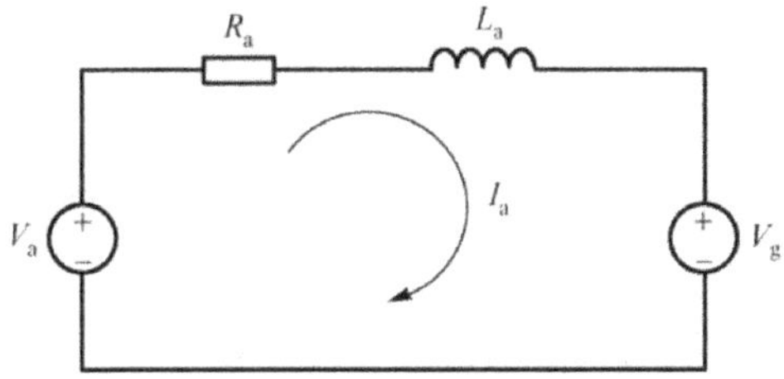

Figure 9.18. Armature circuit.

where V_a and I_a denote armature voltage and current, respectively; R_a and L_a denote armature resistance and inductance coefficient, respectively; V_g denotes the inverse electromotive force, which is proportional to the angular velocity Ω_m through the voltage constant k_v; and k_v is determined by the specific structure of the motor and the coil magnetic flux.

The mechanical balance of a brushless DC servo motor is described by the following equations:

$$\boldsymbol{C}_m = (sI_m + F_m)\Omega_m + \boldsymbol{C}_1 \tag{9.8}$$

$$\boldsymbol{C}_m = \boldsymbol{k}_t I_a \tag{9.9}$$

where $\boldsymbol{C}_m$ and $\boldsymbol{C}_1$ denote the drive torque and load torque, respectively; I_m and F_m denote the moment of inertia and the coefficient of viscous friction on the motor shaft, respectively; and the torque constant $\boldsymbol{k}_t$ is equal to the value of k_v for the compensating motor in standard international units.

In order to quantitatively describe the impact of using a transmission between a servomotor and a driven joint, reference can be made to a mechanical gear pair formed by a pair of cylindrical gears with radii r_m and r. The mechanical gears are shown schematically in Figure 9.19. It is assumed that the mechanical gear pair is ideal (no backlash) and connects the servo motor rotation axis to the corresponding joint axis.

For the servo motor, it is assumed that its rotor is characterized by the moment of inertia I_m and the coefficient of viscous friction F_m around its axis of rotation; I and F denote the moment of inertia and the coefficient of viscous friction of the load, respectively. It is assumed that the moment of inertia of the gear and the coefficient of viscous friction are included in the corresponding parameters of the

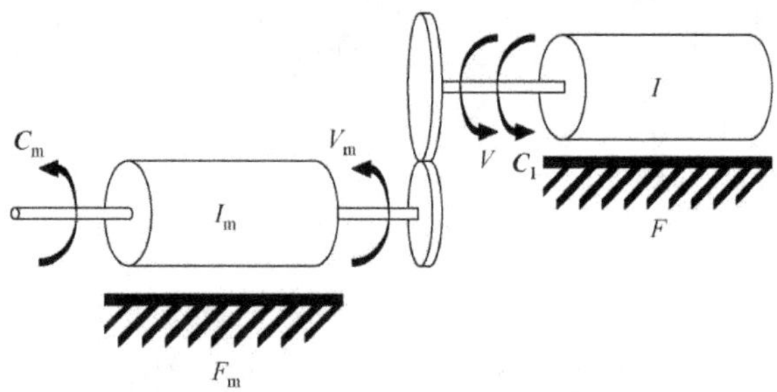

Figure 9.19. Mechanical gear schematic diagram.

servo motor (gear with radius r_m) and the load (gear with radius r). Let $\boldsymbol{C}_m$ denote the drive torque of the servo motor and $\boldsymbol{C}_1$ denote the load torque applied to the load. Similarly, let ω_m and θ_m denote the angular velocity and position of the servo motor shaft, respectively, while ω and θ denote the corresponding quantities of the load. Finally, $\boldsymbol{f}$ denotes the exchange force when the teeth of the two gears bite.

The gear reduction ratio is defined as

$$k_r = \frac{r}{r_m} = \frac{\theta_m}{\theta} = \frac{\omega_m}{\omega} \tag{9.10}$$

Because there is no slippage in the kinematic pair, we have

$$r_m\theta_m = r\theta \tag{9.11}$$

For the case of robotic servomotors coupled to joints, the gear reduction ratio has a value much larger than 1 ($r_m \ll r$), with typical values ranging from tens to hundreds. The exchanged force $\boldsymbol{f}$ between the two gears generates a load moment on the servomotor shaft $\boldsymbol{f}r_m$ and a driving moment on the rotational motion of the load $\boldsymbol{f}r$. The mechanical equilibrium of the servo motor and the load are, respectively,

$$\boldsymbol{C}_m = I_m\dot{\omega} + F_m\omega_m + \boldsymbol{f}r_m \tag{9.12}$$

$$\boldsymbol{f}r = I\dot{\omega} + F\omega + \boldsymbol{C}_1 \tag{9.13}$$

In order to describe the motion about the angular velocity of the servo motor, consider Eq. (9.10) and combine Eqs. (9.12) and (9.13) to give the equations of mechanical equilibrium of the servomotor:

$$\boldsymbol{C}_{\mathrm{m}} = I_{\mathrm{eq}}\dot{\omega}_{\mathrm{m}} + F_{\mathrm{eq}}\omega_{\mathrm{m}} + \frac{\boldsymbol{C}_1}{k_{\mathrm{r}}} \tag{9.14}$$

where

$$I_{\mathrm{eq}} = I_{\mathrm{m}} + \frac{I}{k_{\mathrm{r}}^2} \quad F_{\mathrm{eq}} = F_{\mathrm{m}} + \frac{F}{k_{\mathrm{r}}^2} \tag{9.15}$$

In the case of gears with large reduction ratios, Eqs. (9.14) and (9.15) describe how the rotational moment of inertia and viscous friction coefficient of the load are reflected in the servo motor shaft by the deceleration factor $1/k_{\mathrm{r}}^2$. On the contrary, the load torque decreases by a factor of $1/k_{\mathrm{r}}$. If this torque depends on θ in a nonlinear manner, a high reduction ratio will tend to linearize the dynamic equation.

Disregarding the specific mechanical type of the manipulator, the control action (generalized force of the joint effectors) is realized in the joint space. In such a practical context, a generalized general scheme for joint space control is designed, as shown in Figure 9.20. The scheme's control structure is a closed loop, which can take advantage of all the benefits of feedback, i.e., robustness under model uncertainty, and thus reduce the effects of disturbances. Overall, the joint-space problem is actually quite clear. First, the robot kinematic inverse solution transforms the motion demand x_{d} from the workspace to the corresponding motion θ_{d} in the joint space. Then, the joint space control scheme is designed to track the reference input with the real motion θ. However the disadvantage of this approach is that the joint space control scheme does not affect the workspace variables x_{e} and x_{e} controlled in open-loop form through the mechanical structure of the robot. It is clear that any uncertainty in the structure (e.g., structural tolerances, calibration misses, gear gaps, and elasticity) or any inaccuracy in the information of the end-effector

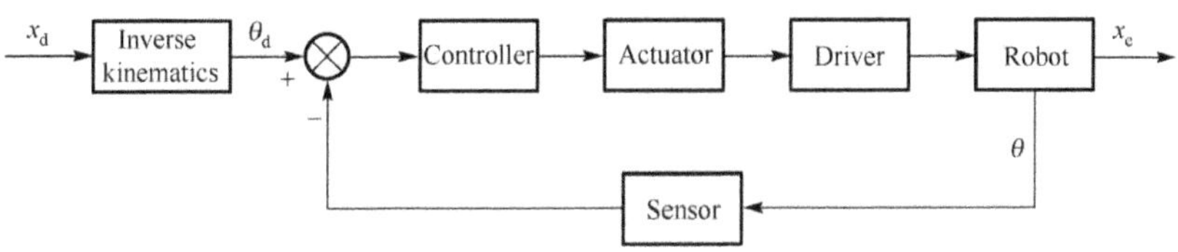

Figure 9.20. General scheme of joint space control.

with respect to the target pose will cause a reduction in the accuracy of the workspace variables.

Because the multi-joint control of the robot platform is a multi-input multi-output (MIMO) problem, modeling is more difficult and needs to use vectors to represent the joint position, velocity, and acceleration; therefore, in engineering practice, the assumption of nonlinear control or model-based control methods is not used. Here, we use the PID controller as a joint space control controller to solve the problem. A PID controller is a feedback control system according to the proportional (P), integral (I), and derivative (D) law regulator, referred to as the PID regulator:

$$P(t) = K_{\mathrm{P}} \left\{ e(t) + \frac{1}{T_{\mathrm{I}}} \int e(t)\mathrm{d}t + T_{\mathrm{D}} \frac{\mathrm{d}e(t)}{\mathrm{d}t} \right\} \tag{9.16}$$

PID parameter tuning provides a reasonable choice of the three parameters of PID. From the stability of the system, response speed, overshoot and steady state error, and other aspects of the problem, the roles of the three PID parameters are as follows.

① The function of the proportional parameter K_{P} is to accelerate the response speed of the system and improve the regulation accuracy of the system. When the value of K_{P} is increased, the response speed of the system is accelerated, the regulation accuracy of the system is increased, but the system is prone to overshooting, and the stability of the system deteriorates. When the value of K_{P} decreases, the response speed of the system slows down, the regulation accuracy of the system decreases, the regulation time becomes longer, and the dynamic and static performance of the system deteriorates.

② One of the main functions of the integral parameter T_{I} is to eliminate the steady-state error of the system. The larger the value of T_{I}, the faster the elimination of the steady-state error of the system, but T_{I} should not be too large, otherwise the integral windup phenomenon will occur in the early stages of the response process. If T_{I} is too small, the steady-state error of the system will be difficult to eliminate, which will affect the regulation accuracy of the system. In addition, as long as there is an integral link in the forward channel of the control system, it is always possible to achieve a steady state without static difference. From a phase point of view, one integral link has a 90° phase delay, perhaps destabilizing the system.

③ The role of the derivative action parameter T_{D} is to improve the dynamic performance of the system, and its main function is to suppress changes in deviation in any direction during the response process and to forecast changes in deviation in advance. However, T_{I} cannot be too large, otherwise it will make the response process brake in advance, prolong the regulation time, and reduce the anti-interference performance of the system. The significance of the three PID parameters is described above, and two PID control algorithms for practical applications are described as follows.

9.3.2.1 *Digital PID position type control algorithm*

$$\int_0^n e(t)\mathrm{d}t = \sum_{j=0}^{n} E(j)\Delta t = T\sum_{j=0}^{n} E(j) \tag{9.17}$$

$$\frac{\mathrm{d}e(t)}{\mathrm{d}t} \approx \frac{E(k)-E(k-1)}{\Delta t} = \frac{E(k)-E(k-1)}{T} \tag{9.18}$$

$$P(k) = K_{\mathrm{P}}\left\{E(k) + \frac{T}{T_{\mathrm{I}}}\sum_{j=0}^{k} E(j) + \frac{T_{\mathrm{D}}}{T}[E(k)-E(k-1)]\right\} \tag{9.19}$$

9.3.2.2 *Digital PID incremental control algorithm*

$$\begin{aligned} P(k-1) = K_{\mathrm{P}}\Bigg\{ & E(k-1) + \frac{T}{T_{\mathrm{I}}}\sum_{j=0}^{k-1} E(j) \\ & + \frac{T_{\mathrm{D}}}{T}[E(k-1)-E(k-2)]\Bigg\} \end{aligned} \tag{9.20}$$

$$\begin{aligned} P(k) = {} & P(k-1) + K_{\mathrm{P}}[E(k)-E(k-1)] + K_{\mathrm{I}}E(k) \\ & + K_{\mathrm{D}}[E(k)-2E(k-1)+E(k-2)] \end{aligned} \tag{9.21}$$

$$\begin{aligned} \Delta P(k) &= P(k) - P(k-1) \\ &= K_{\mathrm{P}}[E(k)-E(k-1)] + K_{\mathrm{I}}E(k) \\ &\quad + K_{\mathrm{D}}[E(k)-2E(k-1)+E(k-2)] \end{aligned} \tag{9.22}$$

The joint space of the working arm is controlled by designing a PID controller, which makes the working arm track the desired kinematic parameters, such as joint angle, angular velocity, and angular acceleration. This improves the joint space control accuracy and response speed and solves the problem of reduced workspace accuracy due to the inaccuracy of the mathematical model of the working arm and the uncertainty of the structure.

9.3.3 *Integrated design of robot workspace control*

In the above control schemes, it is always assumed that desired trajectories expressed as time series of joint positions, velocities, and accelerations are available so that the error of the control scheme can be expressed in joint space. Usually, the motion is specified in the workspace, and the workspace covariates are converted to the corresponding joint space covariates using a kinematic inverse solution algorithm. In addition to the inverse solution of the forward kinematics, first- and second-order differential kinematics inverse solutions are required to convert the desired end-effector positions, velocities, and accelerations to the corresponding joint-level quantities, thus the kinematic inverse solution process increases the computational burden. For this reason, current robot control systems compute joint positions by kinematic inverse solution and then compute velocities and accelerations by numerical differentiation. The control scheme done directly in the joint space is a different approach from this. If the motion is described in the form of workspace variables, the measured joint space variables can be transformed into the corresponding workspace variables according to the positive kinematic relations. Comparing the desired inputs with the reconstructed variables completes the feedback loop design, where the inverse solution of the trajectory is replaced by a suitable coordinate transformation in the feedback loop. Because of the many feedback loop calculations that need to be accomplished (which somehow represent the kinematic inverse solution function), all workspace control schemes impose a considerable computational demand. The burden of the computational requirements relative to the numerical implementation comes mainly from the sampling time, which may cause a degradation of the overall control system performance.

In response to the above discussion, it's essential to outline the overarching framework of workspace control, as illustrated in Figure 9.21. The approach to workspace control aims for a

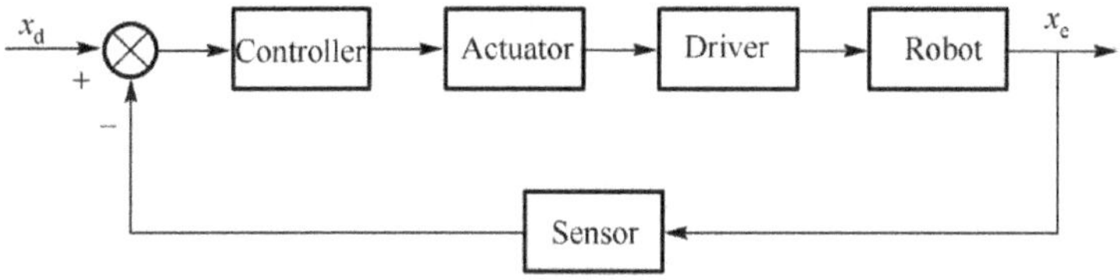

Figure 9.21. General scheme of workspace control.

comprehensive strategy, which, consequently, incurs a higher computational complexity. It's important to note that the solution to the inverse kinematics problem is now integrated into the feedback control loop. From a conceptual standpoint, workspace control offers the advantage of allowing direct intervention on workspace variables. However, this benefit is somewhat theoretical because direct measurement of workspace variables is often challenging. Instead, these measurements are typically derived from solving the kinematic equations for joint-space variables. Such approaches are crucial in scenarios involving interactions between the robot and its environment. In practice, control schemes based on joint space are sufficient for managing motion in unobstructed space. However, when the robot's end-effector encounters environmental constraints (for example, when it contacts an elastic surface), it becomes necessary to control both the position and the contact force. Under these circumstances, the workspace control scheme proves to be more suitable. Consequently, we will present some solutions that are primarily used for motion control but also lay the groundwork for a strategy that manages both force and position.

As mentioned above, the workspace control scheme is based on a direct comparison between the input values of the specified workspace trajectories and the measured values of the corresponding robot outputs. Therefore, the control system can merge some transformation actions from the workspace (where the error is defined) to the joint space (where the control generalized forces are generated). The control scheme that can be employed is called inverse Jacobi matrix control, and the inverse Jacobi matrix control block diagram is shown in Figure 9.22. In this scheme, the orientation x of the end-effector in the workspace is compared with the corresponding desired value x_d to calculate the workspace deviation Δx. Assuming that this deviation is small enough for a good control system, Δx can

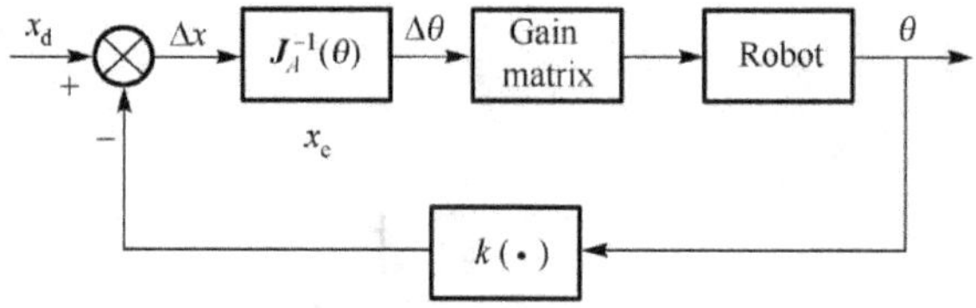

Figure 9.22. Inverse Jacobi matrix control block diagram.

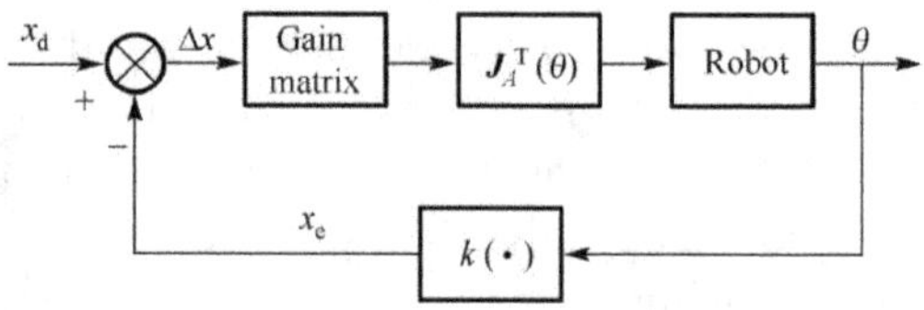

Figure 9.23. Transpose Jacobi matrix control block diagram.

be transformed into the corresponding joint-space deviation $\Delta\theta$ by means of the robot's inverse Jacobi matrix. This way, on top of the amount of the deviation, the appropriate feedback gain matrix can be calculated to generate the generalized forces which can control input. The result is a reduction of $\Delta\theta$ and the corresponding Δx. In other words, the inverse Jacobi matrix control visualizes the whole system as a mechanical system with n generalized spring units in the joint space, whose constant stiffness is determined by the feedback gain matrix. The effect of these systems is to bring the deviation to zero. If the gain matrices are of diagonal form, the generalized spring units are equivalent to n independent elastic elements corresponding to each joint.

A conceptually analogous scheme is the so-called transposed Jacobi matrix control, which is shown in the block diagram of Figure 9.23. In this case, the workspace error is first handled with a gain matrix, and the output of the box can be considered as an elastic force generated by a generalized spring unit, where the function of the elastic force in the workspace is to reduce or eliminate the positional deviation Δx. In other words, the resultant force drives the end-effector to move in the direction of the reduced Δx. This workspace force can be transformed into a generalized force in joint space by transposing the Jacobi matrix to achieve the desired response.

Both the inverse Jacobi matrix control scheme and the transposed Jacobi matrix control scheme are derived in an intuitive manner, and with these two equivalent mathematical solution methods, gravity-compensated PD control also solves the problem that these two schemes are not guaranteed to be effective for system stability and trajectory tracking accuracy.

One of the basic conditions for successfully accomplishing robot control tasks is the ability to handle the interaction between the manipulator and the external environment. A physical quantity that effectively describes the interaction is the contact force of the robot end-effector. It is usually undesirable to have a high value of the contact force because it may cause stress on both the robot end-effector and the target being worked on. Thus, the study of workspace motion control schemes during robot interaction with the external environment is important. For many practical tasks that require the robot end-effector to successfully manipulate a target or perform certain tasks on a flat surface, the control of the robot's interaction with the environment is critical. Typical examples include polishing, burr cleaning, lathe machining, and assembly. In practice, a complete categorization of the tasks that a robot may perform is not feasible because of the variability of possible situations, and such a categorization is of little practical use in finding a general strategy for achieving interactive control of the environment. In interaction, the environment imposes constraints on the geometric paths adopted by the end-effector. The task of interacting with the environment can only be successfully realized through motion control if the task is accurately planned. This sequentially requires accurate models of the robot (kinematics and dynamics) and the environment (geometric properties and mechanical features). Sufficiently accurate modeling of the robot is possible, but a detailed description of the environment is difficult to obtain.

To realize the matching of a mechanical part to a position, it should be ensured that the relative positioning accuracy of the part is one order of magnitude greater than the mechanical tolerance of the part, which is sufficient to understand the importance of the accuracy of the task planning. As long as the absolute position of a component is known exactly, the manipulator can guide the movement of other components with the same accuracy. In practice, planning errors can lead to the appearance of contact forces, which will cause the end-effector to deviate from the desired trajectory. In addition, the control system will attempt to minimize such deviations. This will eventually lead to increased contact forces until the joint effector is saturated or the contacts are damaged. The higher the environmental rigidity and position control accuracy, the more likely the scenario described above will occur. This disadvantage can be overcome if a flexible response can be guaranteed in the interaction.

It is clear from the above discussion that the contact force is the physical quantity that describes the interaction state in its most complete form. For this reason, to improve the performance of interaction control, valid force measurements need to be obtained. Interaction control schemes can be classified into two categories: indirect force control and direct force control. The main difference between the two categories is that the former accomplishes force control through motion control without a force feedback loop closure; on the contrary, the latter is able to control the contact force to the desired value with the help of a force feedback loop closure. The main types of interaction control schemes are flexure control schemes, impedance control schemes, force control schemes, and hybrid force/motion control schemes.

9.4 Integrated Practice of Robot Control

Section 9.3 delved into the comprehensive architecture of the robot's control mechanism, emphasizing the methodology employed in crafting the control system across the robot's driver space, joint space, and operational environment. Upon establishing the framework and parameters of the robot's control system, the implementation phase can commence. This section illustrates the practical application of the control system using the example of a robot platform, serving as a case study for the procedural steps involved.

9.4.1 *Integrated practice of robot drive space control*

In the discussion in the previous two sections, we have given examples of robot joint space control, robot workspace control, and the specific practice process, and we have analyzed target recognition and localization and the inverse kinematics algorithms required for the robot to complete the grasping task. In this section, we test the application of the robotic platform in experimental teaching by taking the actual robot as the experimental object. This section is divided into two parts: ① designing the robot control interface to complete the robot control experiments and verifying the complete functions of the robot; ② writing the robot C/C++ control program and verifying the various motions of the robot by using the ROS node communication method.

First of all, in order to enhance the human–computer interaction and control the robot effectively, the control interface can be designed for it. In this section, Qt 5.12 is used for the design of the robot control interface to control each joint of the robot and view the target recognition results and related images. The local control interface is shown in Figure 9.24, and the key function corresponding to each button of the interface also adopts the method of ROS node for data communication, so that the writing of the core program in the two parts of this section can be accomplished at the same time and the portability of the control program can be enhanced.

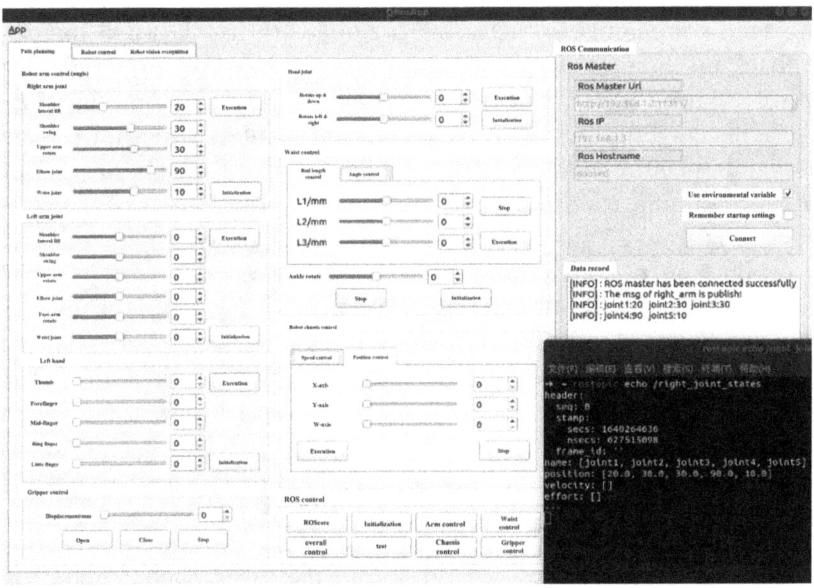

Figure 9.24. Local control interface.

Figure 9.25. Robot's execution effect on a certain control command.

In Figure 9.24, when you need to control some joints of the robot, you have to start the corresponding ROS node and set the angle value of each joint by dragging the slide link. When the "Execute" button is clicked, the local control interface will send out the relevant topic and send commands to each control node through the ROS message callback mechanism. In the terminal interface at the lower-right corner of Figure 9.24, it can be seen that the relevant topic has been issued successfully. At this time, each control node will send the corresponding command packet to each joint motor controller according to the communication protocol, thus realizing the control of the robot joints. Figure 9.25 shows the execution effect of the robot following a certain control command.

The pseudo-code of the drive control program is shown in Figure 9.26. From the pseudo-code, it can be seen that when the local control interface sends out control commands, the corresponding ROS node will continuously send out data frames that can be recognized by the driver; when the driver receives the data frames, it can obtain motion control information, such as the desired position and the desired speed, by unpacking and parsing; finally, the driver adjusts the current and other control units so that the motor/servo will be moved to the corresponding position to complete the control process in the driver's space.

<table>
<tr>
<td>1.
Driver receives frame data</td>
<td>Control parameters.ITEM ← frame data.target index
Control parameters.POS.X ← frame data.target position.X.coordinate
Control parameters.POS.Y ← frame data.target position.Y.coordinate
Control parameters.POS.Z ← frame data.target position.Z.coordinate
Control parameters.SPD.X ← frame data.target velocity.Vx
Control parameters.SPD.Y ← frame data.target velocity.Vy
Control parameters.SPD.Z ← frame data.target velocity.Vz
Control parameters.TIM ← frame data.time interal</td>
<td rowspan="2">3.
Rotational speed control for four motors of trolley</td>
<td rowspan="2">for i ← Current time to i < (Current time + time interval) by i + step time do
If instantaneous position value < (Calculated position value - delt) then
Correct motion direction
Correct function coefficients
i ← i – step time
else
Calculate x-direction position
Calculate x-direction speed
Calculate y-direction position
Calculate y-direction speed
Calculate z-direction position
Calculate z-direction speed
Calculate the synthetic position value
Synthesize the left front wheel motor speed
Synthesize the right front wheel motor speed
Synthesize the left back wheel motor speed
Synthesize the right back wheel motor speed
Send motor rotation speed setting commands
end if
end for</td>
</tr>
<tr>
<td>2.
Calculate coefficients between trajectory and the velocity functions</td>
<td>Ax[] ← Coefficient calculation function (this.POS.X, last.POS.X, this.SPD.X, last.SPD.X, current time, time interval)

Ay[] ← Coefficient calculation function (this.POS.Y, last.POS.Y, this.SPD.Y, last.SPD.Y, current time, time interval)

Az[] ← Coefficient calculation function (this.POS.Z, last.POS.Z, this.SPD.Z, last.SPD.Z, current time, time interval)</td>
</tr>
</table>

Figure 9.26. Driver control program pseudo-code.

9.4.2 *Integrated practice of robotic joint space control*

Through studying the previous chapters, we have gained a basic understanding of the robot's forward and inverse kinematics solutions. This involves constructing the robot's kinematic model and equations to determine the end-effector's position in the workspace (its desired position). Subsequently, we use the inverse kinematics algorithm to calculate the robot joints' angles. By accurately solving for each joint's desired angle, we can control the robot's end-effector to reach its target position steadily, accurately, and swiftly. In controller design, the robot's classical control algorithm, exemplified by PID control, is known for its easy implementation, robustness, and high reliability. Therefore, this section presents and simulates the joint tracking control of the robot using both single-joint PID and *fuzzy* PID. We utilize a simulation tool for the mechanism platform, which allows for the intuitive construction of the robot's 3D model and encapsulates the underlying kinematic algebraic model and dynamic differential equations, making it user-friendly and eliminating the need for complex mechanical analysis. Moreover, the simulation platform includes a plug-in interface for SolidWorks 3D modeling software, enabling the import of SolidWorks 3D robot models into the simulation platform in XML format through relevant plugins.

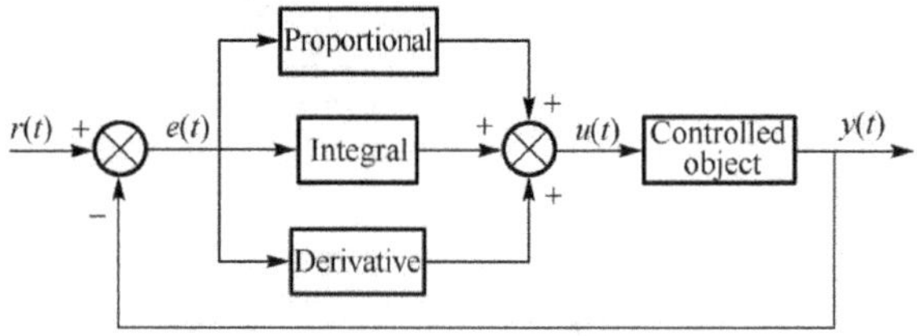

Figure 9.27. Structure of a PID controller □.

The PID control algorithm is easy to implement, highly reliable, and widely used in various fields of control systems. Its working process involves generating a control signal by linearly combining the proportional, integral, and derivative terms of the deviation between the actual measurement and the desired value. This control signal is then used to regulate the controlled object. The structure of the PID controller derived from Eq. (9.21) is shown in Figure 9.27.

In PID control, the proportional component passes the deviation between the actual measured value and the desired value to the control signal proportionally; when the proportional coefficient is larger, the rise time of the system is shorter, but it is easy to make the system overshoot. The integral component eliminates the steady state error of the system, and the integrating effect is weaker when the integrating time constant is larger. The derivative component can reflect the trend of error changes, thereby introducing an effective leading signal in the system, reducing the adjustment time, and also helping to reduce oscillations.

In this section, a single-joint PID control simulation model is built in the simulation platform of the simulation software, as shown in Figure 9.28.

According to the trial-and-error method, we can determine the parameters of the fuzzy PID controller as $\Delta K_{\mathrm{P}} = 10$, $\Delta K_{\mathrm{D}} = 5$, and $\Delta K_{\mathrm{I}} = 0$ and obtain the response curve of the joint's motion trajectory at the input of the desired trajectory and the input of the unit step, which is shown in Figure 9.29.

Due to the nonlinear and coupling characteristics of robot joints, classical PID control cannot meet the demand of such characteristics. Fuzzy PID is one of the more effective methods to meet the demand of such characteristics. In the following, we briefly introduce the fuzzy PID algorithm. Similar to the PID control algorithm, the fuzzy PID

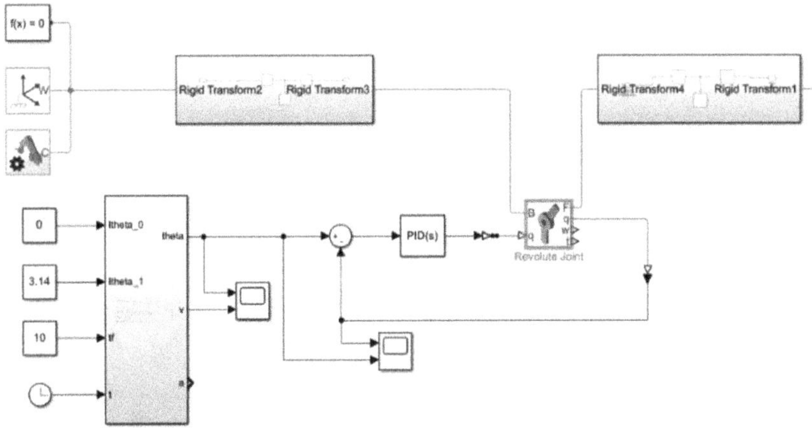

Figure 9.28. Simulation model of single-joint PID control.

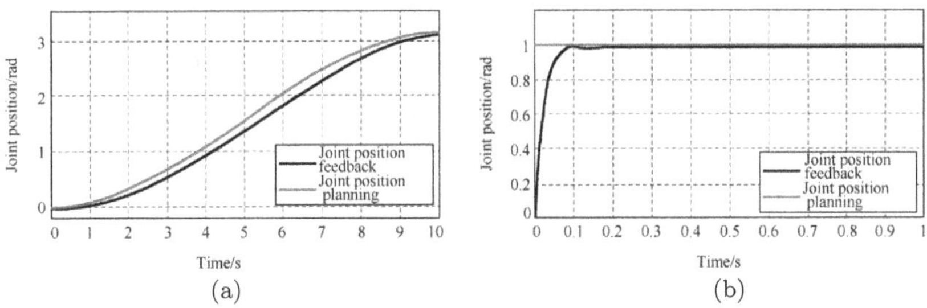

Figure 9.29. Response curve of the motion trajectory of this joint in the PID controller. (a) expected trajectory input and (b) unit step input.

algorithm takes the error e and the error change ec as inputs, which can satisfy the requirements of self-tuning of PID parameters by e and ec at different moments, and modify the PID parameters online by using the fuzzy rules. The structure of the fuzzy PID controller is shown in Figure 9.30.

The fuzzification process converts precise measurements of the input into fuzzy quantities. After the input signal is mapped to a point in the corresponding domain, it is transformed into a fuzzy subset on that domain and defined by its membership function. Fuzzy inference is the core of fuzzy control, primarily based on implication relations and inference rules in fuzzy logic. Subsequently, defuzzification is performed to transform the control quantities obtained

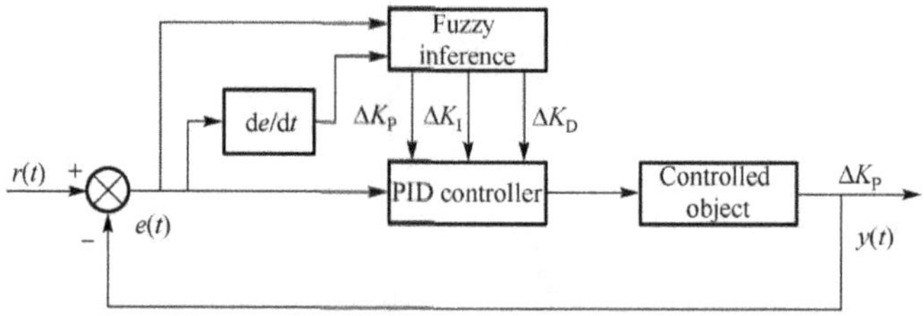

Figure 9.30. Structure of a fuzzy PID controller □.

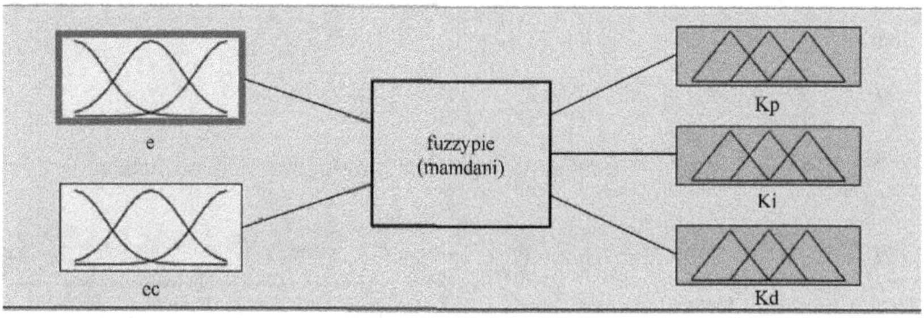

Figure 9.31. Block diagram of the fuzzy inference system.

through fuzzy inference into precise quantities that can be effectively controlled in practice.

In a fuzzy PID controller, *e* and *ec* are taken as inputs, and the regulation quantities ΔK_P, ΔK_I, and ΔK_D of the three parameters are taken as outputs. In the robot control example in this section, *e*, *ec*, and the three regulation quantities ΔK_P, ΔK_I, and ΔK_D are each classified into seven fuzzy levels, i.e., {NM, NB, Z, NS, PS, PB, PM}, which means {Negative Medium, Negative Large, Zero, Negative Small, Positive Small, Positive Large, Positive Medium} . Meanwhile, in the fuzzy module of the simulation software, according to the above fuzzy level division and querying the fuzzy control rule table, the block diagram of the fuzzy inference system can be established, as shown in Figure 9.31.

The error *e* has the same membership function as the error change *ec* with the domain $\{-3, 3\}$, as shown in Figure 9.32.

The membership function of the proportional coefficient K_P is the same as that of the integral coefficient K_I and the derivative coefficient K_D, with the domain $\{-4.5, 4.5\}$, as shown in Figure 9.33.

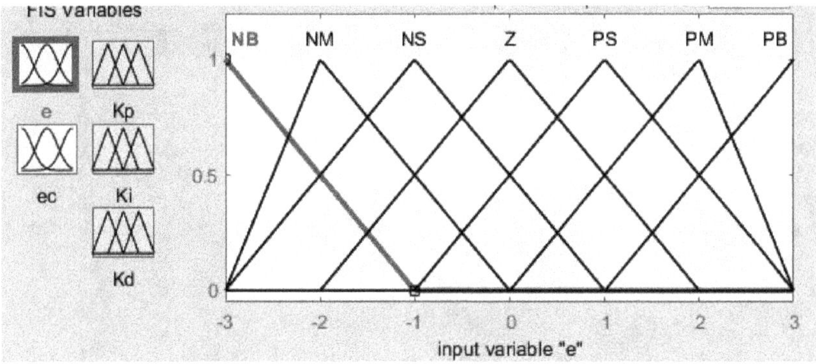

Figure 9.32. Affinity function for error e.

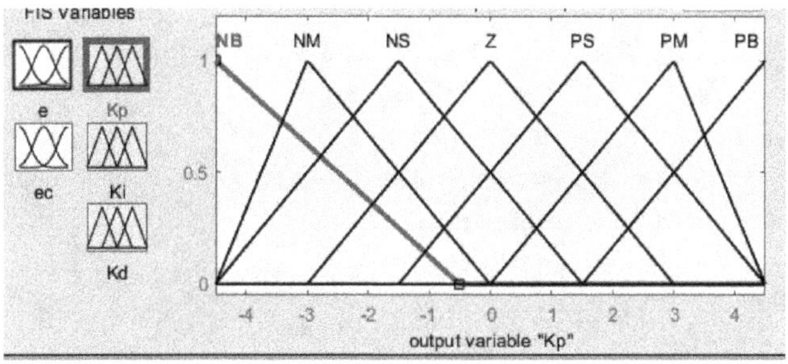

Figure 9.33. Affinity function for scale factor K_P.

Meanwhile, according to the fuzzy rule control table, a total of 49 fuzzy control rules are formulated, as shown in Figure 9.34.

In addition, the simulation model of single-joint fuzzy PID control built is shown in Figure 9.35.

In Figure 9.35, the fuzzy PID control module is shown in Figure 9.36.

Through simulation experiments, the response curves of the motion trajectory of this joint at desired trajectory input and unit step input can be obtained, as shown in Figure 9.37.

As can be seen from the comparison results in Figures 9.29 and 9.37, under the currently given PID parameter settings, for the desired trajectory input, the input trajectory can be well tracked by both the PID control algorithm and the fuzzy PID algorithm. For the unit step input, the response curve of the PID control algorithm

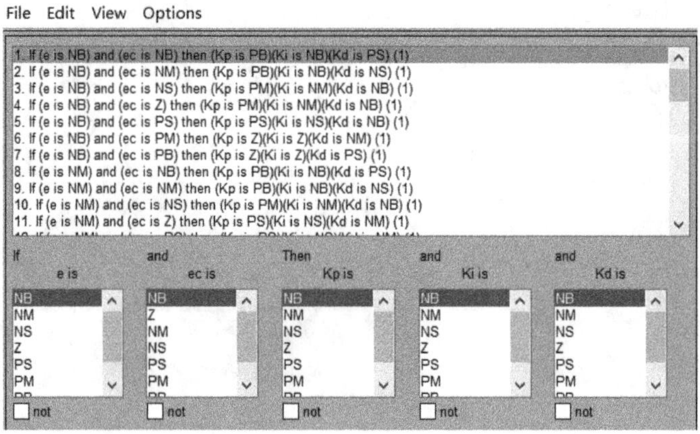

Figure 9.34. Fuzzy control rules.

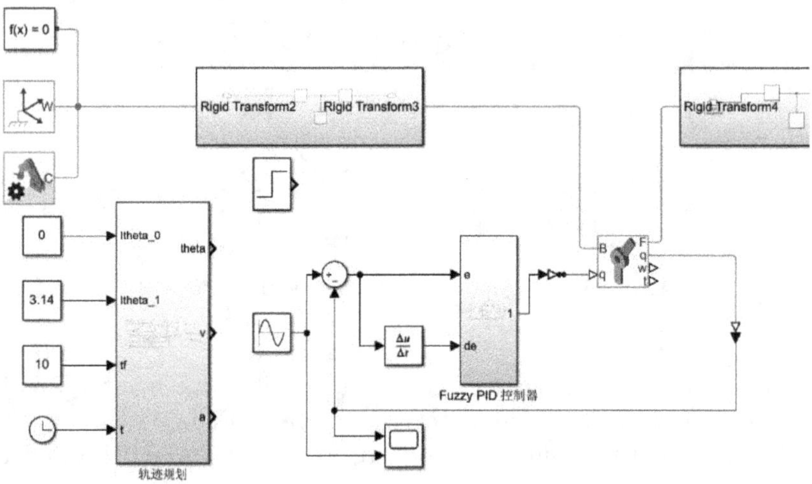

Figure 9.35. Built simulation model of single-joint fuzzy PID control.

is jittery, while the response curve of the fuzzy PID algorithm is relatively smoother. In summary, under the above given PID parameter settings, the fuzzy PID algorithm has some advantages over the PID control algorithm in reducing oscillations.

In practice, the control units encapsulate the PID into the drive control unit, so that the control process is the same as the drive control process: By sending data frames through the ROS node, the control units can self-perform closed-loop control of the position.

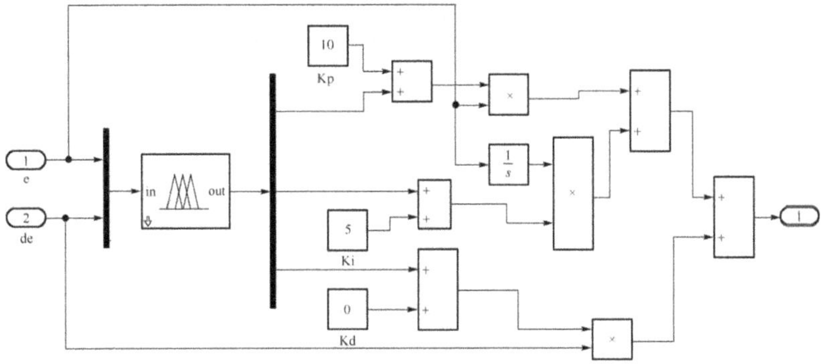

Figure 9.36. Fuzzy PID control module.

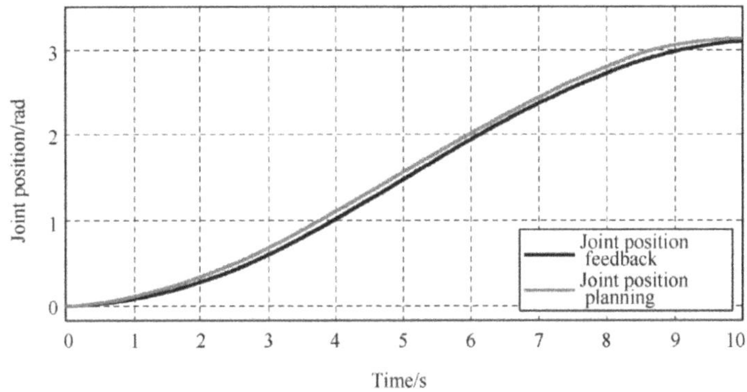

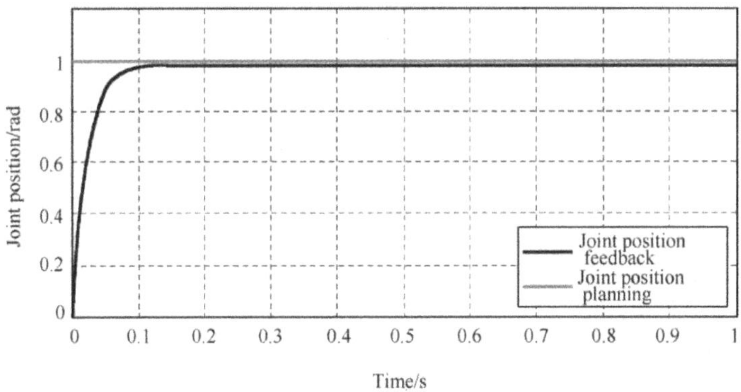

Figure 9.37. Response curve of motion trajectory of this joint in fuzzy PID controller. (a) expected trajectory input and (b) unit step input.

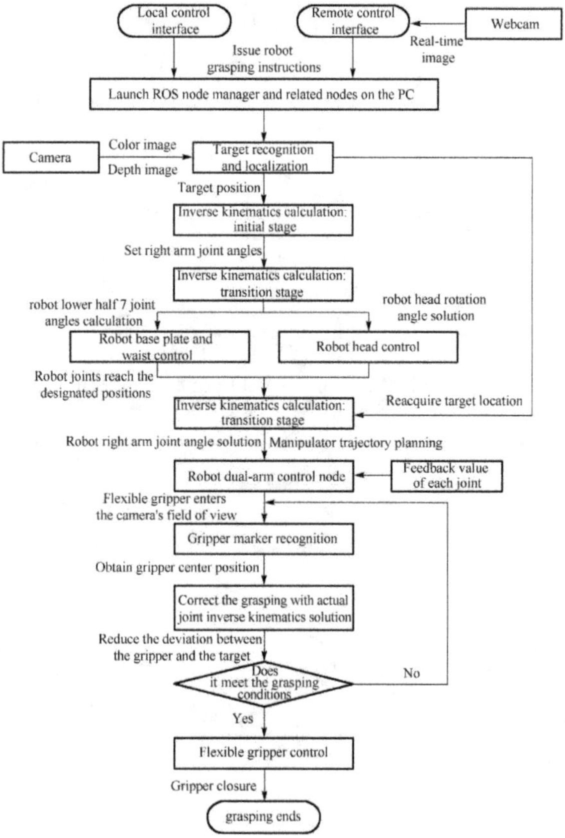

Figure 9.38. Robot autonomous target grabbing process.

9.4.3 *Integrated practice of robot workspace control*

The robot autonomous grasping experiment is a typical case of robot workspace control practice. The whole control system logic process can be illustrated in Figure 9.38. In this robot autonomous grasping process, grasping commands can be sent to the robot through the control interface designed in the previous section. In addition, if the robot is equipped with a wireless communication system, the grasping commands can be sent to the robot remotely through the network control system. In addition, based on the configuration of the node communication characteristics of the ROS system, it is possible to write the relevant control node shown in Figure 9.39. In this

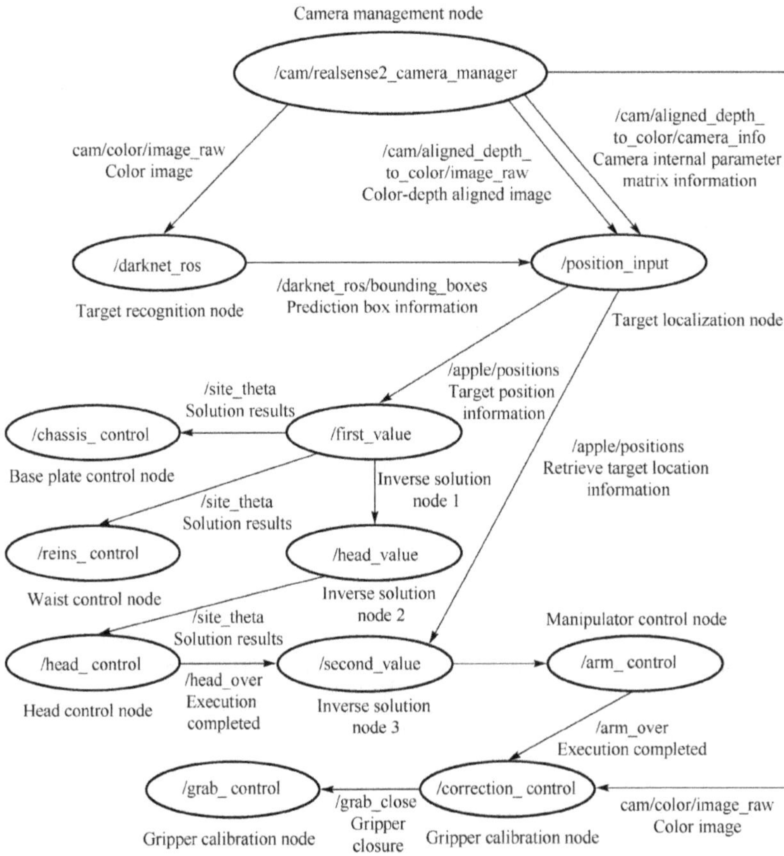

Figure 9.39. Relationship of nodes in the ROS system.

control system node relationship diagram, the circle represents the node, and the arrow represents the process of message transmission between the nodes. Among them, the camera management node is configured by the realsense-ros function package, which can convert the image information acquired by the RGB_D camera into the ROS standard image information, and provide the color and depth image information for object detection and image processing; the target recognition node is configured by the darknet-ros function package, which can receive the ROS standard image information and, through the trained weight file, recognize the target and give the target prediction box information in the image.

The control system of each node in Figure 9.38 can write its own control program, while the target localization node can be determined by the recognition algorithm that has been learned and output the 3D position of the target to be recognized. The inverse solution node 1 can be used by the inverse solution algorithm that has been learned earlier for the initialization and the solution of the various joint variables in the transition phase. The inverse solution node 2 corresponds to the robot's head joint solution algorithm. The inverse solution node 3 corresponds to the operating arm inverse solution algorithm. The robot end-effector correction node corresponds to the robot's inverse solution method. The remaining control nodes accomplish the control of each joint of the robot according to the relevant communication protocols.

The above example validates the robot autonomous recognition of the grasping process through an actual robot grasping experiment.

9.4.3.1 *Acquisition of target position information*

When the robot receives the instruction to grasp the target, it first needs to start the ROS node manager and each related node, including the camera management node, the target recognition/localization node, the inverse kinematics algorithm node, each joint control node, and the hand claw correction node. At this time, it is necessary to keep all the joints of the robot in the initial state; the initial state of the robot and the target recognition results at this time are shown in Figure 9.40.

(a) (b)

Figure 9.40. Robot autonomous gripping experiment (first step): (a) initial state of the robot and (b) target recognition results.

From the target position and orientation acquisition method learned earlier, the position of target A to be grasped under the camera coordinate system is obtained by the binocular camera of the robot as $P_{\mathrm{A}} = (P_{cx}, P_{cy}, P_{cz}) = (52.65, 196.67, 1140.12)$. Combining this positional information with the 50 groups of position and orientation information obtained by binocular visual recognition, we can get the position and orientation of one of the groups of target A under the camera coordinate system ${}^{C}_{\mathrm{A}}\boldsymbol{T}$. Finally, combining with the position and orientation matrix of the mobile cart platform in the coordinate system, the position and orientation of target A in the base coordinate system ${}^{G}_{\mathrm{A}}\boldsymbol{T}$ can be obtained, as shown in Eq. (9.23):

$$ {}^{G}_{\mathrm{A}}\boldsymbol{T} = \begin{bmatrix} 0.5362 & -0.7736 & 0.3377 & 20.15 \\ 0.5831 & 0.05031 & -0.8108 & 1394.10 \\ 0.6102 & 0.6317 & 0.4781 & 1260.08 \\ 0 & 0 & 0 & 1 \end{bmatrix} \tag{9.23} $$

9.4.3.2 *Calculate the angles of each joint in the lower part of the robot*

In this experiment, it is assumed that the target is located below the camera when $P_{cy} = 196.67 > 0$. At this time, the angles of each joint of the robot's main operating arm are set, i.e., let $\theta_1 = 10°$, $\theta_2 = 10°$, $\theta_3 = 30°$, $\theta_4 = 10°$, and $\theta_5 = -60°$. Then, according to the position and orientation of the target in the coordinate system of the robot chest center, the desired position and orientation of the end-effector center in the kinematic model of the robot's upper body, and the coordinate annotation given in Eq. (9.23) and then using the coordinate transformations, we can find out the waist separation point in the geodesic coordinate system of a target position and orientation ${}^{G}_{W}\boldsymbol{T}_{\mathrm{float}}$, as shown in Eq. (9.24):

$$ {}^{G}_{W}\boldsymbol{T}_{\mathrm{float}} = \begin{bmatrix} 0.643 & -0.7655 & 0.02507 & 209.6 \\ 0.7658 & 0.6424 & -0.02955 & 1084.0 \\ 0.006516 & 0.0382 & 10.9992 & 809.3 \\ 0 & 0 & 0 & 1 \end{bmatrix} \tag{9.24} $$

According to the link coordinate system established by the robot and the inverse solution of the robotic arm and Eq. (9.24), 50 sets of solutions for each joint angle of the lower part of the robot can be obtained, and in order to ensure that the robot joints have sufficient workspace, it is necessary to remove the portion of the solutions that do not satisfy the requirements of the joint limits, and then analyze the constraints of each joint angle, as shown in Eq. (9.25):

$$\begin{cases} d_y, , p_{cz}, \ -90^\circ, , \theta_3, , 90^\circ, \ -90^\circ, , \theta_4, , 90^\circ \\ -100\,\text{mm}, , h_z, , 100\,\text{mm}, \ -30^\circ, , \theta_6, , 30^\circ, \ -30^\circ, , \theta_7, , 30^\circ \\ -100\,\text{mm}, , l_1, , 100\,\text{mm}, \ -100\,\text{mm}, , l_2, , 100\,\text{mm}, \\ -100\,\text{mm}, , l_3, , 100\,\text{mm} \end{cases} \tag{9.25}$$

where d_y is the distance moved by the robot chassis in the Y-axis direction; θ_3 and θ_4 are the angle that the robot chassis has turned around itself and the angle that the robot foot rotational joint has turned around, respectively; h_z, θ_6, and θ_7 are the up and down moving distance, the forward and backward bending angle, and the left and right bending angle, respectively, of the parallel structure; and l_1, l_2, and l_3 are the values of the link length change of the parallel mechanism. Between h_z, θ_6, θ_7 and l_1, l_2, l_3, they can be converted to each other based on the inverse kinematics of the parallel structure. In addition, d_x is the distance that the robot chassis moves in the X-axis direction.

At the same time, in order to ensure the stability of the robot during the actual operation, the angles of the robot's front-back bending and left-right bending need to be minimized. Therefore, in this section, among the eight sets of inverse solutions that satisfy the joint constraints, a set of solutions that minimize the difference in the length change of the three links of the parallel mechanism of the robot is selected, i.e., the solution that minimizes the value of $\Delta l = |l_1 - l_2| + |l_1 - l_3| + |l_2 - l_3|$ is selected, from which the values of the joints of the robot's lower part can be obtained, as shown in

Eq. (9.26):

$$\begin{cases} d_x = 209.62\,\text{mm} \\ d_y = 898.48\,\text{mm} \\ \theta_3 = 0^\circ \\ \theta_4 = 49.99^\circ \\ h_z = -55.71 \\ \theta_6 = 2.19^\circ \\ \theta_7 = 0.37^\circ \end{cases} \tag{9.26}$$

In Eq. (9.26), according to the transformation relationship of the parallel mechanism of the robot, h_z, θ_6, and θ_7 can be converted into the values of the link length change of the parallel mechanism, and $l_1 = -52.73$, $l_2 = -59.35$, and $l_3 = -55.06$ can be obtained through calculation. At the same time, according to the data obtained using formula (9.26), we can also control the lower part of the robot joints to move to the corresponding angle. At this time the robot state and the image obtained by the camera are shown in Figure 9.41.

(a) (b)

Figure 9.41. Robot autonomous gripping experiment (step 2): (a) robot state and (b) image acquired by the camera.

9.4.3.3 *Reacquisition of target positions*

In order to re-acquire the target position, according to the coordinates of the target position and orientation in the field of view of the camera and the relative coordinates of the camera in the earth coordinate system, the angle that the two joints of the robot head should move at this time can be derived from the target position, i.e., $\alpha_1 = -22.31°$ (when the head is tilted up, raising the head is positive) and $\alpha_2 = -18.53°$ (when the head is rotated, turning the head counterclockwise is positive); the corresponding angles of rotation of the joints of the robot head can finally be obtained from the results of the calculation. The state of the robot head before and after the rotation is shown in Figure 9.42.

At the same time, the target position information can also be reacquired through the target position output node. At this time, the position of the target A' in the camera coordinate system is $P'_{\mathrm{A}} = (p'_{cx}, p'_{cy}, p'_{cz}) = (-11.74, 44.22, 337.21)$. Since the target position has a greater influence on robot grasping, only the target position is considered here. Then, the position $\boldsymbol{R}$ of the target A' in the camera coordinate system can be set as a unit matrix. At the same time, according to the correspondence between the center of the end-effector and the position and orientation in the base coordinate system, the position and orientation of the target A' in the coordinate system {O} of the center of the robot's chest ${}^{G}_{A'}\boldsymbol{T}$

(a) (b)

Figure 9.42. Robot autonomous gripping experiment (step 3): state of the robot (a) before head rotation and (b) after head rotation.

can also be calculated correspondingly, and the result is shown in Eq. (9.27):

$$
{}^{G}_{A'}\boldsymbol{T} = \begin{bmatrix} 0.1206 & -0.9251 & -0.3599 & -2.277 \\ 0.294 & 0.3796 & -0.8772 & -352.1 \\ 0.9482 & 0 & 0.3178 & 86.99 \\ 0 & 0 & 0 & 1 \end{bmatrix} \tag{9.27}
$$

9.4.3.4 *Solve for the values of each joint in the upper part of the robot*

In order to improve the robot grasping accuracy, the five joint angles of the upper part of the robot can be solved using a numerical method for each joint of the robot. In this example, we set the initial values of the iteration as $\theta_1 = 10°$, $\theta_2 = 10°$, $\theta_3 = 30°$, $\theta_4 = 10°$, and $\theta_5 = -60°$, and set the positioning accuracy of the system as eps = 0.01. Then, the five joint values of the robot can be solved in four iterations, as shown in Eq. (9.28):

$$
\begin{cases} \theta_1 = 41.48° \\ \theta_2 = 0.54° \\ \theta_3 = -27.127° \\ \theta_4 = -7.743° \\ \theta_5 = -28.01° \end{cases} \tag{9.28}
$$

According to the values of the five joints of the robot obtained from Eq. (9.28) and the control method of robot trajectory planning in joint space, trajectory planning is performed for each joint of the robot, so as to control the five joints of the upper part of the robot to move to the corresponding angle. Then, the results of trajectory planning of the joints of the upper part of the robot (the main operating arm) and the images acquired by the camera at this time are shown in Figure 9.43.

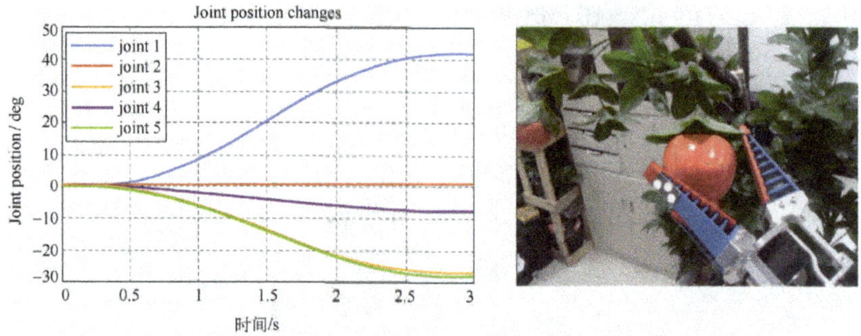

(a) (b)

Figure 9.43. Robot autonomous gripping experiment (step 4). (a) robot upper half trajectory planning and (b) Images captured by the camera.

9.4.3.5 *Robot gripping operation control process*

According to the control flow for the robotic arm's grasping depicted in Figure 9.38, each joint begins to move based on the trajectory planning results. The joint controllers read the feedback values to align each joint's movement closely with the desired trajectory, utilizing the spatial control method for robot joints. As the end-effector approaches the target, the camera captures its image. The next step involves employing the robot grasping control method. This enables the correct identification of the end-effector's center point B within the camera's coordinate system. Consequently, the arm's kinematics are applied to determine the actual position of each joint, enhancing accuracy through the control system, which adjusts errors based on each joint's trajectory planning input. When the deviation between the end-effector's center and the target falls within the grasping criteria, the control process concludes. Finally, the end-effector's fingers are commanded to close, enabling a successful grasp of the target.

In addition, since the average diameter of the simulated target in this case is about 80 mm and the opening diameter of the robot end-effector we designed is about 120 mm, the robot end-effector can still successfully grasp the target even if the center of the end-effector has a small deviation from the target in the experiment. Through multiple experimental verifications of this example, the results show that in the camera coordinate system, when the deviation between the end-effector and the target in each axis is less than 15 mm, it is

(a) (b)

Figure 9.44. Robot autonomous gripping experiment (step 5): (a) robot end-effector calibration process and (b) end-effector closure gripping target state.

not necessary to continue to correct the end-effector position, and then the end-effector can successfully grasp the target. The experimental results show that a certain margin can be reserved for the end-effector in the design of the arm according to the actual size of the object to be grasped and the allowable error of the experimental or control system. For this experiment, the end-effector calibration process of the robot and the end-effector closure to grasp the target are shown in Figure 9.44.

9.4.3.6 *Robot completes capture*

When the robot end-effector fingers are closed and the target object is successfully grasped, the robot will return to the initial state and store the target object that has been successfully grasped so as to carry out the next grasping task. Following the above experimental verification of the robot's grasping process of the target object, it shows that the robot's autonomous grasping function can be realized by the target recognition algorithm and the inverse kinematics algorithm in the laboratory environment.

9.5 Summary

In this chapter, we explore the theoretical foundations of robot control and integrated control design, as well as the practical applications of robot control. Initially, we introduced the basic concepts of control system design, including the overall composition of a robot control system, the hierarchy of system control, and control technology. Subsequently, we delved into the integrated design of robot control, covering the generalized design of robot driver space, joint space, and workspace. This included modeling various common robot structures, designing control systems, and introducing control methods. Finally, we demonstrated the practical application of control in the robot's driver space, joint space, and workspace through target grasping experiments on a robot platform. Through specific control examples, we deepened our understanding of factors influencing the robot control system, such as the success rate and accuracy of grasping control under varying conditions of distance, illumination, and design margins of the operating arm. This chapter not only enhanced our comprehension of the robot control system but also allowed us to synthesize and apply the control theories and methods we have learned.

9.6 Exercises

9.1 What is decomposed motion control? Why is decomposed motion control necessary?

9.2 Design and implement the approach and methods of decomposed motion control.

9.3 What are the objectives of decomposed motion acceleration control? How is it achieved?

9.4 Discuss the similarities and differences between joint space control and workspace control for robots.

9.5 Design and practice a joint-space closed-loop control algorithm and a workspace closed-loop control algorithm for a robot with six degrees of freedom.

Chapter 10

Robot Integrated Practice

10.1 Introduction

This chapter focuses on the practical training of the algorithms covered in this book, which mainly includes four aspects: (1) practical training on performance analysis of visual SLAM 3D environment construction; (2) practical training on target object, environment obstacle recognition, and 3D reconstruction of obstacles; (3) practical training on motion planning of mobile robot in 2D space; (4) practical training on motion planning of operating arm in 3D space.

10.1.1 *Introduction to software and hardware systems*

The comprehensive practice in this chapter is carried out on the platform introduced in Chapter 2, as shown in Figure 10.1. The robot head is loaded with an Intel Realsense D435 camera, and the bottom cart is designed with a McNamee wheel structure, which is more flexible and maneuverable, enabling the robot to realize rotation and translation in any direction and to move flexibly in small spaces, which is conducive to the realization of the path planning of the cart. The cart is equipped with a laptop computer as the core processor; the CPU is Intel core i7, CPU @GHz, and the GPU is GeForce GTX1660s, which has a better performance and is able to

Figure 10.1. A photograph of the robot platform.

process a variety of information quickly. The robot as a whole adopts a series–parallel hybrid structure with many degrees of freedom and a large workspace for the end-effector, which is favorable for target grasping.

10.1.2 *Software system*

The software system used in the experiment has been introduced in Section 2.3. The ROS system is a robotic software development system with strong compatibility, wide versatility, and good open-source support. The system is perfectly compatible with many programming languages such as C++ and Python. It also supports many simulation tools such as RVIZ and gazebo, which are well suited to solve the problems brought by the difficulty of simulation experiments due to code redundancy in the robot control process. At the same time, with more and more researchers and teams developing ROS, most of which are open source in nature, it benefits the practical training of learners of this book. The ROS software system framework structure is shown in Figure 10.2.

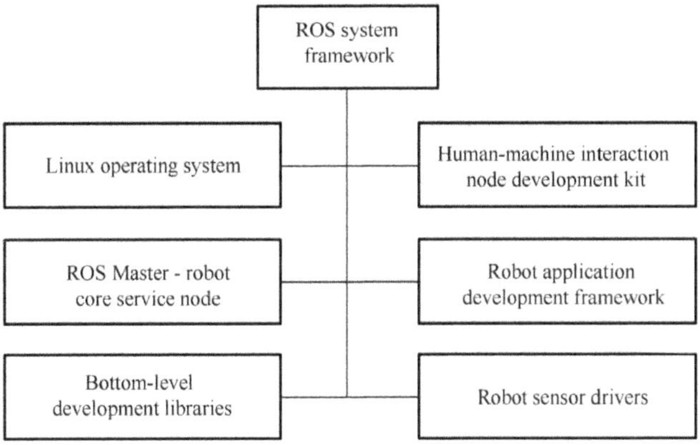

Figure 10.2. ROS software system framework structure.

10.2 SLAM System Performance Analysis

10.2.1 *Tum dataset testing performance analysis*

The Tum dataset is an RGB_D dataset published by Computer Vision Lab in Germany, which includes a variety of texture-rich scenes, such as offices, halls, and factories, and covers a variety of camera motions, such as slow motion and rotation, with a total of 39 packets, making it a very powerful dataset. Meanwhile, this dataset provides Python tools to analyze the experimental results by comparing the ground truth (the real camera pose in the SLAM system captured by eight high-frame-rate tracking cameras, which is considered the absolute pose of the camera here) in this dataset. The following are included in each data package:

(1) rgb.txt and depth.txt, which record the collection time and file name corresponding to color and depth images;
(2) rgb/ and depth/, which are folders for color and depth images;
(3) groundtruth.txt, which contains the real pose of the camera in the SLAM system in the format (time,x,y,z,qx,qy,qz,qw), representing the acquisition timestamp, the x, y, and z positions, and the quaternion of the orientation for each image.

There are two common metrics for SLAM performance, which are Absolute Trajectory Error (ATE) and Relative Pose Error (RPE). Absolute Trajectory Error is a global measure of the difference between the camera pose estimated by the algorithm and the true camera pose, which is usually quantitatively calculated using the root mean square error (RMSE), and the formula for RMSE is shown in Eq. (10.1):

$$\text{ATE}_{\text{RMSE}}(\hat{X}, X) = \sqrt{\frac{1}{n}\sum_{i=1}^{n}\left[\mathbf{trans}(\hat{X}_i) - \mathbf{trans}(X_i)\right]^2} \tag{10.1}$$

where $\hat{X} = \{\hat{X}_1, \hat{X}_2, \ldots, \hat{X}_n\}$ is the estimated motion sequence of the camera, $X = \{X_1, X_2, \ldots, X_n\}$ is the real motion sequence of the camera, and **trans** is the translation vector of the camera pose.

The relative pose error is a local measure of the difference between the camera pose estimated by the algorithm and the camera's true pose, which represents the difference in the amount of pose change at the same two timestamps. Its calculation formula is shown in Eq. (10.2):

$$\text{RPE} = \frac{1}{n}\sum_{ij}(\delta_{ij} - \hat{\delta}_{ij})^2 \tag{10.2}$$

where $\delta_{ij} = x_i - x_j$ is the difference of the camera's true pose at a certain time and $\hat{\delta}_{ij} = \hat{x}_i - \hat{x}_j$ is the difference of the camera's estimated pose at a certain time.

The RPE is applicable to the translation of the system, while the ATE is applicable to the performance evaluation of the whole SLAM system. In this book, the ATE is used as the measurement standard, and the following datasets are selected for error testing, as shown in Figures 10.3–10.6.

(1) **Fr1_xyz:** This dataset contains motion data of the camera slowly translating along the direction of the main axis, which can test the pose estimation performance of the horizontal motion of the camera.

(2) **Fr1_rpy:** This dataset contains the motion data of the camera slowly rotating around the three main axes, which can test the performance of the pose estimation of the camera's rotational motion.

(3) **Fr2_desk and Fr2_large_with_loop:** Both datasets contain information about the start–end point overlap, which can test the closed-loop detection performance of the SLAM system.

In Figure 10.3, the dashed line represents the true trajectory curve of the camera, and the solid line composed of various colors represents the estimated trajectory curve of the camera, with the colors ranging from blue to red, representing the trend of the error from small to large.

Intuitively, from Figures 10.3–10.6, all the estimated camera trajectory curves are almost able to coincide with the trajectory curves of the real camera, which side by side reflects that the SLAM system designed in this book is more reasonable. For a more intuitive illustration, the maximum error, minimum error, average error, and RMSE of the trajectories under each dataset are calculated, as shown in Table 10.1.

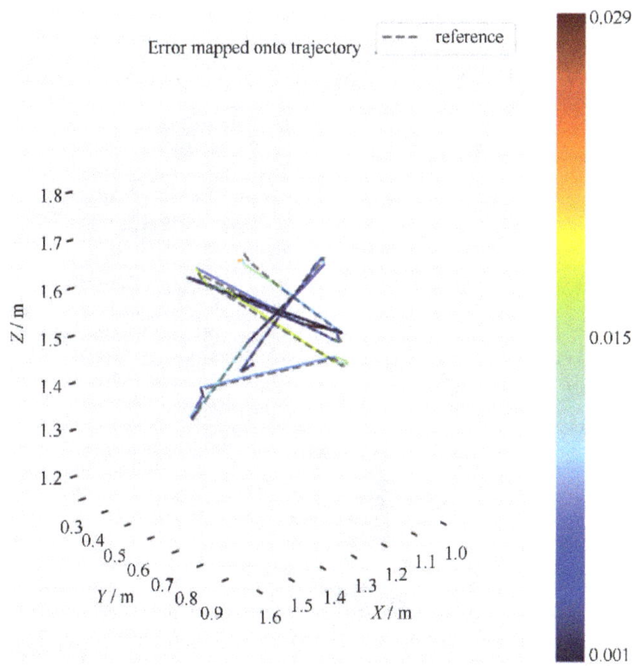

Figure 10.3. Absolute trajectory error curves for the Fr1_xyz dataset.

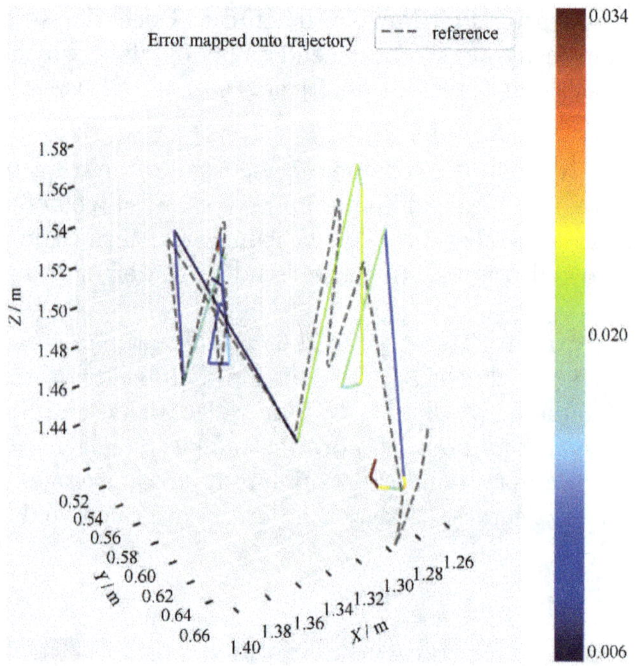

Figure 10.4. Absolute trajectory error curves for the Fr1_rpy dataset.

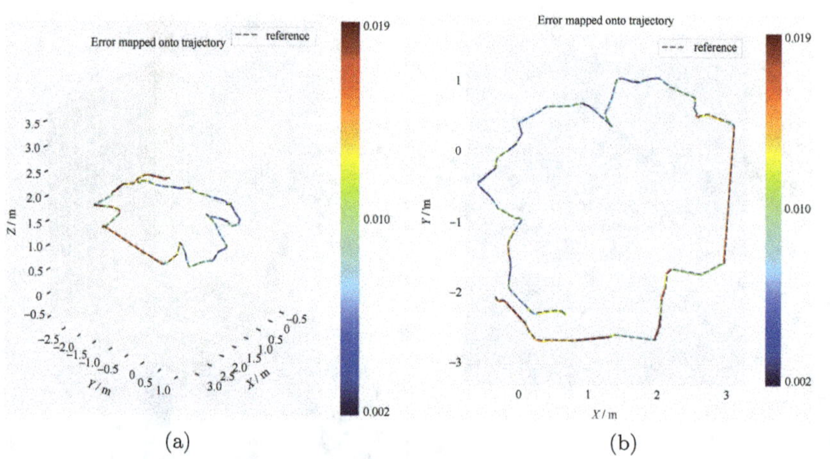

Figure 10.5. Absolute trajectory error curves for the Fr2_desk dataset: (a) 3D trajectory error curve and (b) error curve Projection in XY plane.

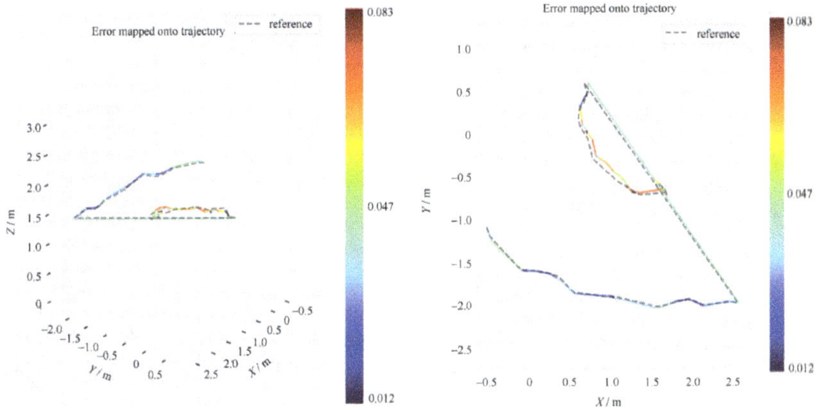

(a) (b)

Figure 10.6. Absolute trajectory error curves for the Fr2_large_with_loop dataset: (a) 3D trajectory error curve and (b) error curve Projection in XY plane.

Table 10.1. Error statistics under four datasets.

Dataset	Fr1_xyz	Fr1_rpy	Fr2_desk	Fr2_large_with_loop
Maximum error/m	0.028541	0.033733	0.018738	0.083014
Minimum error/m	0.001205	0.005759	0.001989	0.011727
Average error/m	0.009597	0.017513	0.010680	0.043901
Root mean square error/m	0.011384	0.019127	0.010436	0.047421

Comparing the values in Table 10.1, we can see that the SLAM system designed in this book shows the following performance: (1) In the smoother translation or rotation scenarios, the maximum error is less than 0.05 m, and the RMSE is below 0.02 m, which meets the requirement of centimeter-level accuracy. (2) In the third smaller closed-loop scenario, both the maximum error and the RMSE are less than 0.02 m. It can also be observed in Figure 10.5 that the closed-loop circuit is clearly measured in this scenario, indicating that the closed-loop section in this book can be effectively implemented. (3) In the Fr2_large_with_loop scenario, the system has a large error, with the maximum error reaching 0.08 m, and the RMSE reaching 0.047 m, which is due to the phenomenon of tracking loss in this scenario and the influence of the large lighting, which has a greater impact on the extraction of feature points. However, in larger scenarios, the accuracy of 0.05 m basically meets the navigation needs of mobile robots.

The validation of the tum dataset concludes that the SLAM system designed in this book achieves a value of 0.02 m in the metric of error accuracy, which meets the experimental requirements.

10.2.2 *Experiment of map establishment*

The 3D map scene is the comprehensive laboratory of intelligent robotics in the School of Automation of Northwestern Polytechnical University, and the environment of the mapping scene, Scene 1, is the main activity area of the robot, as shown in Figure 10.7, with an overall length of 7.5 m and a width of 3.5 m This is used to test the mapping performance of the SLAM system in the indoor open area. Scene 2 mainly utilizes two large experimental platforms to form a small aisle area to test the performance of the SLAM system in a narrow space, as shown in Figure 10.8.

The point cloud maps are constructed for scenes 1 and 2, as shown in Figures 10.9 and 10.10, respectively.

Scene 1 mainly scanned the spatial extent below the height of the robot's head (1.4 m), as shown in Figure 10.9. When conducting this experiment, the mobile robot scanned two circles around the laboratory, taking a total of about 200 s, and obtained 313 keyframe images. It can be seen that the SLAM system in this book is overall able to reconstruct the ground environment and its surroundings

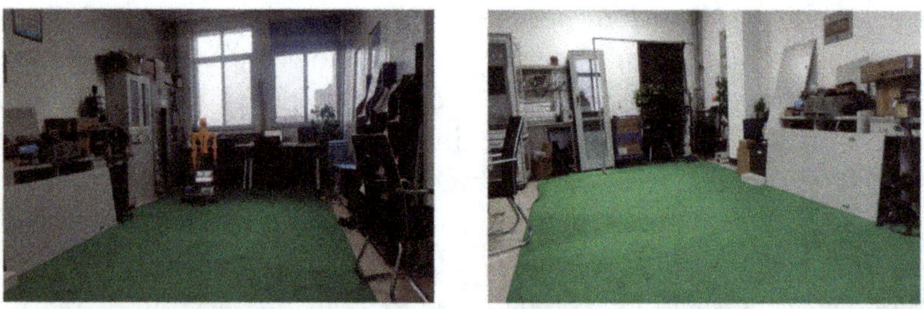

(a) (b)

Figure 10.7. Scene 1 actual picture: (a) laboratory scene taken from south to north and (b) laboratory scene taken from north to south.

(a) (b)

Figure 10.8. Scene 2 actual picture: (a) scene shot on the west side of the lab bench and (b) scene shot on the east side of the lab bench.

Figure 10.9. Point cloud map constructed for scene 1.

Figure 10.10. Point cloud map constructed for scene 2.

quite well. Tables, chairs, experimental platform, etc., in the point cloud image can be clearly seen, and the reconstruction effect is good.

In scene 2, as shown in Figure 10.10, the narrow aisle consisting of two large experimental platform can be clearly seen, which is more obvious, and at the same time, the upper part of the figure is all walls, which makes the reconstruction effect more prominent. However, on the left and right sides of the box range, the point cloud has a tendency to diverge; this is because the area is made of transparent glass, causing infrared light transmission, resulting in inaccurate depth measurement. There are some "empty" areas in the point cloud map, which are caused by the camera not scanning or not being in the keyframe. Overall, the SLAM system in this book is able to adapt to narrow aisle spaces.

The estimated values were measured for individual objects in both scenes using the MeshLab tool, and the data obtained are shown in Table 10.2.

As can be seen from Table 10.2, the relative errors of the SLAM systems in this book are all around 2%, i.e., an absolute error of 2 cm, while the cumulative error increases with the increase in the length of the measured object, but basically also stays around 2%. Such accuracy can meet the requirements of the robot's navigation planning operation.

In the experimental process, due to the point cloud map storage being too large, containing a large amount of redundant information, storage and reading speed will be reduced. This will affect the execution speed of the robot's subsequent path planning, and it cannot meet the real-time requirements of path planning. So, this book utilizes an octree map form to store and update the map. Take scene 1

Table 10.2. Discrepancy between estimated and actual values for objects.

Measurement category	Lab cabinet length	Green carpet width	Narrow aisle width	Sofa width	Sofa height
Estimated value/m	2.341	3.04	1.185	0.915	0.817
Actual value/m	2.30	3.00	1.20	0.90	0.80
Absolute error/m	+0.041	+0.04	−0.015	+0.015	+0.017
Relative error/%	1.7	1.3	1.25	1.6	2.1

Figure 10.11. Octree map with 2 cm resolution corresponding to scene 1.

Figure 10.12. Octree map with 5 cm resolution corresponding to scene 1.

as an example: Set the resolution to 2 cm and 5 cm, carry out the octree map-building analysis, and obtain Figures 10.11 and 10.12.

As can be seen from Figures 10.11 and 10.12, the octree map is also able to show the location relationship of the actual scene better. Several reconstruction experiments were conducted for scene 1, and the storage sizes of the point cloud map and the octree map were obtained, as shown in Table 10.3.

As can be seen from Table 10.3, when using the form of point cloud map, the storage space is above the MB level, but when using the form of octree map, the storage space is at the KB level (the octree map with color is only for visual display and has no practical application value). The results demonstrated that octree maps exhibit moderate storage requirements, which correlates with superior read speeds. Considering that the SLAM system presented in this study exhibits a relative error of approximately 2%, an octree with a resolution of 2 cm was utilized for mapping implementation. The final mapping outcome of Scenario 1 is illustrated in Figure 10.13.

Table 10.3. Map storage sizes for different formats.

Map form	Point cloud map	Colorized octree map	Colorless octree map
Map size for the first experiment	324.5 MB	17.5 MB	648 KB
Map size for the second experiment	274.9 MB	17.7 MB	676 KB
Map size for the third experiment	250.3 MB	17.2 MB	694 KB

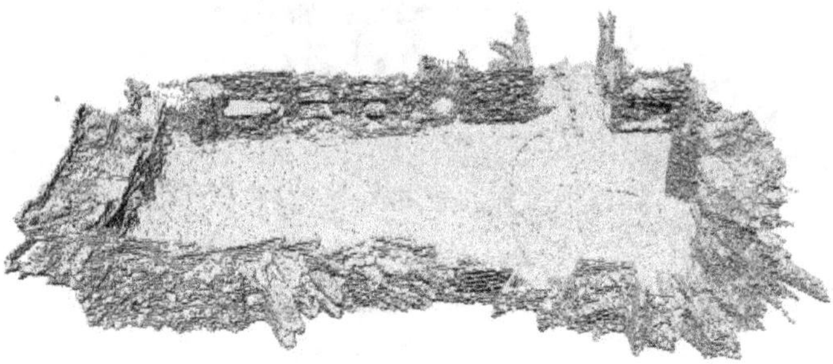

Figure 10.13. Final scene 1 build results.

10.3 Practice of Object Recognition and 3D Environment Reconstruction

10.3.1 *Experimental datasets and related parameters*

10.3.1.1 *Description of the dataset*

Target and environmental obstacle recognition using the I-YOLOv3 algorithm first requires the collection of images as an experimental dataset for training and testing. A total of 1,000 images were collected as source data, of which 325 were collected in normal light scenes, 258 in high-intensity light scenes, 203 in low-light scenes, and 214 in dark light scenes, and the images collected in different scenes are shown in Figure 10.14. Training in different light can enhance the robustness

(a) (b) (c) (d)

Figure 10.14. Images captured in different scenarios: (a) normal lighting scene, (b) high-intensity lighting scene, (c) low lighting scene, and (d) dark lighting scene.

and generalization ability of the neural network. The camera shooting distances are all in the range of 1.0–2.0 m.

10.3.1.2 *Description parameters of the labeling and training process*

After acquiring the dataset, it is necessary to carry out the labeling operation on all the pictures, i.e., to label the location and related category attributes of the target and environmental obstacles in the pictures. In this book, the LabelImg annotation tool is utilized to carry out the annotation work, and different annotation methods are adopted for target objects and environmental obstacles, as shown in Figure 10.15. The category and location information is then saved to the xml folder, which is used as the input file for the neural network.

For the labeling of environmental obstacles, if a large annotation box is directly used for labeling, it will contain a large amount of background information, greatly diminishing the effectiveness of training. For example, in regions (1) and (5), a large annotation box is used to frame the entire environmental obstacles, which is not conducive to training because of the large number of environmental obstacles in such a annotation box. This book adopts small annotation boxes similar to (2), (3), and (4) to label the environmental obstacles. The use of such a annotation box can minimize the influence of peripheral information on the network training. As for the

(a) (b)

Figure 10.15. LabelImg schematic diagrams: (a) environmental obstacle labeling and (b) target labeling.

target objects, the regular rectangular boxes are directly used for labeling.

The overall information of the statistical training set is as follows: (1) Each image contains an average of 3.6 target annotations and 9.5 segments of environmental obstacle annotations, totaling more than 3,600 target samples and 9,500 segments of environmental obstacle samples;. (2) The total number of images in the training set is 774, and the total number of images in the testing set is 226, with the ratio of approximation being 3:1. In order to compare the relevant performances, all the experimental conditions are made identical, synchronously training the YOLOv3 algorithm and the I-YOLOv3 algorithm proposed in this book. The training process parameters are shown in Table 10.4.

10.3.2 *Target object and environmental obstacle recognition effect*

Figures 10.16 and 10.17 show the resulting images of the target and environmental obstacle recognized by the I-YOLOv3 algorithm, respectively, where the box for recognizing the environmental obstacle is not shown in order to highlight the effect of target recognition, and the box for recognizing the target (apple) is not shown in order to highlight the effect of environmental obstacle recognition.

Table 10.4. Training process parameters.

Related configurations	Parameters
Runtime environment and related library versions	CUDA10.1, cuDNN7.6.5, OpenCV3.4.7
Training algorithms	I-YOLOv3, YOLOv3
Modulus of elasticity/GPa	120–145
momentum factor	0.9
Weight decay factor	0.001
Maximum number of training batches	12,000
Steps = 9,600 and 10,800	Scales = 0.1,0.1
Learning rate	0.001

Figure 10.16. Target recognition effect.

Figure 10.17. Environmental obstacle recognition effect.

As can be seen in Figures 10.16 and 10.17, the algorithms used in this book are able to perform target and environmental obstacle recognition well. The following performance tests are conducted using various metrics.

10.3.3 *Neural Network Performance Test*

10.3.3.1 *Loss value analysis*

In neural networks, the loss value is calculated by the loss function, which is used to evaluate whether the predicted value and the real value in the algorithm model are consistent, and can show the fitting effect of the model. The loss function is also the objective function to be optimized in the neural network, and the training of the neural network is actually the process of optimizing the minimum loss function. Usually, the smaller the loss value, the better the training effect of the algorithm, the better the robustness of the algorithm model. Record the log file when the model is training, visualize it, and draw its loss change curve, as shown in Figure 10.18.

I-YOLOv3 algorithm's loss value is able to decrease rapidly within the first 5,000 iterations, from more than 1 to about 0.3; it stabilizes gradually after 5,000 iterations; and stays stable at about 0.1 after 10,000 iterations. YOLOv3 algorithm's loss value is able to be fitted rapidly within the first 4,000 iterations, from about 1.0 to about 0.2. After 4,000 iterations, it gradually converges to about 0.1. In the early stage of iteration (before 5,000 iterations), the loss value of YOLOv3 algorithm decreases significantly faster; in the middle stage of iteration (5,000–10,000 iterations), the oscillation amplitude of the curve of the I-YOLOv3 algorithm is smaller, and the curve is smoother, which indicates that the fit is more stable; in the later stage of iteration (after 10,000 iterations), the loss value of both algorithms is stabilized at 0.1. Overall, the I-YOLOv3 algorithm has achieved the same effect as the YOLOv3 algorithm in the index of loss.

10.3.3.2 *P–R curve analysis*

The *P–R* curve represents the relationship between precision and recall. Generally, the horizontal coordinate of the curve is recall and the vertical coordinate is precision, and the integral of the curve represents the average precision (AP) of a certain type of object recognition. The AP value is an important measure of the performance of the algorithm and is very useful for model localization and segmentation. The mean of average precision (mAP) represents the mean of the average precisions of multi-class objects in target detection.

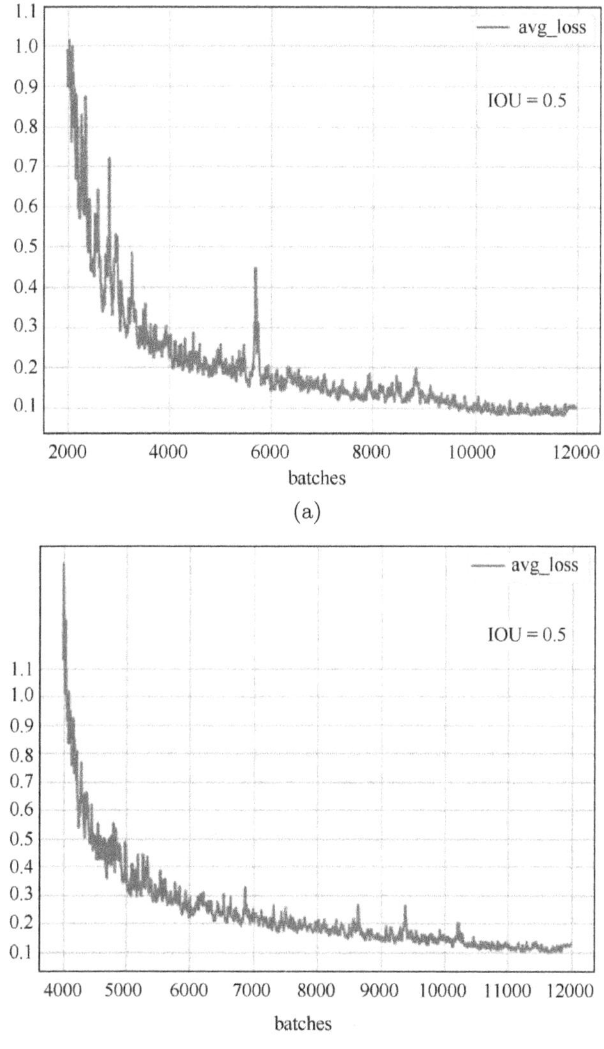

Figure 10.18. Loss curves for the two algorithms: (a) YOLOv3 algorithm and (b) I-YOLOv3 algorithm.

Table 10.5. Confusion matrix.

	Projections	
The real situation	Positive example	Negative example
Positive example	TP	FN
Negative example	FP	TN

First, we introduce the definitions of precision, recall, and AP, as shown in Table 10.5, and when predicting, we denote correctly categorizing positive examples as positive examples as TP; incorrectly categorizing positive examples as negative examples as FN; correctly categorizing negative examples as negative examples as TN; and incorrectly categorizing negative examples as positive examples as FP.

Based on the above matrix, precision and recall can be defined, and the calculation of mAP used in this book is provided, as shown in Eq. (10.3):

$$
\begin{aligned}
P &= \frac{\mathrm{TP}}{\mathrm{TP}+\mathrm{FP}} \\
R &= \frac{\mathrm{TP}}{\mathrm{TP}+\mathrm{FN}} \\
\mathrm{mAP} &= \frac{1}{C}\sum_{k=1}^{N} P(k)\Delta R(k)
\end{aligned}
\tag{10.3}
$$

Each P–R curve corresponds to a threshold value. In this book, the threshold value is selected as 0.5. IOU values greater than 0.5 are referred to as positive examples, and IOU values less than 0.5 are referred to as negative examples. The AP value is obtained by integrating the P–R curve and is used to compare the performance of the I-YOLOv3 and YOLOv3 algorithms. The P–R curves obtained from the training in this book are shown in Figures 10.19 and 10.20.

Integrating each of the four P–R curves in Figures 10.19 and 10.20 yields the results shown in Table 10.6.

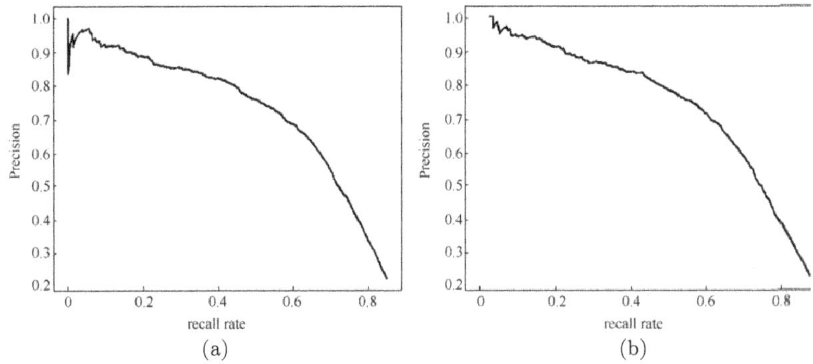

Figure 10.19. *P–R* curves of environmental obstacles for two algorithms: (a) YOLOv3 algorithm and (b) I-YOLOv3 algorithm.

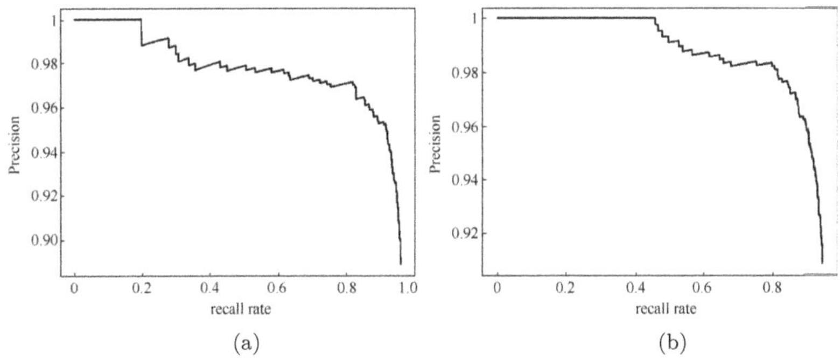

Figure 10.20. *P—R* curves of targets for two algorithms: (a) YOLOv3 algorithm and (b) I-YOLOv3 algorithm.

Table 10.6. Prediction accuracy of two algorithms.

Accuracy	YOLOv3 algorithm (%)	I-YOLOv3 algorithm (%)	Increase
Target recognition accuracy	91.54	95.55	4.01
Environmental obstacle recognition accuracy	62.5	70.2	7.7
mAP	76.02	82.21	6.19

Based on the data in Table 10.6, the I-YOLOv3 algorithm shows a significantly improved accuracy in predicting environmental obstacles and target objects compared to the YOLOv3 algorithm. This

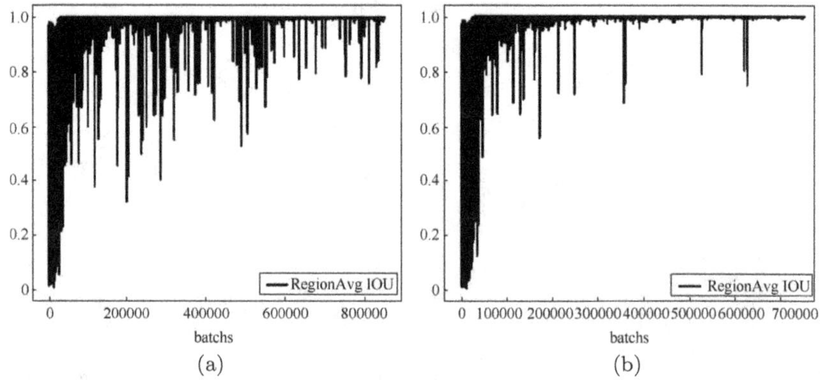

Figure 10.21. IOU curves for the two algorithms: (a) YOLOv3 algorithm and (b) I-YOLOv3 algorithm.

indicates that the improvements made to the YOLOv3 algorithm in this book are correct and can provide precise information for subsequent "target grasping" tasks.

10.3.3.3 *IOU curve analysis*

Observing Figures 10.21, it is observed that the IOU curves of both the I-YOLOv3 algorithm and the YOLOv3 algorithm are able to approximately stabilize the IOU values at around 1 after a certain number of iterations. However, the degree of mutation in the IOU curve of the I-YOLOv3 algorithm is significantly lower than that of the YOLOv3 algorithm, and the curve is more stable. Therefore, the position prediction of the I-YOLOv3 algorithm is more accurate and better.

In summary, the I-YOLOv3 algorithm proposed in this book exceeds the YOLOv3 algorithm in all the indicators and has a more obvious improvement effect.

10.3.4 ***3D reconstruction of environmental obstacles***

10.3.4.1 *Obtaining the minimum outer rectangle*

Taking a picture containing environmental obstacles as an example. The process sequence is shown in Figure 10.22 following the minimum outer rectangle acquisition step in this section.

After a series of image processing, it becomes possible to obtain the minimum outer rectangle for predicting environmental obstacles. In Figure 10.22, a total of 10 prediction boxes are predicted using the I-YOLOv3 algorithm for the figure, and the minimum prediction box size is 30×35. A total of 10 minimum outer bounding rectangles have been obtained through image processing without loss of information about the prediction boxes, which indicates that the thresholds and corrosion expansion coefficients adopted in this example are more reasonable, and each minimum outer bounding rectangle basically follows the edge of the environmental obstacle closely, which indicates that the method implemented in the book is feasible.

10.3.4.2 *Multi-parameter conditional constraint validation and fitting experiments*

To visually illustrate the experimental data, each prediction box is labeled in the minimum outer rectangular prediction plot, and the results are shown in Figure 10.23.

Record the pixel coordinates of the four corners and the center point of each prediction box. Corner order: The bottom-left corner of each prediction box is Rect(1), then clockwise, Rect(2), Rect(3), and Rect(4).

According to the principle of environmental obstacle reconstruction, taking prediction box 1 as the basis, based on the two environmental obstacle constraints, and judging the relationship between

Table 10.7. Environmental obstacle inflection point information.

Prediction box	Rect(1)	Rect(2)	Rect(3)	Rect(4)
1	(643,717)	(610,717)	(610,680)	(643,680)
2	(732,714)	(698,714)	(698,661)	(732,661)
3	(665,646)	(647,643)	(651,608)	(674,610)
4	(628,601)	(607,607)	(582,504)	(602,498)
5	(661,597)	(629,594)	(632,416)	(667,422)
6	(689,584)	(671,578)	(707,494)	(729,507)
7	(566,488)	(541,376)	(564,372)	(586,484)
8	(664,412)	(643,408)	(660,248)	(681,251)
9	(765,413)	(749,409)	(781,320)	(801,324)
10	(539,361)	(526,256)	(545,252)	(557,357)

Figure 10.22. Process sequence for environmental obstacle recognition: (a) the original picture of the obstacle, (b) I-YOLOv3 algorithm recognition effect picture, (c) thresholding diagram (d) corrosion operation diagram, (e) expansion operation diagram, and (f) minimum outer rectangle diagram for environmental obstacles.

prediction box 1 and all the remaining prediction boxes in turn, we get that prediction boxes 3, 4, 5, 6, and 7 satisfy the conditions. Then, based on prediction boxes 3, 4, 5, 6, and 7, according to the constraints, prediction box 6 satisfies the relationship with prediction box 3. Considering that prediction box 6 also has a relationship with prediction box 1, there is no further judgment on prediction box 6 in this step. Prediction boxes 7 and 10 satisfy the relationship with prediction box 4, and prediction box 7 has a relationship with prediction box 1, so prediction box 7 is no longer judged in this step. As prediction box 5 has a relationship with prediction box 9, we now get three arrays: $V_1 = \{1,3,6\}$, $V_2 = \{1,5,9\}$, and $V_3 = \{1,4,7\}$. Then, proceed to the next step of judgment, determining the relationships for prediction boxes 6, 7, and 9. Prediction box 10 is found to satisfy the relationship with prediction box 7, while prediction box 8 is found to satisfy the relationship with prediction box 6. This will give three arrays of $V_1 = \{1,3,6,8\}$, $V_2 = \{1,5,9\}$, and $V_3 = \{1,4,7,10\}$. At this point, all prediction boxes based on prediction box 1 have been fully searched, leaving prediction box 2 as a separate array. Finally, the environmental obstacle fitting is carried out using the cubic polynomial, and the step sizes of V_1, V_2, and V_3 fitting are 39.2, 44.1, and 51.2 (in pixels), respectively, and the final results are shown in Figure 10.24.

10.3.4.3 *Reconstruction performance assessment*

An environmental obstacle reconstruction experiment was performed on 50 images, some of the results of which are shown in Figure 10.25.

As can be seen in Figure 10.25, after the two processes of environmental obstacle recognition and reconstruction, most of the environmental obstacles in the image can be fitted well, but fitting deviations can also occur. The fitting deviation can be roughly

Figure 10.23. Environmental obstacle LabelImg map.

Figure 10.24. Effect of fitting environmental obstacles.

(a) (b) (c)

Figure 10.25. Partial environmental obstacle fit.

divided into two categories: In the first category, as shown in the red box in Figure 10.25(a), the right part of the environmental obstacle is not completely fitted, which is due to two neighboring prediction boxes belonging to the same group of the environmental obstacle are far away from each other, which causes fitting errors, but the overall shape of the environmental obstacle is already fitted, and this kind of error does not have much effect; in the second category, as shown in the blue box in Figure 10.25(c), the environmental obstacle in the blue box is actually a branch of the environmental obstacle in the yellow box, but due to the existence of the environmental obstacle with severe occlusion, it does not satisfy the two constraints of the reconstruction of the environmental obstacle, and there is a case of environmental obstacle compartmentalization, which needs to be solved further in depth.

We can use the method proposed by Yang Changhui *et al.* to define the environmental obstacle accuracy as

$$A_r = \frac{C_{rb}}{A_b} \times 100\% \tag{10.4}$$

where A_r is the accuracy of reconstruction, C_{rb} is the number of correctly reconstructed environmental obstacles, and A_b is the number of actual environmental obstacles.

The final average of the accuracy of multiple sets of environmental obstacles was found to be 82.5%, i.e., the accuracy of environmental obstacle reconstruction in this book is 82.5%.

10.4 Object Grabbing Experiment

10.4.1 *Grab experiment examples*

During the experiment, (p_{Cx}, p_{Cy}, p_{Cz}), $(p'_{Cx}, p'_{Cy}, p'_{Cz})$, and $(p''_{Cx}, p''_{Cy}, p''_{Cz})$ represent the position of the target seen by the binocular camera for the first time with respect to the coordinate system $\{C\}$, the position obtained after the binocular tracking of the target, and the position of the target after the compensation of the moving cart, respectively; (p_{Gx}, p_{Gy}, p_{Gz}) and $(p'_{Gx}, p'_{Gy}, p'_{Gz})$ represent the theoretical position and actual position of the target in the geodetic coordinate system, respectively; (p_{Ox}, p_{Oy}, p_{Oz}) and $(p'_{Ox}, p'_{Oy}, p'_{Oz})$ represent the positions of the secondary tracked target in the chest center coordinate system and the position of the target in the robot chest center coordinate system after the cart compensation, respectively.

10.4.1.1 *First step*

For the robot grasping system platform, at the beginning of the experiment, each joint of the robot is in the initial state, and each rotation angle and translation distance are 0. Therefore, after obtaining the target position and orientation ${}_{A}^{C}\boldsymbol{T}$, the position of the target in the geodetic coordinate system ${}_{A}^{G}\boldsymbol{T}$ can be obtained according to Eq. (3.11), as shown in Eq. (10.5):

$$
\begin{aligned}
{}_{A}^{G}\boldsymbol{T} = {}_{C}^{G}\boldsymbol{T}{}_{A}^{C}\boldsymbol{T} &= \begin{bmatrix} 1 & 0 & 0 & -32.5 \\ 0 & 0 & 1 & 254.5 \\ 0 & -1 & 0 & 1050 \\ 0 & 0 & 0 & 1 \end{bmatrix} \\
&\times \begin{bmatrix} 0.8672 & -0.04662 & -0.4958 & 123.117 \\ -0.4348 & -0.5561 & -0.7083 & 204.16 \\ -0.2427 & 0.8298 & -0.5025 & 1151 \\ 0 & 0 & 0 & 1 \end{bmatrix} \\
&= \begin{bmatrix} 0.8672 & -0.04662 & -0.4958 & 90.62 \\ -0.2427 & 0.8298 & -0.5025 & 1405.5 \\ 0.4348 & 0.5561 & 0.7083 & 845.8 \\ 0 & 0 & 0 & 1 \end{bmatrix}
\end{aligned} \tag{10.5}
$$

According to the above, the target is above the line of sight of the binocular camera as judged by $p_{Cy} = 204.16 > 0$, so the joints from the center of the arm end-effector (hand claw) to the waist model are set to 10°, 10°, 30°, 10°, and −60°. According to Eqs. (4.7) and (10.5), the target position of the waist in the geodetic coordinate system ${}_{W}^{G}\boldsymbol{T}_{\text{float}}$ is obtained, as shown in Eq. (10.6):

$$
{}_{W}^{G}\boldsymbol{T}_{\text{float}} = {}_{A}^{G}\boldsymbol{T}{}_{W}^{A}\boldsymbol{T} = \begin{bmatrix} 0.2441 & -0.9096 & -0.3362 & -203.8 \\ 0.103 & 0.369 & -0.9237 & 1079.0 \\ 0.9643 & 0.1909 & 0.1838 & 573.6 \\ 0 & 0 & 0 & 1 \end{bmatrix} \tag{10.6}
$$

10.4.1.2 *Second step*

According to Eqs. (4.12) and (10.6), the required motion of each joint in the cart-to-waist model is obtained as shown in Eq. (10.7):

$$\begin{cases} d_x = -203.8\,\text{mm} \\ d_y = 892.9\,\text{mm} \\ \theta_3 = 0^\circ \\ \theta_4 = -20^\circ \\ h_z = 28.64\,\text{mm} \\ \theta_6 = -\ 10.59^\circ \\ \theta_7 = -\ 11.2^\circ \end{cases} \tag{10.7}$$

Therefore, the angle values of each joint in the grasping branch are solved under the condition of satisfying the assumption that the main operating arm has successfully grasped the target. In addition, the difference between the selected base coordinate system in the two models of the end-effector of the operating arm to the waist and the waist to the end-effector of the operating arm leads to the fact that the positivity and negativity of the angle values of the last five joints are opposite to the given angle values when verified in the grasping branch, i.e., the angle values of the individual joints of the main operating arm are 60°, −10°, −30°, −10°, and −10°, respectively.

10.4.1.3 *Third step*

The continuous tracking of the target is realized, and the robot's head motion is solved as $\alpha_1 = -7.39^\circ$ (head up is positive) and $\alpha_2 = -19.55^\circ$ (counterclockwise is positive), and then the orientation of the target acquired by the camera is recorded as the square array $\boldsymbol{R}$ of 3×3, and its position is $(p'_{Cx}, p'_{Cy}, p'_{Cz}) = (38.18, 62.87, 248)$. Because of the large influence of the position when grasping the target, only the influence of its position is considered here, and the position coordinates are converted to the robot's chest center coordinate system $\{O\}$ to get the position coordinate value of $(p_{Ox}, p_{Oy}, p_{O_z}) = (129.5, -285.8, 111.3)$, i.e., according to Eq. (3.16), we get ${}^O_A\boldsymbol{T}$ (the pose is that in the ideal case), as shown in Eq. (10.8):

$${}^O_A\boldsymbol{T} = {}^O_C\boldsymbol{T}\,{}^C_A\boldsymbol{T} = \begin{bmatrix} 0.04304 & -0.9917 & -0.1212 & 220.3 \\ 0.3318 & 0.1286 & -0.9345 & -74.8 \\ 0.9423 & 0 & 0.3346 & -7.704 \\ 0 & 0 & 0 & 1 \end{bmatrix}$$

$$\times \begin{bmatrix} & \boldsymbol{R} & & 38.18 \\ & & & 62.87 \\ & & & 248 \\ 0 & 0 & 0 & 1 \end{bmatrix}$$

$$= \begin{bmatrix} 0.5101 & 0.606 & 0.6103 & 129.5 \\ 0.761 & 0.01258 & -0.6486 & -285.8 \\ -0.4007 & 0.7954 & -0.4548 & 111.3 \\ 0 & 0 & 0 & 1 \end{bmatrix} \tag{10.8}$$

The ${}^{O}_{A}\boldsymbol{T}$ in Eq. (10.8) is the desired target position and orientation of the end-effector in Figure 4.6. Based on Eq. (4.22), the numerical method, and Eq. (10.8), two sets of angle values conforming to the limits of each joint of the operating arm are obtained, as shown in Eq. (10.9):

$$\begin{cases} \text{analysis}: (59.57°, -15.12°, -27.38°, -7.05°, -5.74°) \\ \text{iteration}: \ (60.83°, -23.33°, -31.60°, 8.996°, 1.91°) \end{cases} \tag{10.9}$$

Substituting the two sets of solutions into Eq. (4.20), respectively, we obtain the actual position and orientation reached by the end-effector, ${}^{o}_{P}\boldsymbol{T}_{\text{analysis}}$ and ${}^{o}_{P}\boldsymbol{T}_{\text{iteration}}$, as shown in Eqs. (10.10) and (10.11), respectively:

$${}^{o}_{P}\boldsymbol{T}_{\text{analysis}} = \begin{bmatrix} 0.5176 & 0.5685 & 0.6395 & 102.4 \\ 0.7577 & 0.04282 & -0.6513 & -347.7 \\ -0.3976 & 0.8216 & -0.4086 & 125.2 \\ 0 & 0 & 0 & 1 \end{bmatrix} \tag{10.10}$$

$${}^{o}_{P}\boldsymbol{T}_{\text{iteration}} = \begin{bmatrix} 0.3621 & 0.6102 & 0.7046 & 129.5 \\ 0.886 & 0.009612 & -0.4636 & -285.8 \\ -0.2897 & 0.7922 & -0.5372 & 111.3 \\ 0 & 0 & 0 & 1 \end{bmatrix} \tag{10.11}$$

Observing Eqs. (10.10) and Eq. (10.11), it is concluded that although the position and orientation obtained by both sets of solutions do not

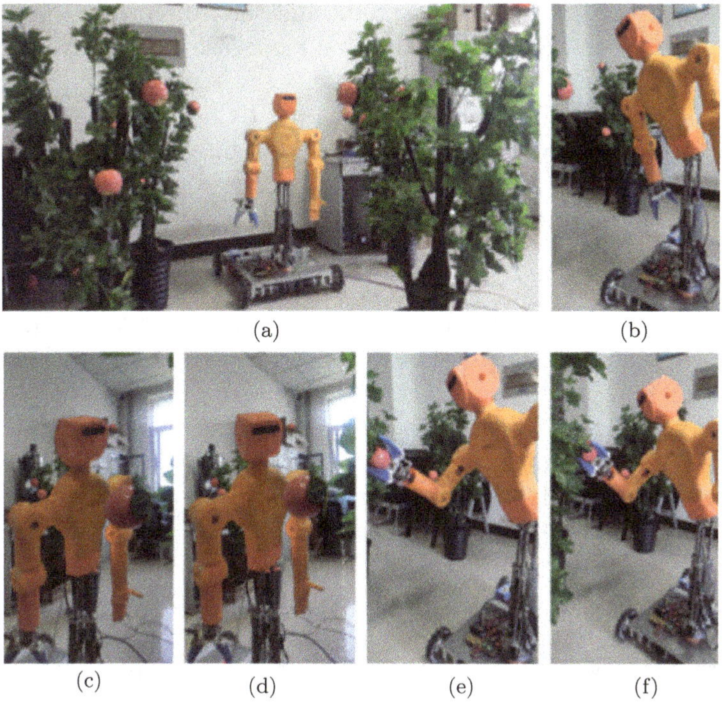

Figure 10.26. Experimental grabbing process: (a) initial state, (b) below-waist movement, (c) before target tracking, (d) after target tracking, (e) analytical solution, and (f) numerical solution.

match the actual values of the target, the most influential factor in the grasping process is the position, and as long as the position is within a reasonable margin of error, the target can be successfully grasped. In this experiment, both sets of solutions can successfully grasp the target, but in order to ensure the accuracy of grasping, this book uses iterative solutions for grasping. The experimental grasping process is shown in Figure 10.26.

Figure 10.26 shows the whole motion process of the robot from the initial state to the grasping to the target object. In the initial state shown in Figure 10.26(a), the center point of the mobile platform is the center point of the large geographic coordinate system $\{G\}$, and the position information of the target is obtained by the vision system and transformed by the perception system, which determines the position and orientation relationship of the target to be grasped in the geographic coordinate system, and then the preset joint angle values of the end-effector of the main operating arm after having grasped the target are given to obtain the motion amount of the joints of the robot's lower body; Figure 10.26(b) shows the robot state when the lower body joint values obtained in the initial state are sent down to the bottom layer and driven. Figure 10.26(b) shows the state of the robot after the lower body joint values obtained in the initial state are sent to the bottom layer and the lower body is driven to the target point; Figure 10.26(c) shows that the target is not visible in the camera's field of view after the lower body is moved; Figure 10.26(d) shows that after the target tracking, the target appears in the camera's field of view again, and the position relation of the target with respect to the robot's chest center coordinate system can be acquired. Figure 10.26(d) shows that after target tracking, the target appears again in the camera field of view and the position information of the target relative to the robot's chest center coordinate system $\{O\}$ can be obtained; Figures 10.26(e) and 10.26(f) show that different inverse solutions are used to obtain the angle values of the joints of the main working arm and send them to the driving servo to complete the grasping task, respectively.

10.4.2 *Error correction*

When the error in the cart motion and the robot's own structure is large, the model separation inverse kinematics algorithm proposed in this book will also fail to find a suitable solution; in this case, in order to successfully grasp the target object, it is necessary to move the cart to compensate for its error, as follows:

(1) Obtain the position and orientation of the target object in the global coordinate system ${}^{G}_{A}\boldsymbol{T}$ for the first time it is observed.
(2) After the cart, robot foot, and waist have moved to the specified position, the camera-tracked position and orientation of the target ${}^{O}_{A}\boldsymbol{T}$ relative to the robot chest center coordinate system $\{O\}$ are acquired.
(3) Find the relationship ${}^{W}_{O}\boldsymbol{T}$ between the center of the robot's chest $\{O\}$ and the coordinate system of the waist end-effector $\{W\}$,

as shown in Eq. (10.12):

$$
{}^{W}_{O}\boldsymbol{T} = \begin{bmatrix} 1 & 0 & 0 & 275 \\ 0 & 0 & -1 & 0 \\ 0 & 1 & 0 & 0 \\ 0 & 0 & 0 & 1 \end{bmatrix} \tag{10.12}
$$

(4) Substituting the motions of the cart, foot, and waist joints of the robot into Eq. (4.9), we get ${}^{G}_{W}\boldsymbol{T}'$. Therefore, the actual position and orientation of the target with respect to the geodetic coordinate system is obtained as ${}^{G}_{W}\boldsymbol{T}'\,{}^{W}_{O}\boldsymbol{T}\,{}^{O}_{A}\boldsymbol{T}$, and it is written as ${}^{G}_{A}\boldsymbol{T}'$.

(5) In summary, it can be seen that ${}^{G}_{A}\boldsymbol{T}$ and ${}^{G}_{A}\boldsymbol{T}'$ are the theoretical and measured positions and orientations of the target in the geodetic coordinate system, respectively, but only the positional relationship between the two is compared here. This is because ${}^{G}_{A}\boldsymbol{T}$ is to rotate the end-effector to be consistent with the large geodetic coordinate system, while ${}^{G}_{A}\boldsymbol{T}'$ is not, so the positions and orientations of the two will be different, but the positional values are theoretically equal. Therefore, the values of the cart after the end of the motion in the directions of x and y that still need to be compensated for are d'_x and d'_y respectively, as shown in Eq. (10.13):

$$
d'_x = {}^{G}_{A}\boldsymbol{T}(1,4) - {}^{G}_{A}\boldsymbol{T}'(1,4) \quad d'_y = {}^{G}_{A}\boldsymbol{T}(2,4) - {}^{G}_{A}\boldsymbol{T}'(2,4) \tag{10.13}
$$

(6) After moving the cart along the directions x and y by distances d'_x and d'_y, respectively, the binocular camera can still observe the target object because it is a trace compensation, so the inverse kinematics can be continued to be solved using Newton's iterative method.

Therefore, when that situation occurs, we will utilize the above method to compensate for the error, thus solving the problem of not finding a suitable solution, which will be experimented with in the following.

At the beginning of the grasping experiment, the position and orientation of the target in the geodetic coordinate system ${}^{G}_{A}\boldsymbol{T}$ are given by Eq. (10.14):

$$ {}^{G}_{A}\boldsymbol{T} = {}^{G}_{C}\boldsymbol{T}\,{}^{C}_{A}\boldsymbol{T} = \begin{bmatrix} 0.9341 & -0.07268 & 0.3494 & 22.3 \\ 0.3527 & 0.3372 & -0.8728 & 1234.0 \\ -0.05439 & 0.9386 & 0.3406 & 1164.0 \\ 0 & 0 & 0 & 1 \end{bmatrix} \tag{10.14} $$

According to Eqs. (4.6) and (10.14), the position and orientation of the waist in the geodetic coordinate system ${}^{G}_{W}\boldsymbol{T}'$ are obtained as shown in Eq. (10.15) when the values of each joint angle in the model of the center of the arm end-effector (hand claw) to the waist are given as 0°, 45°, 20°, 5° and −110°:

$$ {}^{G}_{W}\boldsymbol{T}' = {}^{G}_{A}\boldsymbol{T}\,{}^{A}_{W}\boldsymbol{T} = \begin{bmatrix} 0.3472 & -0.7818 & 0.518 & -153.1 \\ 0.0567 & -0.5338 & -0.8437 & 854.4 \\ 0.9361 & 0.3223 & -0.141 & 611.6 \\ 0 & 0 & 0 & 1 \end{bmatrix} \tag{10.15} $$

According to Eqs. (4.12) and (10.15), the required motion of each joint in the cart-to-waist model is obtained as shown in Eq. (10.16):

$$ \begin{cases} d_x = -153.1\text{mm} \\ d_y = 668.4\text{mm} \\ \theta_3 = 0^\circ \\ \theta_4 = 31.55^\circ \\ h_z = 66.56\text{mm} \\ \theta_6 = 8.106^\circ \\ \theta_7 = -19^\circ \end{cases} \tag{10.16} $$

When the waist reaches the “levitation” state, using the Gauss–Newton algorithm, solving when the robot is continuously tracking the target, the head needs to rotate by the angle values of $\alpha_1 = 20.67^\circ$ (head up is positive) and $\alpha_2 = -31.31^\circ$ (counterclockwise is positive). Furthermore, the coordinates of the target object relative to the coordinate system $\{C\}$ at this moment were obtained as $(p'_{Cx}, p'_{Cy}, p'_{Cz}) = (-42.4878, 0.7516, 232)$, Therefore, based on

Eq. (3.16), O_T^A is obtained as follows, as shown in Eq. (10.17).

$$
{}_A^O\boldsymbol{T} = {}_C^O\boldsymbol{T}\,{}_A^C\boldsymbol{T} = \begin{bmatrix} 0.3907 & 0.2934 & 0.8725 & 333.7 \\ 0.8694 & 0.1939 & -0.4545 & -276.9 \\ -0.3026 & 0.9361 & -0.1793 & 92.09 \\ 0 & 0 & 0 & 1 \end{bmatrix} \tag{10.17}
$$

According to the target position and orientation in Eq. (10.17), no suitable inverse solution can be found using both analytical and iterative methods. Therefore, we utilize the movement of the cart to compensate for errors, as mentioned above. The specific process is as follows:

(1) According to Eq. (10.14), the theoretical target position and orientation of the target object relative to the geodetic coordinate system ${}_A^G\boldsymbol{T}$ are obtained, where the position value is $(p_{Gx}, p_{Gy}, p_{Gz}) = (22.3, 1234, 1164)$.
(2) According to Eqs. (10.12), (10.15), and (10.17), the actual position and orientation of the target in the geodetic coordinate system ${}_A^G\boldsymbol{T}'$ are obtained as shown in Eq. (10.18):

$$
{}_A^G\boldsymbol{T}' = {}_W^G\boldsymbol{T}'\,{}_O^W\boldsymbol{T}\,{}_A^O\boldsymbol{T} \tag{10.18}
$$

The position of ${}_A^G\boldsymbol{T}'$ is $(p'_{Gx}, p'_{Gy}, p'_{Gz}) = (-13.21, 1172, 1191)$. When the cart is moved to compensate for the error, it can only do so in the directions of x and y. Therefore, in order to satisfy the condition that (p_{Gx}, p_{Gy}) and (p'_{Gx}, p'_{Gy}) are equal, the cart needs to move along the directions of x and y by the distances shown in Eq. (10.19):

$$
\begin{aligned} d'_x &= 35.51\,\text{mm} \\ d'_y &= 62\,\text{mm} \end{aligned} \tag{10.19}
$$

When the cart moves to the final specified position, use the camera to read the position of the target as $(p''_{Cx}, p''_{Cy}, p''_{Cz}) = (65.1125, 28.458, 288)$, which is converted into the position in the robot chest center coordinate system $\{O\}$ as $(p'_{Ox}, p'_{Oy}, p'_{Oz}) =$

$(304.9, -279.2, 213.1)$. At this time, the obtained ${}_A^O\boldsymbol{T}'$ is shown in Eq. (10.20):

$$
{}_A^O\boldsymbol{T}' = {}_C^O\boldsymbol{T}\,{}_A^C\boldsymbol{T}' = \begin{bmatrix} 0.3907 & 0.2934 & 0.8725 & 304.9 \\ 0.8694 & 0.1939 & -0.4545 & -279.2 \\ -0.3026 & 0.9361 & -0.1793 & 213.1 \\ 0 & 0 & 0 & 1 \end{bmatrix} \tag{10.20}
$$

Taking ${}_A^O\boldsymbol{T}'$ as the target position and orientation of the end-of-arm effector, the appropriate inverse solution, i.e., each angular value within the joint limits, is obtained by Newton's iterative and analytical methods, as shown in Eq. (10.21):

$$
\begin{cases} \text{analysis: } (106.81^\circ, -21.44^\circ, -29.14^\circ, -35.05^\circ, 17.66^\circ) \\ \text{iteration: } (103.4^\circ, -26.07^\circ, -30.84^\circ, -25.69^\circ, 20.47^\circ) \end{cases} \tag{10.21}
$$

Substituting the two sets of solutions into Eq. (4.20), respectively, we obtain the actual position and orientation reached by the end-effector ${}_P^O\boldsymbol{T}_{\text{analysis}}$, ${}_P^O\boldsymbol{T}_{\text{iteration}}$, as shown in Eqs. (10.22) and (10.23), respectively:

$$
{}_P^O\boldsymbol{T}_{\text{analysis}} = \begin{bmatrix} 0.3891 & 0.2684 & 0.8812 & 313.6 \\ 0.8755 & 0.1898 & -0.4444 & -296.4 \\ -0.2865 & 0.9444 & -0.1611 & 218.6 \\ 0 & 0 & 0 & 1 \end{bmatrix} \tag{10.22}
$$

$$
{}_P^O\boldsymbol{T}_{\text{iteration}} = \begin{bmatrix} 0.3285 & 0.3164 & 0.89 & 305.1 \\ 0.8994 & 0.1829 & -0.397 & -279.0 \\ -0.2883 & 0.9309 & -0.2245 & 213.1 \\ 0 & 0 & 0 & 1 \end{bmatrix} \tag{10.23}
$$

From Eqs. (10.22) and (10.23), it can be seen that the error between the tracking position and actual position of the target is within the permissible range, so both sets of solutions can successfully capture the target.

(a) (b) (c)

Figure 10.27. Experiment of grasping the target after correcting the error: (a) tracking the target, (b) gripping before error compensation, and (c) gripping after error compensation.

Figure 10.27(a) represents the grasping state of the end-effector on the target in the camera's field of view without error correction; Figure 10.27(b) represents the actual state of the robot's end-effector grasping the target without error correction; and Figure 10.27(c) represents the state of the main operating arm successfully grasping the target after correcting the error.

10.4.3 *Dual-arm grasping experiment*

In this experiment, when the main operation system is grasping a target that is obscured by an obstacle, if a collision-free path does not exist or the planned collision-free path is unreasonable, the auxiliary operation arm can be operated to pull the environmental obstacle away, and then the main operation arm can be used to grasp the target. In this regard, in order for the main and auxiliary operation arms to be able to coordinate grasping, the angle of rotation of the robot foot is made as small as possible. At the same time, the target environmental obstacle point of the end-effector of the auxiliary operation arm is also kept at a certain distance from the target to be grasped. The experimental process is shown in Figure 10.28.

Figure 10.28(a) represents the target to be grasped that is blocked by the obstacle, as recognized by the camera; Figures 10.28(b) and 10.28(c) represent the detected state in the camera and the actual grasping state in the field when the main operating arm grasps the target after the auxiliary operating arm clears away the branch and the main operating arm grasps the target, respectively.

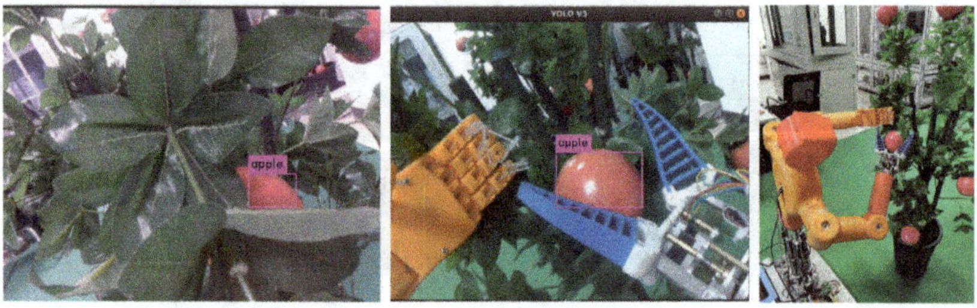

(a) (b) (c)

Figure 10.28. Dual-arm grasping experiment: (a) hidden target, (b) tracked target, and (c) dual-arm grabbing of target.

Therefore, when the target to be grasped is hidden behind an obstacle and the collision-free path planning is unreasonable, the auxiliary operating arm can be used to assist the main operating system to complete the operation task.

The combination of the above experiments yielded an average grasping success rate of 97% for the robotic platform, and the average grasping time from the start of the main operating arm's movement to the successful grasping was 8.2 s.

10.5 Summary

In this chapter, firstly, simulation experiments were carried out on the established model through the MATLAB Robotics Toolbox. Then, according to the control algorithm proposed in Chapter 9, the control program was written under the ROS system, and real-time control experiments were carried out on the robot. The errors of the experimental process were then corrected. Finally, the grasping experiments were carried out using the main and auxiliary operating arms to grasp the target objects hidden behind the obstacles.

10.6 Exercises

10.1 Design and practice target object recognition algorithms.
10.2 Design and practice environmental obstacle recognition algorithms.

10.3 Design and practice environmental obstacle reconstruction algorithms.
10.4 Design and practice closed-loop control algorithms based on binocular vision information.
10.5 Design and practice a map building program.

Index

P

Q

R

S

U

V

W

Y

www.ingramcontent.com/pod-product-compliance
Lightning Source LLC
LaVergne TN
LVHW010330220926
848484LV00004BB/869
9789819810246